W0234716

Encyclopaedia of
Mathematical Sciences

Volume 86

Editor-in-Chief: R. V. Gamkrelidze

Springer
Berlin
Heidelberg
New York
Barcelona
Hong Kong
London
Milan
Paris
Singapore
Tokyo

M. I. Zelikin

Control Theory and Optimization I

Homogeneous Spaces and the Riccati Equation
in the Calculus of Variations

Springer

Title of the Russian edition:
Odnorodnye Prostranstva i uravnenie rikkati v variatsionnom ischislenii
Publisher Faktorial, Moscow 1998

Mathematics Subject Classification (1991): 49-XX, 53-XX

ISSN 0938-0396
ISBN 3-540-66741-5 Springer-Verlag Berlin Heidelberg New York

This work is subject to copyright. All rights are reserved, whether the whole or part of the material is concerned, specifically the rights of translation, reprinting, reuse of illustrations, recitation, broadcasting, reproduction on microfilm or in any other way, and storage in data banks. Duplication of this publication or parts thereof is permitted only under the provisions of the German Copyright Law of September 9, 1965, in its current version, and permission for use must always be obtained from Springer-Verlag. Violations are liable for prosecution under the German Copyright Law.

© Springer-Verlag Berlin Heidelberg 2000
Printed in Germany

Typesetting: By B. Everett using a Springer T$_{\rm E}$X macro package.
Printed on acid-free paper   SPIN 10732015   46/3143LK - 5 4 3 2 1 0

List of Editors, Authors and Translators

Editor-in-Chief

R. V. Gamkrelidze, Russian Academy of Sciences, Steklov Mathematical Institute, ul. Gubkina 8, 117966 Moscow; Institute for Scientific Information (VINITI), ul. Usievicha 20a, 125219 Moscow, Russia; e-mail: gam@ipsun.ras.ru

Author

M. I. Zelikin, Department of Mathematics, MGU, Vorob'evy Gory, 119899 Moscow, Russia

Translator

S. A. Vakhrameev, Department of Mathematics, VINITI, ul. Usievicha 20a, 125219 Moscow, Russia

Preface

This book is devoted to the development of geometric methods for studying and revealing geometric aspects of the theory of differential equations with quadratic right-hand sides (Riccati-type equations), which are closely related to the calculus of variations and optimal control theory.

The book contains the following three parts, to each of which a separate book could be devoted:

1. the classical calculus of variations and the geometric theory of the Riccati equation (Chaps. 1–5),
2. complex Riccati equations as flows on Cartan–Siegel homogeneity domains (Chap. 6), and
3. the minimization problem for multiple integrals and Riccati partial differential equations (Chaps. 7 and 8).

Chapters 1–4 are mainly auxiliary. To make the presentation complete and self-contained, I here review the standard facts (needed in what follows) from the calculus of variations, Lie groups and algebras, and the geometry of Grassmann and Lagrange–Grassmann manifolds. When choosing these facts, I prefer to present not the most general but the simplest assertions. Moreover, I try to organize the presentation so that it is not obscured by formal and technical details and, at the same time, is sufficiently precise.

Other chapters contain my results concerning the matrix double ratio, complex Riccati equations, and also the Riccati partial differential equation, which arises in the minimization problem for a multiple integral.

The book is based on a course of lectures given in the Department of Mechanics and Mathematics of Moscow State University during several years. Therefore, when writing the book, I imagined the ideal readers to be the undergraduate and graduate students in this department, who are familiar with the foundations of calculus, differential equations, differential geometry, and algebra (although, in some cases, I assumed that the reader is familiar with a deeper mathematical technique). However, I hope that a wider audience will find this book interesting. I also hope that the informed reader will tolerate aspects that seem trivial to him while the reader unfamiliar with one or another mathematical object and encountering some difficulties in understanding the text will resolve them using the literature cited and then excuse me for the less-detailed explanation. Always in such cases, an author should strive for a balance between the difficult and the obvious in order to transform the first into the second. The reader can conclude whether I am successful in this.

I am grateful to my friends and colleagues for their indispensible help in preparing this book. I am particularly grateful to the Candidates of Physics and Mathematics V. F. Borisov, A. V. Dombrin, and L. F. Zelikina, to my student R. Hildebrant for the laborious work in improving the text, to Professors A. V. Arutyunov, A. S. Mishchenko, A. N. Parshin, and E. L. Tonkov for their very useful advice and suggestions, to Professor G. Freiling for very valuable bibliographic information, to Professor S. S. Demidov for his valuable help in the history of mathematics, and to E. Yu. Khodan for the careful editing of the manuscript.

The publication of the Russian edition of this book would have been impossible without the financial support of the Russian Foundation for Basic Research (Grant No. 95-01-02867 for publication of the book, Grant No. 96-01-01360 for research, and Grant No. 96-15-96072 for supporting leading scientific schools), for which I am grateful.

I am deeply grateful to Professor R. Gamkrelidze, Professor S. Vakhrameev, B. Everett, and Springer-Verlag for providing the publication of the English translation of this book.

M. I. Zelikin   Moscow, September 1999

Contents

Introduction

The Riccati equations were named in honor of the famous Italian mathematician[1] Count Jacopo Francesco Riccati (1676–1754), who published the paper "Animadversationes in aequationes differentiales secundi gradus" in the journal *Acta Eruditorum* in 1724 (see [90]). This paper was devoted to methods for separating variables and for reducing the order of differential equations. It should be noted that the content of the paper had become known to the mathematical community somewhat earlier [12]. When studying the problem of reconstructing a plane curve using its curvature properties, Riccati had obtained the equation

$$b\frac{dx}{dt} = at^\alpha + x^2. \tag{1}$$

In the same volume of *Acta Eruditorum* where Riccati's paper was published, a paper by 22-year-old Daniel Bernoulli followed it, in which he wrote that both the Nicholas Bernoullis (elder and younger), Johann Bernoulli, and Daniel Bernoulli himself had studied this equation and they all, independently of each other, had found conditions for the parameter α under which this equation admits a separation of variables and is therefore integrable in quadratures. Using a popular manner of that time, D. Bernoulli wrote the answer in the form of an anagram. Riccati answered this note in the next number of the journal. Without regard to the integrability conditions, which were familiar to him,[2] Riccati in fact refused to dispute the priority, saying that he had no intention to challenge the Bernoulli family, which he deeply respected.

D. Bernoulli waited two more years, during which nobody wrote anything on the Riccati equation (called such since that time), and then he published his solution in explicit form: Eq. (1) admits a separation of variables for

$$\alpha = \frac{4n}{-2n+1},$$

where n is an integer. For $n \to \infty$, we obtain the value $\alpha = -2$, for which the equation is also integrable in quadratures, from this formula. In 1841, Liouville[3] proved that Eq. (1) is not integrable in quadratures for any other value of the parameter α.

---

[1] It suffices to note here that a number of universities asked Riccati to be a professor, Peter the Great proposed the presidency of the St. Petersburg Academy of Sciences to him, and the Court of Vienna invited him to be the Imperial Chancellor. Riccati refused all these offers, preferring to live with his family and study mathematics.

[2] These conditions had been discussed in an exchange of letters with the younger Nicholas Bernoulli during five years before the paper was published. The paper itself was sent to Johann Bernoulli, editor of *Acta Eruditorum,* through the mediation of Nicholas Bernoulli, and its content therefore became known to Daniel Bernoulli [40].

[3] Liouville, J. (1841): Remarke nouvelles sur l'équation de Riccati. J. Math. Pures et Appl. **6**, 1–13.

Later, as often happens in mathematics, the name "Riccati equation" came to refer to not only an equation of form (1) but any system of ordinary differential equations with quadratic right-hand sides.

When studying the Riccati equation, one usually uses purely analytic methods, although geometric aspects sometimes appear in them (see [66, 67, 89, 101, etc.]). In the last two decades, there has been an intensive development of geometric methods for studying the Riccati equation. In particular, it was recently observed that Riccati equations define a smooth flow on Lagrange–Grassmann manifolds (see [11, 30, 45, 46, 97, etc.]). However, these ideas, which are mainly presented in journals, have not yet found a sufficient reflection in monographs. Our book is intended to fill this gap. Our goal is to present the known geometric aspects of the theory of differential equations with quadratic right-hand sides and to reveal new ones.

To explain the geometric nature of the Riccati equation, we consider the following second-order linear differential equation in the unknown function $y(t)$ with variable coefficients:

$$\ddot{y} + a(t)\dot{y} + b(t)y = 0. \tag{2}$$

The change of variable $\dot{y} = xy$ reduces this equation to the following first-order equation, whose right-hand side depends quadratically on the unknown function x:

$$\dot{x} + A(t)x^2 + B(t)x + C(t) = 0. \tag{3}$$

In the theory of ordinary differential equations, Eq. (3) is called the *Riccati equation*.

A close connection between the differential Riccati equation and the group of linear-fractional transformations was observed long ago. We here present the classical results describing this connection [39, 101].

Theorem 0.1. *The general solution to Riccati equation (3) is a linear-fractional function of the constant of integration. Conversely, every first-order differential equation having this property is a Riccati equation.*

Proof. Let x_1 be a particular solution to Eq. (3). Then the change of variable $y = x_1 + y$ reduces (3) to the following equation without a free term:

$$\dot{y} + A(t)y^2 + (2A(t)x_1(t) + B(t))y = 0.$$

Using the change of variable $z = 1/y$, we reduce this equation to the linear equation

$$-\dot{z} + A + (2Ax_1 + B)z = 0. \tag{4}$$

The general solution to Eq. (4) has the form $z = C\phi(t) + \psi(t)$, where C is the constant of integration. Consequently,

$$x = x_1 + \frac{1}{C\phi + \psi}$$

is a linear-fractional function of the constant of integration C.

Conversely, let $f(t)$, $g(t)$, $\phi(t)$, and $\psi(t)$ be arbitrary smooth functions, and let

$$x = \frac{Cf(t) + g(t)}{C\phi(t) + \psi(t)}, \tag{5}$$

where C is an arbitrary constant, satisfy a certain first-order differential equation. We solve (5) with respect to the constant C:

$$C = \frac{g - \psi x}{x\phi - f}.$$

Differentiating this relation, we confirm that x satisfies a differential equation with a quadratic right-hand side, i.e., a Riccati equation.

Theorem 0.2. *Let x_1, x_2, x_3, and x_4 be particular solutions to Riccati equation (3) that correspond to the values C_1, C_2, C_3, and C_4 of the constant C. Then*

$$\frac{x_3 - x_1}{x_3 - x_2} : \frac{x_4 - x_1}{x_4 - x_2} = \frac{C_3 - C_1}{C_3 - C_2} : \frac{C_4 - C_1}{C_4 - C_2}$$

for any t.

Proof. By Theorem 0.1, a solution $x(t)$ to Eq. (3) is a linear-fractional function of the constant of integration C for any fixed t. The double ratio of a quadruple of points of the projective line is an invariant of the linear-fractional transformation group. $\square$

Theorem 0.2 means that the double ratio of four particular solutions to the Riccati equation is an integral of this equation. In particular, this implies that if three particular solutions x_1, x_2, and x_3 to the Riccati equation are known, then we can write the formula for any solution to this equation by equating the double ratio of the quadruple of solutions to a constant:

$$\frac{x - x_1}{x - x_2} : \frac{x_3 - x_1}{x_3 - x_2} = \text{const.}$$

The projective properties of the Riccati equations, which appear in Theorems 0.1 and 0.2, are explained as follows. Equation (3) is obtained from the linear homogeneous equation (2), which describes a flow of linear transformations of the phase plane $(\dot{y}, y)$, by passing from the homogeneous coordinates to the affine coordinate $x = \dot{y}/y$. In this case, linear transformations of the plane pass to projective transformations of the projective line, which are linear-fractional transformations of the affine coordinate. This is a deep intrinsic reason for the necessary appearance of the Riccati equation in many fields of mathematics that seem far from each other, e.g., in algebraic geometry (flat structures and vector bundles over Riemann surfaces [106, 111]), conformal mapping theory [22, 98], the theory of completely integrable Hamiltonian systems [73, 119], the theory of automorphic functions [56], and quantum field theory [43, 112]. It is especially important for the subsequent presentation that the Riccati equations arise in the calculus of variations and optimal control theory. When examining minimization problems that lead to this equation, we try to reveal the main geometric aspect of the Riccati equation and its close connection with the geometry of homogeneous spaces.

Instead of a single differential equation, we study a system of differential equations with quadratic right-hand sides. In this case, the role of the projective line is played by its higher-dimensional analogues, the Grassmann and Lagrange–Grassmann manifolds. We need a higher-dimensional analogue of the double ratio, the *matrix double ratio*. It is difficult to indicate a paper where the matrix double ratio first appears; we can mention a number of parallel and often independent publications [60, 65, 89, 92, 110, 116, 118, etc.]. As the most striking, we should consider the fundamental Siegel paper [100], in which a Hermitian metric was introduced on the generalized Siegel upper half-plane using the matrix double ratio. In accordance with the traditional approach, in this book, we use the matrix double ratio not only to study the matrix Riccati equations and search for its integrals but also to study the geometry of Grassmann and Lagrange–Grassmann manifolds. In terms of the matrix double ratio, we obtain theorems that concern isoclinic planes, Clifford parallels, and totally geodesic submanifolds. We show that the formula for seeking the fourth harmonic mean of certain triples of points of Grassmann (or Lagrange–Grassmann) manifolds yields a simple explicit formula for involutive isometries of these manifolds.

The idea of complexification of the Riccati differential equation proves very fruitful. From the algorithmic standpoint, this idea gives us a possibility to halve the dimension of the space in certain cases (by introducing a complex variable instead of a real one). At the same time, this complexification allows us to reveal a surprising intrinsic connection between complex Riccati equations and symmetrical homogeneity domains of spaces of several complex variables; depending on the specific character of the coefficients of the complex Riccati equation, it defines a flow on one or another symmetrical Cartan–Siegel homogeneity domain. In particular, those Riccati equations that arise in the classical calculus of variations define a flow on the generalized Siegel upper half-plane.

One more essential result consists in the construction of an analogue of the differential Riccati equation in the minimization problem for a multiple integral. In this case, the role of the Riccati equation is played by a partial differential equation with a quadratic (in the unknown function) right-hand side. Solutions to this equation define an affine connection on a bundle whose sections play the roles of unknown functions in the minimization problem for the Dirichlet integral. There exists a field of extremals iff the curvature of this connection vanishes.

Because, as already mentioned, the calculus of variations and optimal control theory serve as the main source for obtaining and interpreting the differential Riccati equation, we begin our book with a presentation of the foundations of the calculus of variations.

Chapter 1
Classical Calculus of Variations

§1. Euler Equation

1.1. Brachistochrone Problem. In 1696, J. Bernoulli posed the brachistochrone problem to the mathematical community promising "to give his due to the mathematician who would succeed in its solution." Contemporaries were not unmoved by this, and during a short period, Bernoulli received several letters (one of them by an anonymous author) containing various solutions to this problem. Namely because of this, 1696 is accepted as the year of birth of the calculus of variations. We suggest the readers follow the example of the anonymous author (it was Isaac Newton) and accept the challenge coming from far 1696.

The *brachistochrone problem* is posed as follows. Two points a and b are given in a vertical plane. It is required to determine the shape of a curve such that a body of mass m going down this curve from the point a to the point b under the action of gravity arrives in the minimum time. The absence of friction is assumed.

We introduce a Cartesian coordinate system in this plane taking the point a as the origin and directing the axis Oy downward. Let $y = y(x)$ be the equation of the desired curve. By the energy conservation law, the speed of the body at the point $(x, y(x))$ is the same as in the free-fall motion from the height $y(x)$, i.e., $v = \sqrt{2gy(x)}$. Therefore, integrating along the curve $y = y(x)$, we obtain the time of motion

$$\int \frac{ds}{v} = \int_0^x \frac{\sqrt{1 + (\dot{y}(x))^2}}{\sqrt{2gy(x)}}\, dx. \tag{1.1}$$

Thus, the problem is to choose a function $y = y(x)$ such that the conditions $y(0) = 0$ and $y(x_0) = y_0$ are satisfied and functional (1.1) has the minimum value.

As a rule, to find a solution to a specific problem, mathematicians construct a general theory for solutions of arbitrary problems of the same type. In some cases, when such a theory is already constructed, we find that the initial problem does not satisfy basic assumptions of the theory. Then the process of generalization starts, which often transforms a simple initial construction into a considerably more cumbersome one. Because our goal is the construction of the general theory, not the solution of the brachistochrone problem, we boldly follow this path and consider the general minimization problem for an integral functional under assumptions that are convenient for us. This problem is formulated as follows.

Problem 1. Among all curves $x(\cdot) \in C^1([t_0, t_1], \mathbb{R}^n)$ that have a continuous derivative and satisfy the boundary conditions

$$x(t_0) = a, \qquad x(t_1) = b, \tag{1.2}$$

where a and b are two given points in $\mathbb{R}^n$, find a curve with the minimum value of the functional

$$J(x(\cdot)) = \int_{t_0}^{t_1} f(t, x(t), \dot{x}(t)) \, dt, \tag{1.3}$$

where $f \in C^2([t_0, t_1] \times \mathbb{R}^n \times \mathbb{R}^n)$. By the minimum value, we understand a local minimum in the space $C^1([t_0, t_1])$.

Definition 1.1. We say that a *weak minimum* is attained for a curve $\hat{x}(\cdot)$ if there exists $\delta > 0$ such that for any $x(\cdot) \in C^1([t_0, t_1])$ with $\|x(\cdot) - \hat{x}(\cdot)\| < \delta$, $x(t_0) = a$, and $x(t_1) = b$, the inequality $J(x(\cdot)) \geq J(\hat{x}(\cdot))$ holds.

Remark. The strong minimum is defined below on p. 43.

1.2. Euler Equation. The subspace in the function space $C^1([t_0, t_1], \mathbb{R}^n)$ defined by the conditions

$$h(t_0) = h(t_1) = 0 \tag{1.4}$$

is denoted by $C_0^1([t_0, t_1], \mathbb{R}^n)$ or merely by C_0^1.

Let a curve $\hat{x}(\cdot)$ be a solution to the problem. We consider a one-parameter family of curves $\hat{x}(\cdot) + \lambda h(\cdot)$, where $h(\cdot) \in C_0^1([t_0, t_1], \mathbb{R}^n)$ and $\lambda \in \mathbb{R}$. By condition (1.4), each of the curves of this family satisfies conditions (1.2). We consider a scalar-valued function $\phi(\lambda)$ obtained by substituting curves of our family in functional (1.3):

$$\phi(\lambda) = \int_{t_0}^{t_1} f(t, \hat{x}(t) + \lambda h(t), \dot{\hat{x}}(t) + \lambda \dot{h}(t)) \, dt. \tag{1.5}$$

Because $\phi(0)$ is the value of functional (1.3) at $\hat{x}(\cdot)$, function (1.5) attains a local minimum for $\lambda = 0$. We note that the theorem on the differentiability of an integral with respect to a parameter and the condition $f \in C^2$ imply that the function $\phi(\lambda)$ is twice differentiable in λ (see [76], Vol. 1, Sec. 11.8). Therefore, $\phi'(0) = 0$ and $\phi''(0) \geq 0$. We have

$$\phi'(0) = \int_{t_0}^{t_1} [f_x(t, \hat{x}(t), \dot{\hat{x}}(t)) h(t) + f_{\dot{x}}(t, \hat{x}(t), \dot{\hat{x}}(t)) \dot{h}(t)] \, dt = 0, \tag{1.6}$$

where f_x and $f_{\dot{x}}$ are n-dimensional row vectors and $f_x h$ and $f_{\dot{x}} \dot{h}$ are the products of a row vector and a column vector. We note that Eq. (1.6) holds for any function $h(\cdot) \in C_0^1$.

Lemma 1.1 (Du Bois-Reymond). *Let* $a(t), b(t) \in C([t_0, t_1], \mathbb{R}^n)$, *and let the condition*

$$\int_{t_0}^{t_1} [a(t)h(t) + b(t)\dot{h}(t)] \, dt = 0 \tag{1.7}$$

hold for any function $h(t) \in C_0^1([t_0, t_1], \mathbb{R}^n)$. *Then the function* $b(t)$ *is continuously differentiable and*

$$-\frac{d}{dt} b(t) + a(t) = 0. \tag{1.8}$$

Proof. Let $A(t)$ denote an arbitrary primitive of the function $a(t)$, i.e., let $A(t) = \int_{t_0}^t a(\tau) \, d\tau + K$. Integrating the first summand in (1.7) by parts and taking (1.4) into account, we obtain

$$\int_{t_0}^{t_1} [-A(t) + b(t)] \, \dot{h}(t) \, dt = 0 \tag{1.9}$$

(for any choice of the constant vector K). We now choose a function $h(t)$ such that a full scalar square stands under the integral sign. For this, we set

$$h(t) = \int_{t_0}^t [-A(\tau) + b(\tau)] \, d\tau. \tag{1.10}$$

Then $h(t_0) = 0$, and fulfillment of the condition $h(t_1) = 0$ is ensured by the choice of the constant K. Substituting (1.10) in (1.9), we obtain

$$\int_{t_0}^{t_1} [-A(t) + b(t)]^2 \, dt = 0.$$

Therefore, $A(t) \equiv b(t)$. Consequently, the function $b(t)$ is continuously differentiable, and Eq. (1.8) holds.

Applying the Du Bois-Reymond Lemma to (1.6), we find that the solution $\hat{x}(t)$ to Problem 1 should satisfy the differential equation

$$-\frac{d}{dt} f_{\dot{x}}(t, \hat{x}(t), \dot{\hat{x}}(t)) + f_x(t, \hat{x}(t), \dot{\hat{x}}(t)) = 0, \tag{1.11}$$

which is called the *Euler equation*. The Euler equation is equivalent to the system of n equations

$$-\frac{d}{dt} f_{\dot{x}^i} + f_{x^i} = 0, \quad i = 1, \ldots, n,$$

each of which is an ordinary second-order differential equation. The order of the system equals $2n$, and the general solution hence depends on $2n$ arbitrary constants that should be chosen such that boundary conditions (1.2) are satisfied (their number is also equal to $2n$).

Any solution $x(t)$ to the Euler equation is called an *extremal* of Problem 1.

Example 1.1. We find the extremals of the brachistochrone problem. We have

$$f = \frac{\sqrt{1+\dot{y}^2}}{\sqrt{y}}, \qquad f_y = -\frac{\sqrt{1+\dot{y}^2}}{2\sqrt{y^3}}, \qquad f_{\dot{y}} = \frac{\dot{y}}{\sqrt{1+\dot{y}^2}\sqrt{y}}.$$

Hence,

$$\frac{d}{dx}f_{\dot{y}} = \frac{\ddot{y}}{\sqrt{(1+\dot{y}^2)^3}\sqrt{y}} - \frac{\dot{y}^2}{2\sqrt{1+\dot{y}^2}\sqrt{y^3}}.$$

Consequently, the Euler equation has the form

$$-\frac{\ddot{y}}{\sqrt{y}\sqrt{(1+\dot{y}^2)^3}} + \frac{\dot{y}^2}{2\sqrt{(1+\dot{y}^2)y^3}} - \frac{\sqrt{1+\dot{y}^2}}{2\sqrt{y^3}} = 0$$

or

$$2y\ddot{y} + \dot{y}^2 + 1 = 0.$$

Because this equation does not contain x, its order can be reduced by the change of variable

$$\dot{y} = p(y), \qquad \ddot{y} = p\frac{dp}{dy}.$$

We obtain

$$2yp\frac{dp}{dy} + p^2 + 1 = 0, \qquad \ln(1+p^2) = \ln\frac{C}{y}, \qquad p = \pm\sqrt{\frac{C}{y} - 1}.$$

Choosing the plus sign in the latter formula, we obtain

$$\frac{dy}{\sqrt{C/y - 1}} = dx.$$

Replacing y with $C\sin^2 t$, we obtain

$$x = \int \frac{2C\sin t \cos t \, dt}{\sqrt{1/\sin^2 t - 1}} = C\int 2\sin^2 t \, dt.$$

Therefore,

$$\begin{cases} x + D = Ct - C\dfrac{\sin 2t}{2} = \dfrac{C(2t - \sin 2t)}{2}, \\[2mm] y = C\sin^2 t = \dfrac{C(1 - \cos 2t)}{2}. \end{cases} \tag{1.12}$$

The extremals in the brachistochrone problem are therefore cycloids.

Remark. More precisely, the assumptions for the Euler equation do not hold for the brachistochrone problem, because the integrand has a point of discontinuity for $y = 0$. Therefore, the above calculation can be considered a heuristic deduction. (For a formal proof of the optimality of solution (1.12), see [117], Sec. 5.)

We write the simplest integrals of the Euler equation.

Momentum integral. If the integrand f does not explicitly depend on x^i, then $f_{\dot{x}^i}$ is an integral of the Euler equation because the ith Euler equation in this case has the form

$$\frac{d}{dt} f_{\dot{x}^i} = 0.$$

Energy integral. If the integrand f does not explicitly depend on t, then

$$H(x, \dot{x}) = \sum_{i=1}^{n} f_{\dot{x}^i} \dot{x}^i - f$$

(this formula is briefly written in the form $H = f_{\dot{x}} \dot{x} - f$) is an integral of the Euler equation. Indeed, because

$$\frac{d}{dt} f_{\dot{x}} = f_x,$$

it follows from (1.11) that

$$\frac{d}{dt} H(x, \dot{x}) = \frac{d}{dt}(f_{\dot{x}})\dot{x} + f_{\dot{x}}\ddot{x} - f_x\dot{x} - f_{\dot{x}}\ddot{x} = 0.$$

1.3. Geodesics on a Riemannian Manifold. We recall that a manifold M is called a *Riemannian manifold* of class C^r, $r > 1$, if the structure of a Euclidean space is defined in each tangent space, i.e., if a positive-definite quadratic form that defines an inner product of two tangent vectors is given. In this case, it is assumed that the matrix $g_{ij}(x)$, which defines this form in a certain atlas of the manifold M, is a function of class C^r. The length of a curve $x(t)$, $t \in [t_0, t_1]$, that lies on the manifold M is given by the formula

$$\mathcal{L} = \int_{t_0}^{t_1} \sqrt{\sum_{i,j} g_{ij}(x(t)) \frac{dx^i}{dt} \frac{dx^j}{dt}} \, dt \tag{1.13}$$

(see [28], Sec. 3).

Extremals of functional (1.13) are called geodesics on M. If a manifold is embedded in a Euclidean space, then it admits a Riemannian structure that is induced by the embedding. For example, for a two-dimensional surface in $\mathbb{R}^3$ given by the equation $\mathbf{r} = \mathbf{r}(u, v)$, the metric has the form $g_{11} = \langle \mathbf{r}_u, \mathbf{r}_u \rangle$, $g_{12} = g_{21} = \langle \mathbf{r}_u, \mathbf{r}_v \rangle$, and $g_{22} = \langle \mathbf{r}_v, \mathbf{r}_v \rangle$. The expression

$$ds^2 = \sum_{i,j} g_{ij} \, du_i du_j$$

is called the *first quadratic form of a surface* ([88], Sec. 47).

Example 1.2 (geodesics on the sphere). We consider the unit sphere and pass from the Cartesian coordinates to the spherical ones using the formulas $x = \cos\phi\cos\psi$, $y = \sin\phi\cos\psi$, and $z = \sin\psi$. In these coordinates, the first quadratic form becomes $ds^2 = \cos^2\psi\,d\phi^2 + d\psi^2$. We consider ψ as an independent variable and express ϕ through ψ: $\phi = \phi(\psi)$. Then

$$\mathcal{L} = \int \sqrt{\cos^2\psi\,\dot\phi^2 + 1}\,d\psi.$$

Because the integrand does not depend on ϕ, the corresponding Euler equation has the momentum integral

$$\frac{\dot\phi\,\cos^2\psi}{\sqrt{\dot\phi^2\,\cos^2\psi + 1}} = C;$$

this implies

$$\dot\phi = \frac{\pm C}{\sqrt{\cos^4\psi - C^2\cos^2\psi}}.$$

Integrating this equation, we find ϕ:

$$\phi = \int \frac{\pm C\,d\tan\psi}{\sqrt{1 - C^2(\tan^2\psi + 1)}} = B \pm \arcsin\frac{C\tan\psi}{\sqrt{1 - C^2}}.$$

Therefore, $\sin(\phi - B) = A\tan\psi$. From this, we obtain

$$A_1\sin\phi\cos\psi + A_2\cos\phi\cos\psi + A_3\sin\psi = 0$$

and hence $A_1 x + A_2 y + A_3 z = 0$. Therefore, geodesics are intersections of the sphere with planes passing through the origin, i.e., great circles of the sphere.

Example 1.3 (geodesics on the Lobachevsky plane). In the Klein–Poincaré model, the Lobachevsky plane is described as follows. We consider the upper half-plane $y > 0$ and introduce a Riemannian metric on it defined by the first quadratic form

$$ds^2 = \frac{dx^2 + dy^2}{y^2}.$$

We find the geodesics on this manifold. Taking x as an independent variable, we obtain

$$\mathcal{L} = \int \frac{\sqrt{1 + \dot y^2}}{y}\,dx.$$

Because the integrand does not explicitly depend on x, the Euler equation has the energy integral

$$\frac{\dot y}{y\sqrt{1 + \dot y^2}}\,\dot y - \frac{\sqrt{1 + \dot y^2}}{y} = -\frac{1}{R},$$

i.e., $y\sqrt{1+\dot{y}^2} = R$ or $y\,dx = \sqrt{R^2 - y^2}\,dy$. Integrating this relation, we obtain $(x + D)^2 = R^2 - y^2$. Therefore, geodesics on the Lobachevsky plane in the Klein–Poincaré model are circles orthogonal to the line $y = 0$; in geometry, this line is called the *absolute*. If we take y as an independent variable, then for the integrand

$$\frac{\sqrt{\dot{x}^2 + 1}}{y},$$

we can write the momentum integral

$$\frac{\dot{x}}{y\sqrt{\dot{x}^2 + 1}} = C.$$

In addition to the extremals already obtained, we obtain vertical lines for $C = 0$. Thus, the role of lines in the Lobachevsky plane is played by vertical lines and circles that are orthogonal to the absolute.

Exercises.

1. Prove that for "lines" on the Lobachevsky plane, all the postulates of the Euclidean plane except the parallel postulate hold.

2. Prove that the Gaussian curvature of a manifold with the metric

$$ds^2 = \frac{dx^2 + dy^2}{y^2}$$

is constant and negative.

3. Prove that linear-fractional transformations that transform the upper half-plane into itself preserve ds^2 and define a group of motion of the Lobachevsky plane.

4. Does there exist a linear-fractional transformation that sends the upper half-plane into itself and maps two given points into two other given points?

We return to the general problem of seeking extremals of functional (1.3). We consider curves without singularities, i.e., we assume that $|\dot{x}| \neq 0$.

Because functional (1.13) does not depend on the choice of the parameterization of a curve $x(t)$, we can assume that the arc length is taken as the parameter. In this case, we have

$$\sum_{i,j=1}^{n} g_{ij}(x(t))\,\dot{x}^i \dot{x}^j \equiv 1 \tag{1.14}$$

on the curve considered.

We find the equation for geodesics. Taking (1.14) into account, we obtain

$$f_{\dot{x}_l} = \sum_{j=1}^{n} g_{lj}\dot{x}^j, \qquad f_{x_l} = \frac{1}{2}\sum_{i,j=1}^{n} \frac{\partial g_{ij}}{\partial x_l}\,\dot{x}^i \dot{x}^j.$$

The Euler equation has the form

$$-\sum_{j=1}^{n} g_{lj}\ddot{x}^j - \frac{1}{2}\sum_{i,j=1}^{n}\frac{\partial g_{lj}}{\partial x^i}\dot{x}^i\dot{x}^j$$

$$-\frac{1}{2}\sum_{i,j=1}^{n}\frac{\partial g_{il}}{\partial x^j}\dot{x}^i\dot{x}^j + \frac{1}{2}\sum_{i,j=1}^{n}\frac{\partial g_{ij}}{\partial x_l}\dot{x}^i\dot{x}^j = 0. \tag{1.15}$$

Let g^{kl} denote the inverse matrix of g_{lj} (the matrix g_{lj} is positive definite and therefore nonsingular). We multiply both sides of Eq. (1.15) by the matrix g^{kl} and obtain

$$\ddot{x}^k + \frac{1}{2}\sum_{i,j,l=1}^{n} g^{kl}\left[\frac{\partial g_{lj}}{\partial x^i} + \frac{\partial g_{il}}{x^j} - \frac{\partial g_{ij}}{\partial x_l}\right]\dot{x}^i\dot{x}^j = 0. \tag{1.16}$$

The expressions

$$\frac{1}{2}\sum_{l=1}^{n} g^{kl}\left[\frac{\partial g_{lj}}{\partial x^i} + \frac{\partial g_{il}}{\partial x^j} - \frac{\partial g_{ij}}{\partial x_l}\right]$$

are called the *Christoffel symbols* (or connection coefficients) and are denoted by Γ_{ij}^k. In this notation, Eq. (1.16) becomes

$$\ddot{x}^k + \sum_{i,j=1}^{n}\Gamma_{ij}^k(x)\,\dot{x}^i\dot{x}^j = 0, \quad k = 1,\dots,n.$$

This is just the equation for geodesics.

§2. Hamiltonian Formalism

2.1. Legendre Transform. We consider the class $\mathfrak{G}$ consisting of the functions $f : \mathbb{R} \to \mathbb{R}$ for which $f \in C^2(\mathbb{R})$ and, moreover, the following conditions hold:

1. $f''(u) > 0$, $u \in \mathbb{R}$;
2. the mapping $f' : \mathbb{R} \to \mathbb{R}$, $f' : u \mapsto f'(u)$, is surjective.

Into correspondence with each such function $f \in \mathfrak{G}$, we set a new function $f^*(p)$ defined as

$$f^*(p) = pu - f(u), \tag{1.17}$$

where the new independent variable p is related to u by

$$p = \frac{df(u)}{du}. \tag{1.18}$$

Using conditions 1 and 2, we can easily prove that relation (1.18) defines a one-to-one correspondence between u and p. The mapping $\Lambda : f \to f^*$ is called the *Legendre transform*.

Theorem 1.1. *The Legendre transform Λ has the following properties*:

a. *Λ maps $\mathfrak{G}$ into $\mathfrak{G}$;*
b. *the mapping Λ is involutive, i.e., Λ^2 is the identity transformation.*

Proof. The last two summands in the relation

$$\frac{df^*(p)}{dp} = u + p\frac{du}{dp} - \frac{df}{du}\frac{du}{dp}$$

are mutually annihilated by (1.18). Therefore,

$$\frac{df^*}{dp} = u. \tag{1.19}$$

Differentiating (1.19) with respect to p, we obtain $d^2f^*/dp^2 = du/dp$. To find du/dp, we differentiate (1.18), taking condition 1 into account:

$$1 = \frac{d^2f}{du^2}\frac{du}{dp}, \qquad \frac{du}{dp} = \left(\frac{d^2f}{du^2}\right)^{-1}.$$

Therefore,

$$\frac{d^2f^*}{dp^2} = \left(\frac{df^2f}{du^2}\right)^{-1} > 0. \tag{1.20}$$

Thus, we have proved condition 1 for the function f^*. Condition 2 is now implied by formula (1.19).

We now prove statement b. Indeed, formula (1.19) defines an independent variable of the function $(f^*)^*$. Moreover,

$$(f^*)^*(u) = up - f^*(p) = up - [pu - f(u)] = f(u).$$

The geometric sense of the Legendre transform can be explained as follows. We consider the graph of the function $z = f(u)$. Relation (1.18) means that the tangent to the graph of the function $z = f(u)$ at the point u_0 has the slope p. Formula (1.17) implies that $f^*(p)$ is the value by which the line $z = pu$ should be lowered for it to become the tangent to the graph of the function $f(u)$ (see Fig. 1.1). The function $f^*(p)$ thus defines a set of tangents to the graph of the function $z = f(u)$.

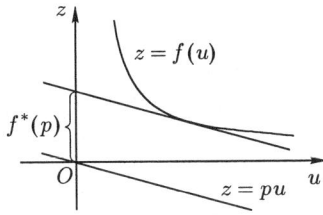

Fig. 1.1

The definition of the Legendre transform is easily extended to the class $\mathfrak{G}$ consisting of functions of several variables, $f : \mathbb{R}^n \to \mathbb{R}$. In this case,

$$f^*(p) = \sum_{i=1}^{n} p_i u_i - f(u), \quad p \in (\mathbb{R}^n)^*,$$

where $p_i = \partial f / \partial u_i$, $i = 1, \ldots, n$. Conditions 1 and 2 become

$1'$. $(f_{u_i u_j}) > 0$, i.e., the Hessian of the function f is a positive-definite matrix;

$2'$. the gradient mapping $u \mapsto f_u(u)$ is surjective.

The proof of properties a and b is similar to the proof of Theorem 1.1, but all the formulas should be understood in the matrix sense. The geometric sense of $f^*(p)$ remains the same. This is the value by which the plane $z = \langle p, u \rangle$ should be lowered for this plane to become the tangent plane to the graph of the function $z = f(u)$.

For nonsmooth and nonconvex functions, instead of the Legendre transform, we should consider the Legendre–Young–Fenchel transform, which is defined as follows. Let $f \in \mathfrak{G}$. We show that $f^*(p)$ satisfies the relation

$$f^*(p) = \sup_u \left[\langle p, u \rangle - f(u) \right]. \tag{1.21}$$

Indeed, condition $1'$ implies that the function $\langle p, u \rangle - f(u)$ is strictly concave, and condition $2'$ implies that it has a unique stationary point $p = f_u(u)$ for every fixed p. Consequently, the maximum value is attained at this point; this proves (1.21).

Formula (1.21) is just the definition of the Legendre–Young–Fenchel transform for nonsmooth and nonconvex functions. This transform is also denoted by f^*.

Formula (1.21) implies the Young inequality

$$f(u) + f^*(p) \geq \langle p, u \rangle. \tag{1.22}$$

Definition 1.2. A function f is called *closed* if its epigraph is a closed set.

The following theorem is an analogue of the involutiveness property of the Legendre transform.

Theorem 1.2 (duality theorem). *If f is a convex and closed function, then $(f^*)^* = f$.*

The proof of this theorem can be found in [1], Sec. 2.6.

We note that in the case where a function f is not closed and convex, $(f^*)^*$ is the convex closure, i.e., the maximum function among all closed convex functions that do not exceed f.

Example 1.4. We find the Legendre transform of the function $f = u^2/2$. In this case, $p = u$ and $f^*(p) = pu - u^2/2 = p^2/2$.

Example 1.5. We find the Legendre–Young–Fenchel transform of the function $f = \gamma\sqrt{1 + |u|^2}$, where $\gamma > 0$ and $|u|^2 = \sum_{i=1}^{n} u_i^2$. It is easy to verify that

$$p_j = \frac{\gamma u_j}{\sqrt{1 + |u|^2}}. \tag{1.23}$$

Squaring and summing over j, we obtain $|p|^2(1 + |u|^2) = \gamma^2|u|^2$. This implies $|p| < \gamma$ and $|u|^2 = |p|^2/(\gamma^2 - |p|^2)$. Substituting this in (1.23), we have

$$\gamma u^j = p^j\sqrt{1 + \frac{|p|^2}{\gamma^2 - |p|^2}} = \frac{\gamma p^j}{\sqrt{\gamma^2 - |p|^2}}, \qquad u_j = \frac{p^j}{\sqrt{\gamma^2 - |p|^2}}.$$

Therefore, the following relations hold for $|p| < \gamma$:

$$f^*(p) = \langle p, u \rangle - f(u) = \frac{|p|^2}{\sqrt{\gamma^2 - |p|^2}} - \frac{\gamma^2}{\sqrt{\gamma^2 - |p|^2}} = -\sqrt{\gamma^2 - |p|^2}.$$

If $|p| = \gamma$, then $\sup_u [\langle p, u \rangle - f(u)] = 0$; if $|p| > \gamma$, then $\sup_u [\langle p, u \rangle - f(u)] = +\infty$. Therefore,

$$f^*(p) = \begin{cases} +\infty & \text{for } |p| > \gamma, \\ -\sqrt{\gamma^2 - |p|^2} & \text{for } |p| \leq \gamma. \end{cases}$$

Exercise. Find the Legendre–Young–Fenchel transform of the function $f = |u|^\alpha/\alpha$ for $\alpha > 1$, where $u \in \mathbb{R}^n$ and $|u| = \sqrt{\sum_{i=1}^{n} u_i^2}$.

2.2. Canonical Variables. As already mentioned, the system of Euler equations

$$-\frac{d}{dt} f_{\dot{x}^i} + f_{x^i} = 0, \quad i = 1, \ldots, n, \tag{1.24}$$

is of order $2n$. To reduce this system to the normal form, it is natural to require

$$\det\left(\frac{\partial^2 f}{\partial \dot{x}^i \partial \dot{x}^j}\right) \neq 0,$$

which is called the *Hilbert condition*. If this condition holds, then system (1.24) can be locally solved with respect to the second derivative $\ddot{x}(t)$, and we can take the first derivatives as new unknown functions. But there is another way, suggested by the English mathematician W. R. Hamilton.

For simplicity, we assume that the function f (as a function of $\dot{x}$ for any fixed values of t and x) satisfies conditions $1'$ and $2'$. Then, the inequality $(\partial^2 f/(\partial \dot{x}^i \partial \dot{x}^j)) > 0$ holds. The Legendre transform of the function f (as a function of $\dot{x}$ for fixed values of t and x) is denoted by $H(t, x, p)$, i.e.,

$$p = f_{\dot{x}}, \qquad H(t, x, p) = \langle p, \dot{x} \rangle - f(t, x, \dot{x}). \tag{1.25}$$

We write system (1.24) in these variables. Because

$$dH = \langle dp, \dot{x} \rangle + \langle p, d\dot{x} \rangle - f_t \, dt - f_x \, dx - f_{\dot{x}} \, d\dot{x}$$

and $\langle p, d\dot{x} \rangle = f_{\dot{x}} \, d\dot{x}$, we have $dH = \langle dp, \dot{x} \rangle - f_t \, dt - f_x \, dx$. By the theorem on the invariance of the first differential, we obtain

$$\frac{\partial H}{\partial p} = \dot{x}, \qquad \frac{\partial H}{\partial t} = -f_t, \qquad \frac{\partial H}{\partial x} = -f_x. \tag{1.26}$$

Using these formulas, we write system (1.24) in the form

$$\begin{cases} \dot{x} = H_p(t, x, p), \\ \dot{p} = -H_x(t, x, p). \end{cases} \tag{1.27}$$

System (1.27) is called the *canonical system* of ordinary differential equations. It has a number of remarkable properties (e.g., see [3], Sec. 35); we use some of them in what follows. In the case where the function f does not satisfy conditions $1'$ and $2'$, the canonical variables are introduced using the Legendre–Young–Fenchel transform.

2.3. Mechanical Meaning of the Canonical Variables. We consider a system of N material points with the masses m_i and the Cartesian coordinates $r_i = (x_i, y_i, z_i)$, $i = 1, \ldots, N$. We assume that the potential energy of this system is given by a function $U(r_1, \ldots, r_N)$. The kinetic energy of this system is $T = \frac{1}{2} \sum_{i=1}^{N} m_i \dot{r}_i^2$. In accordance with the minimum action principle, the motion of the system is realized along extremals of the action functional

$$\mathcal{L} = \int_{t_0}^{t_1} L(r, \dot{r}) \, dt, \tag{1.28}$$

where $L(r, \dot{r}) = T - U$. Extremals of this functional are determined by the Euler equation

$$-\frac{d}{dt} L_{\dot{r}} + L_r = 0 \tag{1.29}$$

for functional (1.28).

System (1.29) consists of $3N$ second-order equations. We write these equations in the canonical form. Formulas (1.25) imply $p_i = L_{\dot{r}_i} = m_i \dot{r}_i$, i.e., p_i in the Cartesian coordinate system coincides with the momentum of a material point. Furthermore, $H = \sum_i \dot{r}_i(m_i \dot{r}_i) - T + U = T + U$, i.e., the Hamiltonian H coincides with the total energy of the system. Because L does not explicitly depend on t, we have $H = \text{const}$, i.e., the total energy of the system is conserved during its motion. Equations (1.29) in the canonical form become

$$\begin{cases} \dot{r}_i = H_{p_i} = \dfrac{p_i}{m_i}, \\[2mm] \dot{p}_i = -H_{r_i} = -\dfrac{\partial U}{r_i}, \quad i = 1, \ldots, N. \end{cases}$$

In the canonical variables, the equations for the extremals are precisely the definitions of the momentum and Newton's second law ($-\partial U / \partial r_i$ is the force acting on the ith point).

2.4. Variation Formula for a Functional with Movable Endpoints.
The formula just introduced plays a crucial role everywhere in what follows.
We consider the integral functional

$$J = \int_{t_0}^{t_1} f(t, x(t), \dot{x}(t)) dt$$

defined on a one-parameter family of curves $x(t, \alpha)$, $\alpha \in \mathbb{R}$, $x \in C^2(\mathbb{R} \times \mathbb{R})$, whose ends vary when α varies. Values referring to the left endpoint are marked by the subscript 0, and to the right, by the subscript 1.

Let $t = t_0(\alpha)$, $x = x_0(\alpha)$ be a smooth curve along which the left endpoint moves, and let $t = t_1(\alpha)$, $x = x_1(\alpha)$ be a smooth curve along which the right endpoint moves. This means that

$$x(t_1(\alpha), \alpha) = x_1(\alpha), \qquad x(t_0(\alpha), \alpha) = x_0(\alpha). \tag{1.30}$$

We have

$$J(\alpha) = \int_{t_0(\alpha)}^{t_1(\alpha)} f(t, x(t, \alpha), \dot{x}(t, \alpha)) dt.$$

We compute $dJ(0)/d\alpha$ (a dot over the variable always denotes differentiation with respect to t):

$$x(t, 0) = \hat{x}(t), \qquad \dot{x}(t, 0) = \hat{\dot{x}}(t), \qquad t_1(0) = \hat{t}_1, \qquad t_0(0) = \hat{t}_0,$$

$$\frac{\partial x}{\partial \alpha}(t, 0) = h(t), \qquad \frac{\partial \dot{x}}{\partial \alpha}(t, 0) = \hat{\dot{h}}(t).$$

The latter formula is implied by the theorem that the partial derivatives are independent of the order of differentiation. Then

$$\frac{dJ(\alpha)}{d\alpha} = f(t_1(\alpha), x(t_1(\alpha), \alpha), \dot{x}(t_1(\alpha), \alpha)) \frac{dt_1}{d\alpha}$$

$$- f(t_0(\alpha), x(t_0(\alpha), \alpha), \dot{x}(t_0(\alpha), \alpha)) \frac{dt_0}{d\alpha}$$

$$+ \int_{t_0(\alpha)}^{t_1(\alpha)} \left(f_x \frac{\partial x}{\partial \alpha} + f_{\dot{x}} \frac{\partial \dot{x}}{\partial \alpha} \right) dt. \tag{1.31}$$

We introduce an abbreviating notation: $\hat{f}(t)$, or merely $\hat{f}$, denotes the function $f(t, \hat{x}(t), \hat{\dot{x}}(t))$; $\hat{f}_0$ denotes the value of this function at the point t_0, i.e., $f(t_0, \hat{x}(t_0), \hat{\dot{x}}(t_0))$; similarly, $\hat{f}_1 = f(t_1, \hat{x}(t_1), \hat{\dot{x}}(t_1))$. The symbols $\hat{f}_x$, $\hat{f}_{\dot{x}}$, $(\hat{f}_{\dot{x}})_0$, etc., are introduced similarly.

Substituting $\alpha = 0$ in (1.31), we obtain

$$\frac{dJ(0)}{d\alpha} = \hat{f}_1 \frac{dt_1}{d\alpha} - \hat{f}_0 \frac{dt_0}{d\alpha} + \int_{\hat{t}_0}^{\hat{t}_1} (\hat{f}_x h + \hat{f}_{\dot{x}} \dot{h}) \, dt. \tag{1.32}$$

In the ensuing deduction of formula (1.35), we assume that $\hat{f}_{\dot{x}}(\cdot) \in C^1([t_0, t_1])$.

We integrate the last summand in (1.32) by parts:

$$\frac{dJ(0)}{d\alpha} = \hat{f}_1 \frac{dt_1}{d\alpha} - \hat{f}_0 \frac{dt_0}{d\alpha} + \hat{f}_{\dot{x}} h|_{\hat{t}_0}^{\hat{t}_0} + \int_{\hat{t}_0}^{\hat{t}_1} \left(\hat{f}_x - \frac{d}{dt} \hat{f}_{\dot{x}} \right) h \, dt. \qquad (1.33)$$

We transform the term outside the integral. For this, we differentiate identities (1.30) with respect to α for $\alpha = 0$:

$$\dot{\hat{x}}(t_1) \frac{dt_1(0)}{d\alpha} + h(\hat{t}_1) = \frac{dx_1(0)}{d\alpha}, \qquad \dot{\hat{x}}(t_0) \frac{dt_0(0)}{d\alpha} + h(\hat{t}_0) = \frac{dx_0(0)}{d\alpha}.$$

Substituting $h(t_i)$, $i = 0, 1$, in (1.33), we obtain

$$\frac{dJ(0)}{d\alpha} = [\hat{f}_1 - (\hat{f}_{\dot{x}})_1 \dot{\hat{x}}(t_1)] \frac{dt_1}{d\alpha} - [\hat{f}_0 - (\hat{f}_{\dot{x}})_0 \dot{\hat{x}}(t_0)] \frac{dt_0}{d\alpha}$$

$$+ (\hat{f}_{\dot{x}})_1 \frac{dx_1(0)}{d\alpha} - (\hat{f}_{\dot{x}})_0 \frac{dx_0(0)}{d\alpha} + \int_{\hat{t}_0}^{\hat{t}_1} \left(\hat{f}_x - \frac{d}{dt} \hat{f}_{\dot{x}} \right) h \, dt. \qquad (1.34)$$

It is convenient to write formula (1.34) in the symbolic form

$$\frac{dJ(0)}{d\alpha} = (p dx - H dt) \Big|_{\hat{t}_0}^{\hat{t}_1} + \int_{\hat{t}_0}^{\hat{t}_1} \left(\hat{f}_x - \frac{d}{dt} \hat{f}_{\dot{x}} \right) h dt. \qquad (1.35)$$

In this notation, substituting the upper limit $\hat{t}_1$ in the expression $p dx - H dt$ means not only substituting $\hat{t}_1$ in the canonical variables p and H but also replacing (dt, dx) with the vector $(dt_1/d\alpha, dx_1/d\alpha)$ tangent to the curve along which the right endpoint of the family moves; a similar remark applies to substituting $\hat{t}_0$.

Remark. In what follows, formula (1.35) is used in only those cases where $\hat{x}(\cdot)$ is an optimal solution; by the Du Bois-Reymond Lemma, the above assumption on the smoothness of the function $\hat{f}_{\dot{x}}$ should hence be valid. We therefore use formula (1.35) without noting special stipulations.

The differential form $p dx - H dt$ is called the *Poincaré–Cartan form*. It plays the fundamental role in the canonical formalism and is repeatedly used in the further presentation.

2.5. Transversality Conditions in the Problem with Movable Endpoints. We replace the boundary conditions in Problem 1 by the more general conditions

$$\Phi(t_0, x(t_0)) = 0, \qquad \Psi(t_1, x(t_1)) = 0, \qquad (1.36)$$

where $\Phi : \mathbb{R}^{n+1} \to \mathbb{R}^k$ and $\Psi : \mathbb{R}^{n+1} \to \mathbb{R}^l$.

Problem 2. Among all curves $x(\cdot)$ of class C^1 that satisfy conditions (1.36), find a curve for which the functional J attains its minimum value.

Let $\hat{x}(t)$ with $t \in [\hat{t}_0, \hat{t}_1]$ be a solution to Problem 2. Then the same curve is a solution to Problem 1 with the fixed endpoints $\hat{x}(\hat{t}_0)$ and $\hat{x}(\hat{t}_1)$; hence, Euler equation (1.11) is satisfied for it. However, in order to find $\hat{t}_0$, $\hat{t}_1$, and also $2n$

parameters that characterize the desired extremal (i.e., in order to find $2n + 2$ parameters), we still have only $k + l$ boundary conditions $\Phi = 0$ and $\Psi = 0$.

To find additional boundary conditions that should hold on an optimal trajectory, we consider an arbitrary tangent vector $(\delta t_1, \delta x_1)$ to the manifold $\Psi = 0$ at the point $(\hat{t}_1, \hat{x}(\hat{t}_1))$. By the definition of a tangent vector (see [1], Sec. 2.3), there exists a smooth curve $\gamma : \alpha \mapsto (t_1(\alpha), x_1(\alpha))$ on the manifold Ψ whose tangent coincides with this vector, i.e., $(t_1(0), x_1(0)) = (\hat{t}_1, \hat{x}(\hat{t}_1))$ and $(dt_1(0)/d\alpha, dx_1(0)/d\alpha) = (\delta t_1, \delta x_1)$. Let $x(t, \alpha)$ denote a smooth, one-parameter family of curves that contains $\hat{x}(t)$ for $\alpha = 0$, the curve γ for $t = t_1(\alpha)$, and the point $\hat{x}(\hat{t}_0)$ for $t = t_0(\alpha)$, i.e., $x(t, 0) = \hat{x}(t)$, $x(t_1(\alpha), \alpha) = x_1(\alpha)$, and $x(\hat{t}_0, \alpha) = \hat{x}(\hat{t}_0)$.

The possibility of constructing such a family (which is sufficiently obvious geometrically) is formally justified in [6], Sec. 61. Substituting this family in the functional, taking into account that the function J attains its maximum value for $\alpha = 0$, and using (1.35), we obtain

$$(p dx - H dt)\Big|_{\hat{t}_0}^{\hat{t}_1} + \int_{\hat{t}_0}^{\hat{t}_1} \left(\hat{f}_x - \frac{d}{dt} \hat{f}_{\dot{x}} \right) h dt = 0.$$

Because the Euler equation is satisfied for $\hat{x}(t)$, the integral term vanishes. Also, the result of substituting $\hat{t}_0$ in the term outside the integral vanishes because all curves of the family pass through the point $(\hat{t}_0, \hat{x}(\hat{t}_0))$; hence, $dt_0/d\alpha = 0$ and $dx_0/d\alpha = 0$. Therefore,

$$p_1 \delta x_1 - H_1 \delta t_1 = 0 \qquad (1.37)$$

for any vector $(\delta t_1, \delta x_1)$ that is tangent to the manifold $\Psi = 0$ at the point $(\hat{t}_1, \hat{x}(\hat{t}_1))$.

In full analogy, we obtain

$$p_0 \delta x_0 - H_0 \delta t_0 = 0 \qquad (1.38)$$

for any vector $(\delta t_0, \delta x_0)$ tangent to the manifold $\Phi = 0$ at the point $(\hat{t}_0, \hat{x}(\hat{t}_0))$. Relations (1.37) and (1.38) are called the *transversality conditions*.

We verify that relations (1.36)–(1.38) give a complete set of conditions. The dimension of the manifold $\Psi = 0$ (in the regular case where the gradients of equations of the system $\Psi = 0$ are linearly independent) equals $n + 1 - l$, and relations (1.37) yield $n + 1 - l$ independent equations. Adding the l equations $\Psi = 0$ to them, we obtain $n + 1$ equations at the right endpoint (as well as at the left one). This yields the $2n + 2$ relations we need.

Example 1.6. We consider the minimization problem for the functional $\int_{t_0}^{t_1} \gamma(t, x)\sqrt{1 + \dot{x}^2}\, dt$, $x \in \mathbb{R}$, on the set of curves going from the point (t_0, x_0) to the terminal manifold $x = \phi(t)$. Let $(\delta t, \delta x)$ be a vector tangent to the terminal manifold. We write relations (1.37) for this problem. Using the result, which was obtained in Example 1.5, we have

$$p = \frac{\gamma \dot{x}}{\sqrt{1 + \dot{x}^2}}, \qquad H = \frac{-\gamma}{\sqrt{1 + \dot{x}^2}},$$

$$(p\delta x - H\delta t) = \frac{\gamma}{\sqrt{1 + \dot{x}^2}}(\dot{x}\delta x + \delta t) = 0. \qquad (1.39)$$

But $\delta x/\delta t = \phi'(t)$ is the tangent of the inclination angle of the tangent line. Therefore, condition (1.39), which has the form $\dot{x}(\delta x/\delta t) = -1$, coincides with the orthogonality condition for the extremal $x(\cdot)$ and the terminal manifold $x = \phi(t)$.

Example 1.7. We consider the problem of seeking a curve of minimum length on a Riemannian manifold M that joins a point $x_0 \in M$ with a submanifold $N \subset M$.

The length of a curve on M is given by formula (1.13):

$$\mathcal{L} = \int_{t_0}^{t_1} \sqrt{\sum_{i,j=1}^{n} g_{ij}(x(t))\frac{dx^i}{dt}\frac{dx^j}{dt}}\, dt,$$

where g_{ij} is the metric tensor. The canonical variables have the following forms for this functional:

$$p_j = \sum_{j=1}^{n} g_{ij}(x)\frac{dx^j}{dt}, \qquad H \equiv 0.$$

Let δx be an arbitrary vector tangent to the submanifold N at the endpoint $x(t_1)$ of the extremal. Transversality condition (1.37) becomes

$$\sum_{i=1}^{n} p_i \delta x^i = 0,$$

i.e., $\sum_{i,j=1}^{n} g_{ij}(x)(dx^j/dt)\delta x^i = 0$. This means that the vector dx/dt should be orthogonal (in the sense of the metric g_{ij}) to any vector δx that is tangent to N at the point where the extremal $x(\cdot)$ arrives at N. Therefore, the extremal should approach N at a right angle (in the metric g_{ij}).

2.6. Weierstrass–Erdmann Conditions. We begin with the following example.

Example 1.8 (minimum-area surface of revolution). We now find a curve $y = y(x)$ that joins two points $(x_0, y_0) = a$ and $(x_1, y_1) = b$ and gives the minimum area for the surface of revolution around the x axis.

The area of the surface of revolution is equal to

$$2\pi \int y\, ds = 2\pi \int_{x_0}^{x_1} y\sqrt{1 + \dot{y}^2}\, dx; \qquad (1.49)$$

therefore, we must minimize the functional $\int y\sqrt{1 + \dot{y}^2}\, dx$ under the conditions $y(x_0) = y_0$ and $y(x_1) = y_1$.

The energy functional has the form

$$-y\sqrt{1+\dot{y}^2} + \frac{y\dot{y}^2}{\sqrt{1+\dot{y}^2}} = C.$$

Elementary transformations yield $y^2 = C^2 + C^2\dot{y}^2$ or

$$\int \frac{Cdy}{\sqrt{y^2 - C^2}} = \int dx.$$

After the change of the variable $y = C\cosh\tau$, we obtain

$$y = C\cosh\frac{x+D}{C}. \tag{1.41}$$

If we rewrite the functional being minimized in the form

$$\int y\,ds = \int y(t)\sqrt{\dot{x}(t)^2 + \dot{y}(t)^2}\,dt,$$

then along with the already obtained solutions, the momentum integral with respect to x, i.e., $y\dot{x}/\sqrt{\dot{x}^2 + \dot{y}^2} = C$, gives vertical lines $x = \mathrm{const}$ for $C = 0$.

Curves (1.41) are called *catenaries* in mechanics, because exactly this form is taken by a heavy, flexible, homogeneous, inextensible cable suspended by its ends. To make this fact clearer, it suffices to note that formula (1.40) defines the coordinate of the center of gravity of a homogeneous curve and the cable tends to occupy a position for which this coordinate is minimum. However, this functional should be minimized on the set of curves of a given length (the cable is inextensible). A complete solution to this problem requires applying the Lagrange multiplier rule, which can be found in [1].

The surface of revolution of a catenary is called the *catenoid*. We reveal whether it is always possible to draw curve (1.41) joining the points a and b. For example, let $a = (q, 1)$ and $b = (-q, 1)$. Then (because the boundary conditions are symmetrical with respect to the origin), $D = 0$, and C is found as a solution to the equation $C\cosh(q/C) = 1$. We let $1/C = z$ and solve the equation $\cosh qz = z$.

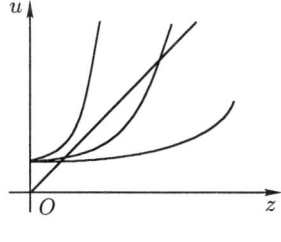

Fig 1.2

Figure 1.2 depicts the graphs of the curves $u = z$ and $u = \cosh qz$ for various values of q. For a sufficiently large q, these curves do not intersect, i.e., in this case, there is no solution to the Euler equation that passes through a and b.

The problem of seeking a surface of minimum area is a particular case of the Plateau problem, which is formulated as follows: among all surfaces that have a given contour K as their boundary, find a surface having the minimum area. From standpoint of the natural sciences, we here speak of seeking the shape of a soap film that covers the contour K (the soap film tends to occupy the position having the minimum surface area because of the strong surface tension). The surface of revolution occurs in the case where the contour K consists of two circles obtained by rotating the points a and b around the x axis. As we revealed, the catenoid serves as a solution. But how can we explain the absence of a solution for large values of q?

To answer this question, we turn to the general theory. We consider the class of admissible curves in Problem 1 and seek the minimum of the functional not in the class of absolutely continuous curves but in the class $\mathcal{P}C^1([t_0, t_1])$ of piecewise smooth curves.

Definition 1.3. The space $\mathcal{P}C^1([t_0, t_1])$ is the set of all functions $x(t)$ that are continuous on $[t_0, t_1]$ and such that for each $x(t)$, there exist finitely many points $\tau_1, \ldots, \tau_n \in [t_0, t_1]$, such that the restriction of $x(t)$ to any of the closed intervals $[t_0, \tau_1], \ldots, [\tau_i, \tau_{i+1}], \ldots, [\tau_n, t_1]$ is a smooth function on this closed interval. This set is endowed with the topology of a subspace of the space $C([t_0, t_1])$.

Exercise. Prove that the space $\mathcal{P}C^1([t_0, t_1])$ endowed with the metric induced by $C([t_0, t_1])$ is not a Banach space.

Obviously, for a solution $\hat{x}(\cdot) \in \mathcal{P}C^1([t_0, t_1])$ to Problem 1, each smooth piece of $\hat{x}(t)$, $t \in [\tau_i, \tau_{i+1}]$, should satisfy the Euler equation because it should give the minimum value to the functional J with the boundary conditions $(\tau_i, \hat{x}(\tau_i))$, $(\tau_{i+1}, \hat{x}(\tau_{i+1}))$. However, most such extremals do not guarantee the minimum value for the functional J. In fact, the following additional condition should hold at each corner point.

Weierstrass–Erdmann Condition. *The canonical variables should be continuous at a corner point.*

This condition is nontrivial because the expression defining the canonical variables contains $\dot{x}$, which has a discontinuity at the corner point.

Proof. Obviously, it suffices to consider the case of a single corner point. Let $\hat{x}(\cdot)$ have the breakpoint $(\tau, \hat{x}(\tau))$. We draw an arbitrary line l through this point with the directing vector $(\delta t, \delta x)$. As above, we consider a smooth, one-parameter family of curves $x(t, \alpha)$ such that $x(t, 0) = \hat{x}(t)$, $x(t_0, \alpha) = \hat{x}(t_0)$, and $x(t_1, \alpha) = \hat{x}(t_1)$ and such that for a fixed α, the function $x(t, \alpha)$ is smooth everywhere except at the intersection $(\tau(\alpha), x(\tau(\alpha), \alpha))$ with the line l.

We represent the integral $J(\alpha)$ as the sum of integrals over segments lying before and after the intersection of the curve $x(t, \alpha)$ with the line l:

$$J(\alpha) = \int_{t_0}^{\tau(\alpha)} f(t, x(t, \alpha), \dot{x}(t, \alpha)) dt + \int_{\tau(\alpha)}^{t_1} f(t, x(t, \alpha), \dot{x}(t, \alpha)) dt.$$

We apply formula (1.35). Because the Euler equation is satisfied on the curve $\hat{x}(t) = x(t, 0)$ and the substitution of t_0 and t_1 in the Poincaré–Cartan form also yields zero (all the curves $x(t, \alpha)$ pass through the points $(t_0, \hat{x}(t_0))$ and $(t_1, \hat{x}(t_1))$), we obtain

$$\frac{dJ(0)}{d\alpha} = [p(\tau - 0)\delta x - H(\tau - 0)\delta t] - [p(\tau + 0)\delta x - H(\tau + 0)\delta t]$$

$$= [p(\tau - 0) - p(\tau + 0)]\delta x - [H(\tau - 0) - H(\tau + 0)]\delta t = 0. \quad (1.42)$$

Because δt and δx are arbitrary, it follows from (1.42) that the canonical variables are continuous at the point τ.

We now return to Example 1.8 and seek piecewise smooth solutions to the problem of the minimal surface of revolution. For this, we write the canonical variables:

$$p = \frac{y\dot{y}}{\sqrt{1 + \dot{y}^2}}, \qquad H = \frac{-y}{\sqrt{1 + \dot{y}^2}}.$$

It can be easily shown that p and H can remain continuous for a jump of $\dot{y}$ iff $y = 0$. But none of the lines (1.41) arrives at the x-axis. Therefore, broken extremals are only lines that consist of a segment of the x axis and the two vertical segments $x = $ const. Consequently, the answer to the question of what happens to the soap film when the two circles forming the contour K go away from each another (i.e., the distance $2q$ between them increases) is very simple: the soap bubble blows up and the film covers just the two circles. This is the surface of revolution of the found broken extremal. The segment of the axis of revolution joining the two disks has zero area and serves only as a reminder of the connectivity of the catenoid. In the general Plateau problem, such lower-dimensional cells play a very important role (see [32]).

2.7. Hamilton–Jacobi Equation. We now consider the minimization problem for functional (1.3) with boundary conditions (1.2) from a slightly different viewpoint. We fix an initial point $(t_0, x_0) = a$ and introduce a function $S(t_1, x_1)$ that is equal to the minimum value of the functional J on trajectories connecting the points $a = (t_0, x_0)$ and $b = (t_1, x_1)$. We assume that in a certain domain of the space of the variables t_1 and x_1, this minimum value is attained at curves that depend smoothly on t_1 and x_1. We find a partial differential equation that the function S satisfies.

We consider an arbitrary, smooth, one-parameter family of trajectories $x(t, \alpha)$ with a fixed left endpoint and a movable right endpoint such that $x(t, 0) = \hat{x}(t)$ (we recall that $\hat{x}(t)$ denotes the solution of the variation problem considered). Applying the variation formula for the functional with movable boundary conditions to this family, we obtain the equation $\partial S / \partial \alpha =$

$p(\partial x_1/\partial \alpha) - H(\partial t_1/\partial \alpha)$. Taking into account that the curve along which the right endpoint moves is arbitrarily chosen, we obtain the relation $dS = p\,dx_1 - H\,dt_1$.

The differential of the function $S(t,x)$ equals the Poincaré–Cartan form. We have thus obtained

$$\frac{\partial S(t,x)}{\partial t} + H\left(t, x, \frac{\partial S(t,x)}{\partial x}\right) = 0. \tag{1.43}$$

This is just the Hamilton–Jacobi equation.

Example 1.9. We find the Hamilton–Jacobi equation for the functional

$$\int_{t_0}^{t_1} \gamma(t,x)\sqrt{1 + \sum_{i=1}^{n}(\dot{x}^i)^2}\,dt.$$

The Hamiltonian of this functional was found in Example 1.5:

$$H(t,x,p) = -\sqrt{\gamma^2 - \sum_{i=1}^{n} p_i^2}.$$

The Hamilton–Jacobi equation has the form

$$\frac{\partial S}{\partial t} = \sqrt{\gamma^2 - \sum_{i=1}^{n}\left(\frac{\partial S}{\partial t^i}\right)^2}$$

or

$$\left(\frac{\partial S}{\partial t}\right)^2 + \sum_{i=1}^{n}\left(\frac{\partial S}{\partial x^i}\right)^2 = \gamma^2(t,x).$$

This equation is called the *eikonal equation for an inhomogeneous medium*.

Exercise. Find the Hamilton–Jacobi equation for the action functional

$$\int \sum_{i,j=1}^{n} g_{ij}(x)\frac{dx^i}{dt}\frac{dx^j}{dt}\,dt$$

on a Riemannian manifold.

§3. Theory of the Second Variation

3.1. Problem of the Second Variation. We return to the deduction of the Euler equation (see Sec. 1.1). A solution $\hat{x}(t)$ to the minimization problem for functional (1.3),

$$J(x(\cdot)) = \int_{t_0}^{t_1} f(t, x(t), \dot{x}(t)) \, dt,$$

was included in a smooth, one-parameter family of curves $\hat{x}(t) + \lambda h(t)$, $\lambda \in \mathbb{R}$, $h \in C_0^1([t_0, t_1])$ (because t_0 and t_1 are fixed in what follows, we write merely C_0^1 instead of $C_0^1([t_0, t_1])$). Computing the value of functional (1.3) on curves of this family, we obtained the function

$$\phi(\lambda) = \int_{t_0}^{t_1} f(t, \hat{x}(t) + \lambda h(t), \dot{\hat{x}}(t) + \lambda \dot{h}(t)) \, dt,$$

which has a local minimum for $\lambda = 0$. Until now, we have considered the condition $\phi'(0) = 0$. We now turn to the condition $\phi''(0) \geq 0$. If $f \in C^2$, then the value $\phi''(0)$ can be found by differentiating[1] the integrand:

$$\phi''(0) = \int_{t_0}^{t_1} \sum_{i,j=1}^{n} \left(\hat{f}_{\dot{x}^i \dot{x}^j} \dot{h}_i \dot{h}_j + 2 \hat{f}_{x^i \dot{x}^j} \dot{h}_j h_i + \hat{f}_{x^i x^j} h_i h_j \right) dt. \qquad (1.44)$$

Expression (1.44) is called the *second variation of the functional J* at the point $\hat{x}(t)$ and is denoted by $\delta^2 J$. The matrix $\hat{f}_{\dot{x}^i \dot{x}^j}$ is denoted by $A(t)$, the matrix $\hat{f}_{x^i \dot{x}^j}$ by $C(t)$, and the matrix $\hat{f}_{x^i x^j}$ by $B(t)$. The matrices $A(t)$ and $B(t)$ are symmetrical according to the theorem that mixed derivatives are independent of the order of differentiation. We note that the matrix $C(t)$ can be asymmetrical in general. In this notation, functional (1.44) is written as

$$\delta^2 J = \int_{t_0}^{t_1} \left(\langle A\dot{h}, \dot{h} \rangle + 2 \langle C\dot{h}, h \rangle + \langle Bh, h \rangle \right) dt. \qquad (1.45)$$

We say that the functional $\delta^2 J$ is *positive semidefinite* $(\delta^2 J \geq 0)$ if it does not assume negative values for functions $h \in C_0^1$. The functional $\delta^2 J$ is *positive definite* $(\delta^2 J > 0)$ if it is positive semidefinite and assumes a zero value only for the function $h(t) = 0$.

Because $\phi''(0) \geq 0$, the positive semidefiniteness of the functional of the second variation $\delta^2 J$ is a necessary minimality condition for functional (1.3).

---

[1] As before, $\hat{f}_{x^i x^j}$ denotes the expression $f_{x^i x^j}(t, \hat{x}(t), \dot{\hat{x}}(t))$. The sense of the symbols $\hat{f}_{\dot{x}^i x^j}$ and $\hat{f}_{\dot{x}^i \dot{x}^j}$ is analogous.

3.2. Legendre Necessary Condition.

Definition 1.4. The functional $\delta^2 J$ satisfies the *Legendre condition* if for all $t \in [t_0, t_1]$, the matrix $A(t)$ is positive semidefinite $(A(t) \geq 0)$.

Theorem 1.3. *The Legendre condition is a necessary condition for the positive semidefiniteness of the functional $\delta^2 J$.*

Proof. We suppose that the Legendre condition is violated at a point $\tau \in (t_0, t_1)$, i.e., there exists a vector $\xi \in \mathbb{R}^n$ such that $\langle A\xi, \xi \rangle < 0$. We choose a number σ such that $t_0 < \tau - \sigma < \tau + \sigma < t_1$ and consider the function

$$\chi(t) = \begin{cases} \sigma - |t - \tau| & \text{for } |t - \tau| < \sigma, \\ 0 & \text{for } |t - \tau| \geq \sigma. \end{cases}$$

Computing the value of the functional $\delta^2 J$ at the function $h(t) = \chi(t)\xi$, we obtain

$$\delta^2 J(\chi(t)\xi) = \int_{\tau-\sigma}^{\tau+\sigma} (\langle A\xi, \xi \rangle + 2\, \mathrm{sgn}(\tau - t)\langle C\xi, \xi \rangle \chi(t)$$
$$+ \langle B\xi, \xi \rangle \chi^2(t))\, dt. \tag{1.46}$$

We consider the first summand in (1.46). Because $A(t)$ is continuous and $\langle A\xi, \xi \rangle < 0$, there exists k such that $\langle A(t)\xi, \xi \rangle < -k$ for $t \in [\tau - \sigma, \tau + \sigma]$ when σ is sufficiently small. The other summands in (1.46) tend uniformly to zero as σ tends to zero, and hence $\delta^2 J < 0$ for a sufficiently small σ.

Because the functional $\delta^2 J$ is negative not at a smooth but at a piecewise smooth function, it suffices to use the lemma on rounding angles (see [1], Sec. 1.4) to complete the proof. This lemma states that the value of an integral functional for any piecewise smooth function can be approximated with arbitrary accuracy by its values on smooth functions satisfying the same boundary conditions. Therefore, if $\delta^2 J$ assumes a negative value for a certain piecewise smooth function, there exists a smooth function for which this functional is negative; this contradicts the positive semidefiniteness of $\delta^2 J$.

We have thus proved that the matrix $A(t)$ is nonnegative at interior points of the closed interval $[t_0, t_1]$. The fact that it is nonnegative at the ends of this closed interval is a consequence of its continuity. $\qquad \blacksquare$

As a consequence of the theorem just proved, we obtain the following necessary minimality condition for functional (1.3).

Theorem 1.4 (Legendre necessary condition). *If a local minimum of the functional J is attained at an extremal $\hat{x}(t)$, then for all $t \in [t_0, t_1]$, the matrix $\hat{f}_{\dot{x}^i \dot{x}^j}(t)$ is positive semidefinite.*

Definition 1.5. The functional $\delta^2 J$ satisfies the *strengthened Legendre condition* if for all $t \in [t_0, t_1]$, the matrix $A(t)$ is positive definite.

In connection with the strengthened Legendre condition, we present a *mistaken* argument due to Lagrange who thought that the Legendre condition is sufficient for the positive semidefiniteness of the functional $\delta^2 J$.

Lagrange's argument. We consider the case of a scalar-valued function x. Then

$$\int_{t_0}^{t_1} 2C\dot{h}h\,dt = -\int_{t_0}^{t_1} \dot{C}h^2\,dt,$$

and the functional $\delta^2 J$ (see (1.45)) can be written in the form

$$\delta^2 J = \int_{t_0}^{t_1} [p(t)\dot{h}^2(t) + q(t)h^2(t)]\,dt.$$

We let $p(t) > 0$, i.e., we assume that the strengthened Legendre condition holds. Adding the summand

$$\frac{d}{dt}[w(t)h^2(t)]$$

to the integrand does not change the value of $\delta^2 J$, because $h(t) \in C_0^1$, and hence

$$\int_{t_0}^{t_1} \frac{d}{dt}[w(t)h^2(t)]\,dt = w(t)h^2(t)\Big|_{t_0}^{t_1} = 0.$$

We choose a function $w(t)$ such that the integrand $p\dot{h}^2 + 2w\dot{h}h + (\dot{w} + q)h^2$ becomes a full square, i.e., find w from the equation

$$\dot{w} + q = p^{-1}w^2. \tag{1.47}$$

Then

$$\delta^2 J = \int_{t_0}^{t_1} (p^{1/2}\dot{h} + p^{-1/2}wh)^2\,dt \geq 0.$$

The flaw is that contrary to the implicit assumption in this argument, Riccati equation (1.47) may have no solution that is continuous on the whole closed interval $[t_0, t_1]$. The theorem on the continuation of a solution does not hold for this equation, and the solution can blow up at a finite time as the following example shows.

Indeed, let

$$\delta^2 J = \int_{t_0}^{t_1} (\dot{h}^2 - h^2)\,dt.$$

Then the Riccati equation has the form $\dot{w} - 1 = w^2$. This implies $\arctan w = t + C$, i.e., $w = \tan(t + C)$. We can see that if the length of the closed interval $[t_0, t_1]$ is greater than π, then $w(t)$ has a point of discontinuity on $[t_0, t_1]$.

3.3. The Associated Problem and the Definition of a Conjugate Point. We use the above results to examine the positive semidefiniteness of the quadratic functional $\delta^2 J$. We note that negative definiteness of the functional $\delta^2 J$ is equivalent to the condition that $h(\cdot) = 0$ yields the minimum value of the functional $\delta^2 J$. Therefore, we consider the following *associated extremal problem*:

$$\delta^2 J = \int_{t_0}^{t_1} [\langle A\dot{h}, \dot{h}\rangle + 2\langle C\dot{h}, h\rangle + \langle Bh, h\rangle]dt \to \inf, \quad h(\cdot) \in C_0^1.$$

We write the Euler equation for the associated problem:

$$-\frac{d}{dt}[A\dot{h} + C^T h] + [C\dot{h} + Bh] = 0. \tag{1.48}$$

Equation (1.48) is called the Jacobi equation for initial Problem 1.

Exercise. Show that the Jacobi equation is the variational equation for Euler equation (1.11).

Lemma 1.2. *The functional $\delta^2 J$ vanishes for any solution $h(\cdot) \in C_0^1$ to Eq. (1.48).*

Proof. We write $\delta^2 J$ in the form

$$\delta^2 J = \int_{t_0}^{t_1} [\langle (A\dot{h} + C^T h), \dot{h}\rangle + \langle (C\dot{h} + Bh), h\rangle]dt.$$

Integrating the first term by parts and taking into account that the term outside the integral vanishes (because $h(\cdot) \in C_0^1([t_0, t_1])$), we obtain

$$\int_{t_0}^{t_1} \left\langle \left[-\frac{d}{dt}(A\dot{h} + C^T h) + (C\dot{h} + Bh) \right], h \right\rangle dt = 0;$$

moreover, the latter relation is implied by the fact that $h(\cdot)$ is a solution to the Jacobi equation.

We consider a matrix solution $U(t)$ to Eq. (1.48):

$$-\frac{d}{dt}[AU' + C^T U] + [CU' + BU] = 0, \tag{1.49}$$

where $U(t)$ is an $n \times n$ matrix satisfying the initial conditions

$$U(t_0) = 0, \qquad U'(t_0) = 1. \tag{1.50}$$

This means that the ith column of the matrix $U(t)$ is a solution $h_i(t)$ to Eq. (1.48) that satisfies the conditions $h_i(t_0) = 0$ and $\dot{h}_i(t_0) = e_i$, where e_i is a unit basis vector.

Definition 1.6. A point $\tau > t_0$ is called a *conjugate point to t_0* if there exists a nontrivial solution $\tilde{h}(t)$ to Eq. (1.48) such that $\tilde{h}(t_0) = \tilde{h}(\tau) = 0$.

Proposition 1.1. *A point $\tau > t_0$ is conjugate to t_0 iff* $\det U(\tau) = 0$.

Proof. Let $\det U(\tau) = 0$. Then there exists a nontrivial n-tuple of constants c_i, $i = 1, \ldots, n$, such that the linear combination of the columns $h_i(\tau)$ of the matrix $U(\tau)$ with the coefficients c_i vanishes: $\sum c_i h_i(\tau) = 0$. We consider the function $\tilde{h}(\tau) = \sum c_i h_i(t)$, which is a solution to system (1.48). This solution is nontrivial because its derivative is different from zero at the point t_0 and this solution vanishes at $t = t_0$ as well as at $t = \tau$.

Now, let τ be a conjugate point. Then the solution $\tilde{h}(\cdot)$ can be represented as a linear combination of solutions $h_i(\cdot)$ by expanding $\dot{\tilde{h}}_i(t_0)$ in the basis e_i: $\tilde{h}(t) = \sum c_i h_i(t)$; moreover, the n-tuple c_i is nontrivial. Because $\tilde{h}(\tau) = 0$, we have a nontrivial linear combination of columns of the matrix $U(\tau)$ that yields a zero column. Consequently, $\det U(\tau) = 0$.

3.4. Necessary Conditions for the Positive Semidefiniteness of $\delta^2 J$.

Theorem 1.5. *Let the strengthened Legendre condition $(A(t) > 0)$ hold. If the quadratic functional $\delta^2 J$ is positive semidefinite, then the interval (t_0, t_1) contains no points conjugate to the point t_0.*

Proof. We suppose the contrary. Let $\tau \in (t_0, t_1)$ be a point conjugate to the point t_0. Then, by definition, there exists a nontrivial solution $\tilde{h}(t)$ to Jacobi equation (1.48) that vanishes for $t = t_0$ and $t = \tau$: $\tilde{h}(t_0) = \tilde{h}(\tau) = 0$. We extend this function to the closed interval $[\tau, t_1]$ by zero:

$$\overline{h}(t) = \begin{cases} \tilde{h}(t) & \text{for } t \in [t_0, \tau], \\ 0 & \text{for } t \in [\tau, t_1]. \end{cases}$$

Applying Lemma 1.2 to the function $\tilde{h}(t)$ and the closed interval $[t_0, \tau]$, we can easily verify that $\delta^2 J(\overline{h}(\cdot)) = 0$. Because $\delta^2 J \geq 0$, we find that $\overline{h}(\cdot)$ is a solution to the associated problem that has τ as a breakpoint. Consequently, the Weierstrass–Erdmann condition should hold at this point. We write this condition. The momentum in the associated problem is equal to $2A\dot{h} + 2C^\top h$. Because $\tilde{h}(\tau) = 0$ and the matrix A is nonsingular, the continuity of the momentum implies continuity of $\dot{\overline{h}}$. But $\dot{\overline{h}}(\tau + 0) = 0$. Consequently, $\dot{\overline{h}}(\tau - 0) = 0$. The function $\tilde{h}(\cdot)$ on the closed interval $[t_0, \tau]$ is a solution to the second-order equation (1.48) and satisfies the conditions $\tilde{h}(\tau) = 0$ and $\dot{\tilde{h}}(\tau) = 0$ at the point τ. Consequently, $\tilde{h}(t) \equiv 0$, which contradicts the definition of a conjugate point.

As a direct consequence of Theorem 1.5, we obtain the following theorem.

Theorem 1.6 (Jacobi necessary condition). *Let $\hat{x}(\cdot)$ be a solution to Problem 1. If the strengthened Legendre condition $(\hat{f}_{\dot{x}\dot{x}} > 0)$ holds for $\hat{x}(\cdot)$, the interval (t_0, t_1) contains no points conjugate to t_0.*

Definition 1.7. We say that the *Jacobi condition holds* for the functional $\delta^2 J$ if the interval (t_0, t_1) contains no points conjugate to t_0. We say that the *strengthened Jacobi condition holds* if a semiopen interval $(t_0, t_1]$ does not contain points conjugate to t_0.

§4. Riccati Equation

4.1. Sufficient Conditions for the Positive Definiteness of $\delta^2 J$.

Theorem 1.7. *Let the strengthened Legendre condition and the strengthened Jacobi condition hold for the functional $\delta^2 J$. Then $\delta^2 J > 0$.*

Proof. We consider the quadratic form $\langle W(t)h, h \rangle$ with the symmetrical matrix $W(t)$. Our goal is to find $W(t)$ satisfying the following conditions:

a. adding the summand $d\langle W(t)h, h \rangle/dt$ to the integrand does not change $\delta^2 J$, i.e.,

$$\int_{t_0}^{t_1} \frac{d}{dt} \langle W(t)h, h \rangle dt = 0;$$

b. adding $d\langle W(t)h, h \rangle/dt$ to the integrand transforms it into a full (scalar) square.

Condition a is satisfied, e.g., if $W(t)$ is a smooth function defined on the entire closed interval $[t_1, t_1]$ and $h \in C_0^1$. We use a more general sufficient condition below (see p. 34).

We find the condition that should be imposed on the matrix W in order to satisfy condition b. After the addition of $d\langle W(t)h, h \rangle/dt$, the integrand becomes

$$\begin{aligned}
\langle A\dot{h}, \dot{h} \rangle &+ 2\langle C\dot{h}, h \rangle + \langle Bh, h \rangle \\
&+ 2\langle W\dot{h}, h \rangle + \langle \dot{W}h, h \rangle = [A^{1/2}\dot{h} + A^{-1/2}(C^{\mathsf{T}} + W)h]^2 \\
&\qquad\qquad - [A^{-1/2}(C^{\mathsf{T}} + W)h]^2 \\
&\qquad\qquad + \langle (B + \dot{W})h, h \rangle, \qquad (1.51)
\end{aligned}$$

where $A^{1/2}$ is a symmetrical positive-definite matrix whose square is equal to A. (To verify that such a matrix exists, it suffices to reduce the matrix A to the diagonal form, extract the arithmetic root from the eigenvalues on the principal diagonal, and return to the original basis.)

Expression (1.51) is a full square if

$$\begin{aligned}
\langle (\dot{W} + B)h, h \rangle &= \langle A^{-1/2}(C^{\mathsf{T}} + W)h, A^{-1/2}(C^{\mathsf{T}} + W)h \rangle \\
&= \langle (C + W)A^{-1}(C^{\mathsf{T}} + W)h, h \rangle,
\end{aligned}$$

i.e., the matrix W should be a solution to the equation

$$\dot{W} + B = (C + W)A^{-1}(C^\mathsf{T} + W). \tag{1.52}$$

Equation (1.52) is called the (matrix) *Riccati equation* for the functional $\delta^2 J$. We must find a symmetrical matrix $W(t)$ that is a solution to Riccati equation (1.52) on the closed interval $[t_0, t_1]$ and satisfies condition a.

We will return to the proof of Theorem 1.7 after we have proved several auxiliary assertions. As explained in the introduction, the scalar Riccati equation arises when the order of a linear second-order differential equation is reduced. A similar relation between solutions to matrix Jacobi equation (1.49) and matrix Riccati equation (1.52) holds. In Lemmas 1.3–1.5, we show that only one solution to Riccati equation (1.52) can be constructed using a solution $U(t)$ to Jacobi equation (1.49) satisfying initial conditions (1.50). (This solution has a singularity at the point t_0.)

Lemma 1.3. *Let $U(t)$ be a matrix solution to Jacobi equation (1.49) that does not degenerate on the semiopen interval $(t_0, t_1]$. Then*

$$W(t) = -(AU' + C^\mathsf{T}U)U^{-1} \tag{1.53}$$

is a matrix solution to Riccati equation (1.52).

Proof. The proof is a direct verification. The right-hand side of (1.52) has the form

$$(C - AU'U^{-1} - C^\mathsf{T})A^{-1}(C^\mathsf{T} - AU'U^{-1} - C^\mathsf{T}) = -(C - AU'U^{-1} - C^\mathsf{T})U'U^{-1}.$$

Substituting (1.53) in the left-hand side of (1.52), we obtain

$$B - (AU' + C^\mathsf{T}U)'U^{-1} + (AU' + C^\mathsf{T}U)U^{-1}U'U^{-1}.$$

Using Jacobi equation (1.49) to transform the second summand, we obtain

$$B - (CU' + BU)U^{-1} + AU'U^{-1}U'U^{-1} + C^\mathsf{T}U'U^{-1} =$$
$$= -CU'U^{-1} + AU'U^{-1}U'U^{-1} + C^\mathsf{T}U'U^{-1},$$

which coincides with the right-hand side of (1.52). The lemma is proved.

The solution $U(t)$ to the Jacobi equation, which is used in the definition of a conjugate point, is nondegenerate only on the semiopen interval $(t_0, t_1]$. At the point t_0 itself, we have $U(t_0) = 0$ and $U'(t_0) = I$ (see (1.50)). Therefore, $W(t)$ constructed using this matrix $U(t)$ has a singularity at the point t_0. We reveal its form.

Lemma 1.4. *The matrix $W(t)$ constructed according to formula (1.53) using the solution $U(t)$ to Eq. (1.49) with initial conditions (1.50) satisfies the relation $W(t) = -A(t_0)/(t - t_0) + \Phi(t)$ in a neighborhood of the point t_0 where the matrix $\Phi(t)$ is continuous in the neighborhood of the point t_0.*

Proof. We use conditions (1.50) to expand the matrices $U(t)$, $U^{-1}(t)$, and $U'(t)$ by the Taylor formula in a neighborhood of the point t_0. We have

$$U'(t) = I + (t - t_0)R_1(t), \qquad U(t) = (t - t_0)[I + (t - t_0)R_2(t)],$$

and hence $U^{-1}(t) = [I + (t - t_0)R_3(t)]/(t - t_0)$. Here, the matrices $R_i(t)$, $i = 1, 2, 3$, are continuous in a neighborhood of the point t_0. Substituting these expansions in formula (1.53) completes the proof of the lemma.

Therefore, $W(t)$ has a simple pole with the residue $(-A(t_0))$ at the point t_0. We prove one more auxiliary result from the theory of ordinary differential equations.

Lemma 1.5. *Let*

$$\dot{x} = x + f(x), \tag{1.54}$$

where $x \in \mathbb{R}^n$, $f \in C^3(V)$, and V is a domain in $\mathbb{R}^n$ that contains the point 0, be a system of ordinary differential equations. Let there exist a constant $K > 0$ such that $|f(x)| \leq K|x|^2$ for all $x \in V$. Then for any $l \in \mathbb{R}^n$, $|l| = 1$, there exists a unique phase trajectory $x(t)$ of system (1.54) such that $x(t) \to 0$ and $\dot{x}(t)/|\dot{x}(t)| \to l$ as $t \to -\infty$.

Proof. We introduce the spherical coordinates $x = r\theta$, $r \in \mathbb{R}^1_+$ and $\theta \in S^{n-1}$, where S^{n-1} is the $(n-1)$-dimensional unit sphere. System (1.54) takes the form

$$\dot{r}\theta + r\dot{\theta} = r\theta + f(r\theta). \tag{1.55}$$

The estimate $|f| \leq K|x|^2$ and the condition $f \in C^3(V)$ allow us to write the function f in the form $f(r, \theta) = r^2\phi(r, \theta)$, where $\phi \in C^1(V)$. Taking the inner product of Eq. (1.55) with θ and taking the relations $\langle \theta, \theta \rangle = 1$ and $\langle \theta, \dot{\theta} \rangle = 0$ into account, we obtain

$$\dot{r} = r + r^2 \langle \theta, \phi(r, \theta) \rangle. \tag{1.56}$$

Substituting (1.56) in (1.55) yields

$$\dot{\theta} = -r\theta \langle \theta, \phi(r, \theta) \rangle + r\phi(r, \theta).$$

We obtain the system

$$\begin{cases} \dot{r} = r + r^2 \langle \theta, \phi(r, \theta) \rangle, \\ \dot{\theta} = -r\theta \langle \theta, \phi(r, \theta) \rangle + r\phi(r, \theta). \end{cases} \tag{1.57}$$

It follows from the first equation of this system that $\dot{r} > 0$ for a sufficiently small r, i.e., in a sufficiently small neighborhood $\widetilde{V}$ of the origin, r tends to zero as $t \to -\infty$. Therefore, we can take r as an independent variable in the domain $\widetilde{V}$. As a result, we have

$$\frac{d\theta}{dr} = \frac{-\theta \langle \theta, \phi(r, \theta) \rangle + \phi(r, \theta)}{1 + r \langle \theta, \phi(r, \theta) \rangle}. \tag{1.58}$$

The right-hand side of Eq. (1.58) is a continuous function of r and θ in the neighborhood $\widetilde{V}$. Let $\chi(r,\theta)$ denote this function. The condition $\theta(0) = l$ corresponds to the boundary conditions $\dot{x}(t)/|\dot{x}(t)| \to l$ as $t \to -\infty$. We have

$$\frac{d\theta}{dr} = \chi(r,\theta), \qquad \theta(0) = l. \tag{1.59}$$

The assertion of the lemma is now implied by the existence and uniqueness theorem for the solution to Cauchy problem (1.59).

Exercise. Explain why a similar proof is not applicable to the system $\dot{x} = Kx + f(x)$ where $K \neq I$.

The meaning of Lemma 1.5 can be explained as follows. The linear system of equations $\dot{x} = x$ has rays emanating from the origin as its trajectories. The addition of a small summand $f(x)$ to the right-hand side of this equation distorts these rays but does not change the qualitative picture in a neighborhood of the origin in the phase space.

We now state the inversion of Lemmas 1.3 and 1.4 by constructing a solution to the Jacobi equation using a solution to the Riccati equation.

Lemma 1.6. *Let $W(t)$ be a solution to Riccati equation (1.52) satisfying the condition*

$$W(t) = -\frac{A(t_0)}{(t - t_0)} + \Phi(t), \tag{1.60}$$

where $\Phi(t)$ is continuous for $t \in [t_0, t_1]$. Then

1. *there exists a unique matrix $U(t)$ satisfying the equation*

$$AU' = -C^\top U - WU \tag{1.61}$$

   and initial conditions (1.50), i.e., $U(t_0) = 0$ and $U'(t_0) = I$, and
2. *the function $U(t)$ is a solution to Jacobi equation (1.49).*

Proof. The fact that statement 1 implies statement 2 is verified by a direct computation, which in fact repeats the proof of Lemma 1.3.

Exercise. Perform the necessary computation.

We prove statement 1. Equation (1.61) is a linear differential equation with respect to the matrix $U(t)$. The coefficients of this equation are discontinuous for $t = t_0$ (exactly because of this, not only the initial conditions $U(t_0)$ but also $U'(t_0)$ are given for $t = t_0$). To remove the discontinuity, we introduce a new phase coordinate $v = t - t_0$ and a new independent variable $\tau = \ln(t - t_0)$. Then (1.61) takes the form

$$\frac{dU}{d\tau} = -(A^{-1}WU)v - (A^{-1}C^\top U)v, \qquad \frac{dv}{d\tau} = v.$$

Using (1.60), we obtain

$$\frac{dU}{d\tau} = U + R(v)Uv, \qquad \frac{dv}{d\tau} = v,$$

where $R(v)$ is a continuous matrix. In the new coordinates, the initial values $U \to 0$ and $v \to 0$ as $\tau \to -\infty$ in the direction l defined by the pair $(I, 1)$ in the (U, v) phase space correspond to the initial values $U(t_0) = 0$ and $U'(t_0) = I$.

The existence and uniqueness of a solution to Eq. (1.61) with the initial conditions $U(t_0) = 0$ and $U'(t_0) = I$ therefore follow from Lemma 1.5.

We now return to the proof of Theorem 1.7. Using the function $U(t)$, which is a matrix solution to Jacobi equation (1.49), we construct the matrix $W(t)$, which is a solution to Riccati equation (1.52) by Lemma 1.3. This matrix is continuous on $(t_0, t_1]$ and has a simple pole at the point t_0. In addition, because $h(\cdot) \in C_0^1$, we have $h(t) = (t - t_0)\rho(t)$, where $\rho(t)$ is continuous. Using both these facts, we obtain

$$\int_{t_0}^{t_1} \frac{d}{dt} \langle W(t)h(t), h(t) \rangle dt = \langle W(t_1)h(t_1), h(t_1) \rangle - \lim_{t \to t_0} \langle W(t)h(t), h(t) \rangle = 0.$$

This means that the value of $\delta^2 J$ is in fact unchanged when $\frac{d}{dt}\langle Wh, h \rangle$ is added to the integrand, and we have proved that condition a is satisfied.

We show that the matrix $W(t)$ thus constructed is symmetrical. Transposing both parts of Eq. (1.52) and using the symmetry of the matrices A and B, we obtain

$$(W^\top)' + B = (C + W^\top)A^{-1}(C^\top + W^\top).$$

Therefore, the matrix $W^\top$ is also a solution to Riccati equation (1.52). Transposing formula (1.60), we find (by the symmetry of A) that $W^\top(t)$ as well as $W(t)$ has a simple pole with the residue $(-A(t_0))$ at the point t_0.

Exactly here we need Lemma 1.6, which allows us to construct a solution to the Jacobi equation using a solution to the Riccati equation. By Lemma 1.6, the unique matrix $\widetilde{U}(t)$ that is a solution to the equation $A\widetilde{U}' = -C^\top \widetilde{U} - W^\top \widetilde{U}$ corresponds to the matrix $W^\top$. In addition, $\widetilde{U}$ is a solution to Jacobi equation (1.49) with the initial values $\widetilde{U}(t_0) = 0$ and $\widetilde{U}'(t_0) = I$. Consequently, by the existence and uniqueness theorem for Eq. (1.49), $\widetilde{U}(t) = U(t)$. Comparing the relations $W^\top U = -AU' - C^\top U$ and $WU = -AU' - C^\top U$, we conclude that $(W - W^\top)U = 0$. Because the matrix U is nonsingular for $t > t_0$, we have $W(t) = W^\top(t)$ for $t > t_0$, i.e., the matrix W is symmetrical. We have thus proved the formula

$$\delta^2 J = \int_{t_0}^{t_1} [A^{1/2}\dot{h} + A^{-1/2}(C^\top + W)h]^2 dt. \tag{1.62}$$

It follows from this formula that $\delta^2 J \geq 0$. We now prove the positive definiteness of $\delta^2 J$.

Let $\delta^2 J(\widetilde{h}(\cdot)) = 0$. Then $A^{1/2}\dot{\widetilde{h}} + A^{-1/2}(C^\top + W)\widetilde{h} \equiv 0$. But $\widetilde{h}(t_0) = 0$, and hence $A^{1/2}(t_0)\dot{\widetilde{h}}(t_0) = 0$. We note that $\widetilde{h}$ yields the minimum of the functional $\delta^2 J$, and it is therefore a solution to Jacobi equation (1.48). The initial conditions $\widetilde{h}(t_0) = 0$ and $\dot{\widetilde{h}}(t_0) = 0$ mean that $\widetilde{h}(t) \equiv 0$. Therefore, $\delta^2 J > 0$.

Exercise. Prove the following theorem:

Theorem 1.8. *Let the strengthened Legendre condition ($A(t) > 0$) hold for the quadratic functional $\delta^2 J$, let the interval (t_0, t_1) contain no points conjugate to the point t_0, and let the point t_1 be conjugate to the point t_0. Then $\delta^2 J \geq 0$ (in this case, $\delta^2 J$ is not positive definite).*

Hint. Let $\delta^2 J(\widetilde{h}(\cdot)) < 0$. Approximate $\widetilde{y}(t)$ in the metric of C^1 with a sequence of functions that vanish at $t = t_0$ and $t = t_1 - 1/n$. Then apply Theorem 1.7.

§5. Morse Index

In this section, we discuss the Morse index theorem for a quadratic functional. The rank of a matrix G is denoted by $\mathrm{rk}\, G$.

Definition 1.8. The *multiplicity of a point* τ that is conjugate to the point t_0 is the number $n - \mathrm{rk}\, U(\tau)$.

Exercise. Prove that the multiplicity of a point τ coincides with the number of linearly independent solutions $\widetilde{h}(t)$ to system (1.48) that vanish at the points t_0 and τ.

Definition 1.9. The *index of a quadratic functional* is the maximum dimension of the subspace on which this functional is negative definite.

Theorem 1.9 (Morse). *Let the strengthened Legendre condition ($A > 0$) hold for the functional $\delta^2 J$. Then the index λ of this functional is equal to the number of points that are conjugate to the point t_0 on the interval (t_0, t_1) if we count each conjugate point together with its multiplicity.*

The proof of this theorem can be found in [70], Sec. 15.

We now give a geometric interpretation of the Morse theorem. We imagine that the infinite-dimensional space on which the functional $\delta^2 J$ is defined is the plane (x, y) and the values of the functional are on the z axis. We examine how the surface $z = z(x, y)$ evolves as the point t_1 moves away from the point t_0.

While the point t_0 is sufficiently close to t_1 such that there are no points that are conjugate to t_0 on the semiopen interval $(t_0, t_1]$, the functional $\delta^2 J$ is positive definite, and the surface is an "elliptic parabaloid." When t_1 attains the first point τ conjugate to t_0, the functional $\delta^2 J$ becomes positive semidefinite. To be more precise, by Lemma 1.2, there arises a "zero subspace" of this functional that consists of the functions $\tilde{h}(t)$ satisfying the definition of a conjugate point (see p. 28). The surface becomes an "elliptic cylinder" whose "ruling" has the dimension equal to the multiplicity of the conjugate point τ.

When t_1 has passed through the point τ, this ruling goes below the "plane (x, y)," and the surface becomes a "hyperbolic paraboloid" such that the dimension of its negative subspace is equal to the multiplicity of the point τ. When t_1 increases further, a new subspace of dimension equal to the multiplicity of conjugate points that are passed by the point t_1 split from the positive subspace (whose dimension is always infinite).

Example 1.10 (harmonic oscillator). We consider the functional

$$J = \int_0^T (\dot{x}^2(t) - \omega^2 x^2(t))\, dt$$

(which defines the action function for the harmonic oscillator) and the extremal $\hat{x}(t) \equiv 0$. The functional J is quadratic; therefore, it coincides with its second variation $\delta^2 J$. The Euler equation (which is also the Jacobi equation) has the form

$$\ddot{x} + \omega^2 x = 0. \tag{1.63}$$

A solution to Eq. (1.63) with the initial conditions $x(0) = 0$ and $\dot{x}(0) = 1$ defines a matrix $U(t)$ (which consists of a single entry because x is one-dimensional):

$$U(t) = \frac{1}{\omega} \sin \omega t.$$

The zeros of this function are the points $t = \pi k/\omega$, $k \in \mathbb{Z}$. Consequently, the extremal $\hat{x}(\cdot)$

1. yields the strict minimum of J for $T < \pi/\omega$,
2. yields the nonstrict minimum of J for $T = \pi/\omega$, and
3. does not yield a minimum of J for $T > \pi/\omega$.

Example 1.11 (geodesics on a cylinder). We define the cylinder by the equation $x = \cos u$, $y = \sin u$, $z = v$. The first quadratic form is $ds^2 = du^2 + dv^2$. The length functional is

$$\int \sqrt{u'^2 + 1}\, dv$$

for the independent variable v. The Euler equation

$$\frac{d}{dt}\left(\frac{u'}{\sqrt{1 + u'^2}} \right) = 0$$

or $u'' = 0$ has the solutions $u = C_1 v + C_2$, which are spirals. If we take u as an independent variable, then $v = b_1 u + b_2$, and there arise rectilinear rulings $v = \text{const}$ in addition to the just found extremals. The Jacobi equation is the variational equation for the Euler equation, i.e., $h'' = 0$; a linear function is its solution with the initial conditions $h(u_0) = 0$ and $h'(u_0) = 1$. It cannot vanish two times. Therefore, there are no conjugate points.

Example 1.12 (continuation of Example 1.2). In the spherical coordinates, the first quadratic form is $ds^2 = \cos^2 \psi d\phi^2 + d\psi^2$. The Euler equation has the form

$$-\frac{d}{dt}\left[\frac{\dot\psi}{\sqrt{\cos^2 \psi + \dot\psi^2}}\right] - \frac{\cos\psi \sin\psi}{\sqrt{\cos^2 \psi + \dot\psi^2}} = 0. \qquad (1.64)$$

As seen in Example 1.2, the solutions to this equation are great circles of the sphere. Because any two great circles can be transformed into one another by rotation of the sphere, it suffices to consider the geodesics $\widehat\psi(\phi) \equiv 0$. We obtain the Jacobi equation if we write the variational equation for (1.64) in a neighborhood of $\widehat\psi(\cdot)$. It is easy to verify that we obtain the equation $\ddot h + h = 0$.

Exercise. Reconstruct the omitted calculation.

In accordance with Example 1.10, the diametrically opposite point of the sphere is a conjugate point to the initial one. By the Jacobi necessary condition, a geodesic prolonged beyond the diametrically opposite point loses the minimality property. All semicircles that connect diametrically opposite points have the same length. That the conjugate point is here the point of intersection of all extremals emanating from one and the same point is a degenerate phenomenon, which is related to the symmetry of the sphere. In Example 1.14 below, we meet a more general situation, that is, the existence of an envelope of a set of extremals emanating from one and the same point.

Example 1.13 (continuation of Example 1.3). By examining this example, we demonstrate another method for finding conjugate points. In Example 1.3, we showed that the general solution to the Euler equation for our problem has the form $y = \sqrt{R^2 - (x + D)^2}$. The arbitrary constants R and D here correspond to the initial data for the geodesic. The Jacobi equation is the variational equation for the Euler equation. Therefore, the derivatives with respect to the initial data of solutions to the Euler equation or, equivalently, the derivatives with respect to R and D are solutions to the Jacobi equation:

$$h_1 = \frac{\partial y}{\partial R} = \frac{R}{\sqrt{R^2 - (x + D)^2}}, \qquad h_2 = \frac{\partial y}{\partial D} = \frac{-(x + D)}{\sqrt{R^2 - (x + D)^2}}.$$

We fix an arbitrary geodesic $\widehat R$, $\widehat D$. Then h_1 and h_2 are linearly independent and hence form a fundamental system of solutions to the Jacobi equation. The general solution to the Jacobi equation is its linear combination

$$C_1 h_1 + C_2 h_2 = \frac{C_1 \widehat R - C_2 (x + \widehat D)}{\sqrt{\widehat R^2 - (x + \widehat D)^2}}.$$

If we take a nonzero solution that vanishes for $x = x_0$, this solution has no other zeros, because the numerator is a linear function of x. Consequently, there are no conjugate points.

Example 1.14 (ballistics problem). We consider

$$\int_0^l \sqrt{y+k}\sqrt{1+\dot y^2}\,dx \to \inf, \qquad y(0) = 0, \qquad y(l) = a > -k.$$

The energy integral yields

$$\sqrt{\frac{y+k}{1+\dot y^2}} = C.$$

The general solution to this equation has the form $(x - A)^2 = 4C^2(y+k-C^2)$. For the solution passing through the point $(0,0)$, we obtain $A^2 = 4C^2(k-C^2)$.

As a parameter that defines the extremal passing through the point $(0,0)$, we choose $y'(0) = \alpha$. Then $\alpha = -A/(2C^2)$ and $A^2 = 4C^2(k - C^2)$. Excluding A and C, we obtain

$$y = \frac{(1+\alpha^2)x^2}{4k} + x\alpha.$$

We find the envelop of this set of parabolas

$$y = \frac{1+\alpha^2}{4k}x^2 + x\alpha, \qquad 0 = \frac{\alpha x^2}{2k} + x.$$

Therefore,

$$y = \frac{x^2}{4k} - k.$$

In ballistics, this curve is called the *safety parabola*. The point of tangency of the extremal with the safety parabola defines a conjugate point.

Exercise. Prove this fact.

§6. Jacobi Envelope Theorem

As already mentioned, the general situation for the appearance of a conjugate point on an extremal is its tangency to the envelope of a one-parameter family of extremals that emanate from one and the same point. Such envelopes were found by Jacobi when he studied the geodesic flow on a triaxial ellipsoid.

We here present a theorem that characterizes a remarkable geometric property of envelopes of a one-parameter family of extremals. In this case, instead of a problem with fixed endpoints where all extremals emanate from one point, we consider a more general problem with a movable left endpoint, which does not complicate the calculations practically.

We consider the functional

$$J = \int_{t_0}^{t_1} f(t, x(t), \dot x(t))\,dt \tag{1.65}$$

with the boundary condition $\Phi(t_0, x(t_0)) = 0$. Here, $x \in \mathbb{R}$, $f : \mathbb{R} \times \mathbb{R} \times \mathbb{R} \to \mathbb{R}$, $f \in C^1(D)$, $\Phi : \mathbb{R} \times \mathbb{R} \to \mathbb{R}$, $\Phi \in C^1(D)$, and $\Phi' \neq 0$; D is a neighborhood of the graph of the curve $x = \hat{x}(t)$, $\dot{x} = \hat{\dot{x}}(t)$, $t \in [\hat{t}_0, \hat{t}_1]$, where $\hat{x}(\cdot)$ is an extremal of the functional J that satisfies transversality condition (1.38) at the point $(\hat{t}_0, \hat{x}(\hat{t}_0))$.

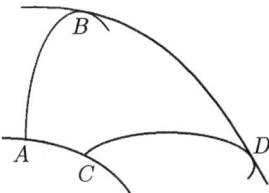

Fig. 1.3

We consider a one-parameter family of extremals $x(t, \alpha)$ satisfying transversality condition (1.38) such that $x(t, 0) = \hat{x}(t)$. We assume that this family has an envelope, such as the arc BD in Fig. 1.3, where AC is the curve $\Phi(t, x) = 0$ and AB and CD are extremals corresponding to α_1 and α_2. The instant of time when the extremal $x(t, \alpha)$ is tangent to the envelope is denoted by $t_1(\alpha)$.

Theorem 1.10 (Jacobi envelope theorem). *If the parameterization of the envelope BD is chosen such that the speed of motion along BD at each point is equal to the speed of motion along the extremal that is tangent to BD at this point, then the difference of values of the functional J on the extremals AB and CD is equal to the value of J on the segment of the envelope BD.*

Proof. Substituting $x(t, \alpha)$ in the integrand of the functional J, we obtain a scalar-valued function $J(\alpha)$. By (1.35), we have

$$\frac{dJ}{d\alpha} = \int_{t_0(\alpha)}^{t_1(\alpha)} \left[f_x - \frac{d}{dt} f_{\dot{x}} \right] \frac{\partial x}{\partial \alpha} dt + [p\delta x - H\delta t] \Big|_{t_0(\alpha)}^{t_1(\alpha)}.$$

The integral term in this formula vanishes because $x(t, \alpha)$ is an extremal, and by transversality condition (1.38) at the points $(t_0(\alpha), x(t_0(\alpha), \alpha))$,

$$[p\delta x - H\delta t]|_{t_0(\alpha)} = 0.$$

Therefore,

$$\frac{dJ}{d\alpha} = f_{\dot{x}} \frac{\partial x}{\partial \alpha}(t_1(\alpha), \alpha) - [-f + f_{\dot{x}}\dot{x}] \frac{\partial t_1}{\partial \alpha}. \tag{1.66}$$

Because BD is an envelope, the tangent to it, which is given by the vector

$$\left(\frac{dt_1}{d\alpha}, \frac{dx}{d\alpha}(t_1(\alpha), \alpha) \right),$$

coincides with the tangent to the extremal, given by the vector $(1, \dot{x}(t_1(\alpha), \alpha))$, i.e.,

$$\dot{x}(t_1(\alpha), \alpha)\frac{dt_1}{d\alpha} = \frac{\partial x}{\partial \alpha}(t_1(\alpha), \alpha).$$

Substituting this relation in (1.66), we obtain

$$\frac{dJ}{d\alpha} = f(t_1(\alpha), x(t_1(\alpha), \alpha), \dot{x}(t_1(\alpha), \alpha)). \tag{1.67}$$

Integrating (1.67) from α_1 to α_2, we have

$$J(\alpha_2) - J(\alpha_1) = \int_{\alpha_1}^{\alpha_2} f(t_1(\alpha), x(t_1(\alpha), \alpha), \dot{x}(t_1(\alpha), \alpha))d\alpha. \tag{1.68}$$

The right-hand side of (1.68) is the value of the functional J on BD, which enters the conditions of the theorem.

The theorem just proved generalizes the well-known property of the evolute of a plane curve (see [88], Sec. 29). We recall that the *evolute* of a plane curve $y = \phi(x)$ is the locus of its centers of curvature and is the envelope of the set of normals to the initial curve $y = \phi(x)$. On the set of plane curves $x = x(t)$, $y = y(t)$, $t_0 \leq t \leq t_1$, satisfying the boundary conditions $y(t_0) = \phi(x(t_0))$, we consider the functional

$$\int_{t_0}^{t_1} \sqrt{\dot{x}^2 + \dot{y}^2}dt,$$

which defines the length of the curve. Straight lines are extremals of this functional; normals satisfy the transversality condition (see Example 1.6). The evolute is the envelope of the set of normals. We now apply the Jacobi theorem.

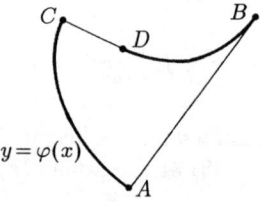

Fig. 1.4

If we draw two normals AB and CD at two points A and C of the curve $y = \phi(x)$ (see Fig. 1.4) and consider the distances to the points B and D of tangency to the evolute along them, then $|AB| - |CD|$ is equal to the length of the evolute arc BD.

Example 1.15. On the set of plane curves $x = x(t)$, $y = y(t)$, $0 \leq t \leq 1$, satisfying the boundary conditions $x(0) = \xi$, $y(0) = \eta$, and $y(1) = x^2(1)$, we consider the functional

$$\int_0^1 \sqrt{\dot{x}^2 + \dot{y}^2}\, dt,$$

which defines the length of the curve. The minimization problem for this functional corresponds to the search for the extremal distance from a point of the plane (ξ, η) to the parabola $y = x^2$.

This problem is one of the variants of the problem of Apollonii, who, in fact, found the evolute of the ellipse when studying the distance to it. A generalization of this problem is the problem of finding extremals of the distance to a given submanifold of a Riemannian manifold, which served as a base for the creation of the Morse theory.

As in the above case, normals serve as extremals that satisfy the transversality condition. The equation of the normal at the point (a, a^2) has the form

$$y - a^2 = -\frac{1}{2a}(x - a).$$

The envelope of this family satisfies the set of equations $2ay - 2a^3 = a - x$, $2y - 6a^2 = 1$. Excluding the parameter a, we obtain the equation of a semicubic parabola (see Fig. 1.5).

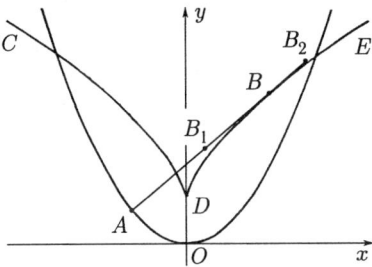

Fig. 1.5

At a certain fixed point A of the parabola $y = x^2$, we consider the normal, which is tangent to the evolute at the point B. Then $AB = R$ is the radius of curvature of the parabola at the point A. If the point (ξ, η) lies on this normal between the points A and B (the point B_1 in Fig. 1.5), then the circle of radius (AB_1) centered at the point B_1 has a curvature greater than the curvature of the parabola at the point A. Consequently, this circle is tangent to the parabola "from within," i.e., in a neighborhood of the point A, it lies in the domain $y \geq x^2$. Thus, a local minimimum is attained at AB_1.

If $|AB_2| > R$, then the circle of radius $|AB_2|$ centered at the point B_2 has a curvature less than the curvature of the parabola at the point A and is tangent to the parabola "from the outside," i.e., in a neighborhood of the point A, this circle lies in the domain $y \le x^2$. A local maximum is attained at AB_2. The osculating circle for the point B passes from one side of the parabola to the other, and there is no extremum.

Therefore, the envelope of the one-parameter family of normals (the arc CDE in Fig 1.5) plays the same role in Example 1.15 as conjugate points in problems with fixed endpoints: the extremal is locally minimal up to the point of tangency to the envelope, and after the tangency, it is no longer locally minimal. The points of the curve CDE are called *focal points of the parabola*.

To explain the meaning of this concept, we recall a visual definition of a one-parametric family of lines on the plane. The *envelope* is the intersection locus of infinitely close lines. In other words, if one of the lines tends to the other one, then the point of their intersection tends to a point that lies on the envelope.

If we imagine that light rays are emitted by a glowing thread having the form of a parabola (we here speak about the problem of light propagation in a homogeneous isotropic medium in a plane), then the points of the curve CDE are the most brightly illuminated. If the medium is inhomogeneous and nonisotropic, then the role of lines is played by extremals of the corresponding functional.

From this standpoint, there is no principal distinction between the concepts of *conjugate* and *focal points*. In the first case, the family of extremals emanates from one common point in all directions (its singularities are conjugate points); in the second case, they emanate from points of the manifolds in directions that satisfy the transversality condition (its singularities are focal points).

The concept *focal point* is close in its sense to the concept of *caustic*, which is used in the theory of partial differential equations. Incidentally, both concepts have a common etymology, *focus* is the fireplace or hearth in Latin; καυστικος (causticos) is burnt or seared in Greek (compare with "caustic soda"). These names are related to optics and characterize points of space where an increase in the density of the energy of light flow is caused by the convergence of light rays.

§7. Strong Minimum

7.1. Weierstrass Necessary Condition. We now consider the following problem: among all curves $x(\cdot) \in C^1([t_0, t_1], \mathbb{R}^n)$ satisfying the boundary conditions
$$x(t_0) = a, \qquad x(t_1) = b,$$
where a and b are two given points in the space $\mathbb{R}^n$, find a curve that gives the minimum value of the functional
$$J(x(\cdot)) = \int_{t_0}^{t_1} f(t, x(t), \dot{x}(t))\, dt,$$

where $f \in C^2([t_0, t_1] \times \mathbb{R}^n \times \mathbb{R}^n)$.

Definition 1.10. We say that a curve $\hat{x}(\cdot)$ realizes the *strong minimum* of the problem considered above if there exists a neighborhood V of $\hat{x}(\cdot)$ in the space $\mathcal{PC}^1([t_0, t_1])$ such that for any curve $x(\cdot) \in V$, the inequality $J(x(\cdot)) \geq J(\hat{x}(\cdot))$ holds.

Obviously, any strong minimum is a weak one, and sufficient conditions for the strong minimum are therefore sufficient conditions for the weak one, and necessary conditions for the weak minimum are necessary conditions for the strong one. However, not every weak minimum is strong. It is therefore natural to pose the problem of necessary conditions for the strong minimum that are not necessary conditions for the weak one.

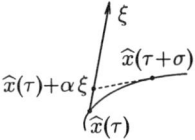

Fig. 1.6

We fix $\tau \in [t_0, t_1)$, a vector $\xi \in \mathbb{R}^n$, and a number $\sigma > 0$. We consider the following family of curves (see Fig. 1.6):

$$
x(t, \alpha) = \begin{cases}
\hat{x}(t), & t \notin [\tau, \tau + \sigma], \\
\hat{x}(\tau) + \xi(t - \tau), & t \in [\tau, \tau + \alpha], \\
\hat{x}(t) + \dfrac{\hat{x}(\tau) + \xi\alpha - \hat{x}(\tau + \alpha)}{\sigma - \alpha}(\tau + \sigma - t), & t \in [\tau + \alpha, \tau + \sigma].
\end{cases}
$$

For a sufficiently small α, the curves $x(t, \alpha)$ lie in an arbitrary small C-neighborhood (but not a C^1-neighborhood) of the curve $\hat{x}(\cdot)$, $x(t, 0) = \hat{x}(t)$. Substituting $x(t, \alpha)$ in the integrand of the functional J, we obtain a function $J(\alpha)$ that is defined for $\alpha \geq 0$ and attains its minimum value for $\alpha = 0$. Consequently, $dJ(0)/d\alpha \geq 0$. We have

$$
J(\alpha) - J(0) = \int_\tau^{\tau+\alpha} f(t, \hat{x}(\tau) + \xi(t - \tau), \xi)\, dt - \int_\tau^{\tau+\sigma} f(t, \hat{x}, \dot{\hat{x}})\, dt
$$

$$
+ \int_{\tau+\alpha}^{\tau+\sigma} f(t, x(t, \alpha), \dot{x}(t, \alpha))\, dt.
$$

Applying (1.35), we obtain

$$
\frac{dJ}{d\alpha} = f(\tau + \alpha, \hat{x}(\tau) + \xi\alpha, \xi) - [p\delta x - H\,\delta t]\Big|_{\tau+\alpha}
$$

$$+ \int_{\tau+\alpha}^{\tau+\sigma} \left[f_x - \frac{d}{dt} f_{\dot{x}} \right] \frac{\partial x}{\partial \alpha} \, dt.$$

Because the Euler equation holds on $\hat{x}(\cdot)$, the integral term vanishes for $\alpha = 0$. Because the left endpoint of the family $x(t, \alpha)$ moves along the line $t = \tau + \alpha$, $x = \hat{x}(\tau) + \alpha\xi$ with the directing vector $(1, \xi)$, it is necessary to substitute the vector $(1, \xi)$ instead of the vector $(\delta t, \delta x)$. As a result, we obtain

$$f(\tau, \hat{x}(\tau), \xi) - f(\tau, \hat{x}(\tau), \dot{\hat{x}}(\tau)) -$$

$$- f_{\dot{x}}(\tau, \hat{x}(\tau), \dot{\hat{x}}(\tau))(\xi - \dot{\hat{x}}(\tau)) \geq 0.$$

The function

$$\mathcal{E}(\tau, x, \dot{x}, \xi) = f(\tau, x, \xi) - f(\tau, x, \dot{x}) - f_{\dot{x}}(\tau, x, \dot{x})(\xi - \dot{x})$$

is called the *Weierstrass function*.

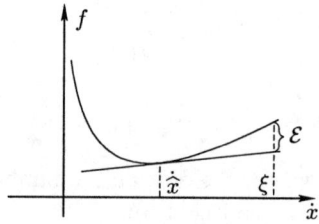

Fig. 1.7

We have proved the following theorem.

Theorem 1.11 (Weierstrass necessary condition). *Let the extremal $\hat{x}(\cdot)$ yield the strong minimum of the functional J. Then $\mathcal{E}(\tau, \hat{x}(\tau), \dot{\hat{x}}(\tau), \xi) \geq 0$ for all $\xi \in \mathbb{R}^n$ and $\tau \in [t_0, t_1]$.*

We now discuss the geometric meaning of this theorem. We fix τ_0, x_0, and $\dot{x}_0$ and consider $y = f(\tau_0, x_0, \xi)$ as a function of the variable ξ. We can then interpret the Weierstrass function as the deviation of the surface $y = f(\tau_0, x_0, \xi)$ from the tangent plane drawn at the point $\dot{x}_0$. The Weierstrass condition thus means that for all $\tau \in [t_0, t_1]$, the surface $y = f(\tau, \hat{x}(\tau), \xi)$ lies entirely over the tangent plane drawn at the point $\dot{\hat{x}}(\tau)$ (see Fig. 1.7).

Remark. Applying the Taylor formula

$$f(t, \hat{x}(t), \xi) = f(t, \hat{x}(t), \dot{\hat{x}}(t)) + f_{\dot{x}}(t, \hat{x}(t), \dot{\hat{x}}(t))(\xi - \dot{\hat{x}}(t))$$

$$+ \frac{1}{2} \langle f_{\dot{x}\dot{x}}(t, \hat{x}(t), \dot{\hat{x}}(t) + \theta(\xi - \dot{\hat{x}}(t)))(\xi - \dot{\hat{x}}(t), (\xi - \dot{\hat{x}}(t)) \rangle,$$

where $0 \leq \theta \leq 1$, we can easily show that the Weierstrass condition implies the Legendre condition $f_{\dot{x}\dot{x}} \geq 0$.

Exercise. Show that the weak minimum of the functional

$$J = \int_0^1 \dot{x}^3 \, dt$$

with the boundary conditions $x(0) = 0$ and $x(1) = 1$ is not a strong minimum.

§8. Poincaré–Cartan Integral Invariant

8.1. Exterior Differential Forms. Let M be a smooth n-dimensional manifold, and let $T_x M$ be the n-dimensional tangent plane to M at the point x, i.e., the linear space of first-order differential operators that act on functions $\phi : M \to \mathbb{R}$. If $x^1, \ldots, x^n$ are local coordinates on M, then the operators of differentiation

$$\frac{\partial}{\partial x^1}, \ldots, \frac{\partial}{\partial x^n}$$

form a basis of the space $T_x M$. A *differential 1-form* is a linear functional on $T_x M$, more precisely, a smooth family of linear functionals parameterized by points of the manifold M.

Linear functionals on $T_x M$ at a point x form the linear space $T_x^* M$ dual to $T_x M$. It is called the *cotangent space*. As a basis of the space $T_x^* M$, we can take the basis dual to

$$\frac{\partial}{\partial x^1}, \ldots, \frac{\partial}{\partial x^n},$$

i.e., a basis such that its ith element (denoted by dx^i) is a linear functional that assumes the value 1 at the vector $\partial/\partial x^i$ and the value 0 at other vectors $\partial/\partial x^j$, $j \neq i$. A general element of $T_x^* M$ has the form

$$\sum_{i=1}^n f_i dx^i.$$

A differential 1-form $\omega^{(1)}$ is a family of elements

$$\omega^{(1)} = \sum_{i=1}^n f_i(x) dx^i$$

depending on x such that the functions f_i smoothly depend on x and are transformed as coefficients of covariant vectors when we pass from one chart to another one. The simplest example of a 1-form is the differential of a function $F : M \to \mathbb{R}$,

$$dF = \sum_{i=1}^n \frac{\partial F}{\partial x^i} dx^i.$$

We can integrate a differential 1-form over curves l lying on M. For this, we divide a curve l into pieces l^α each of which lies in one local coordinate system. Further, we substitute the equation $l_\alpha : x = x_\alpha(t)$, $t_{\alpha 0} \leq t \leq t_{\alpha 1}$, under the integral sign,

$$\int_{l_\alpha} \omega^{(1)} = \int_{t_{\alpha 0}}^{t_{\alpha 1}} \sum_{i=1}^n f_i(x_\alpha(t)) \dot{x}_\alpha^i(t) \, dt,$$

and perform the summation over α.

Differential 2-forms are bilinear skew-symmetrical forms on $T_x M \times T_x M$, i.e., functionals defined on pairs of tangent vectors, linear in each of the arguments, and changing the sign when the arguments are interchanged. The simplest example of such a functional is the determinant

$$\begin{vmatrix} \xi_1 & \eta_1 \\ \xi_2 & \eta_2 \end{vmatrix} = \xi_1 \eta_2 - \xi_2 \eta_1 \tag{1.69}$$

constructed on pairs of two-dimensional vectors.

This example is typical: we can decompose any skew-symmetrical form into a direct sum of forms like (1.69) (see [62], Chap. 14, Sec. 9). The general form of a differential 2-form is

$$\omega^{(2)} = \sum_{i,j=1}^n f_{ij}(x) \, dx^i \wedge dx^j, \tag{1.70}$$

where $f_{ij} = -f_{ji}$ are smooth functions that are transformed as coordinates of a covariant tensor of the second rank when we pass from one coordinate system to another one. The differential form $dx^i \wedge dx^j$ is a form that assumes the value $\xi_i \eta_j - \xi_j \eta_i$ on a pair of tangent vectors with coordinates $\xi_1, \ldots, \xi_n$, $\eta_1, \ldots, \eta_n$. This value is equal to the area of the parallelogram (taken with the sign $+$ or $-$ depending on the orientation) constructed on the projections of the vectors ξ and η onto the (i, j)th coordinate plane in $T_x M$.

To each bilinear skew-symmetrical form $\omega^{(2)}$ in $T_x M$, we can uniquely associate a skew-symmetrical linear operator A_x that sets a 1-form $A_x \xi \in T_x^* M$ in correspondence to each vector $\xi \in T_x M$ such that $\omega_x^{(2)}(\xi, \eta) = A_x \xi(\eta)$. In the basis

$$\frac{\partial}{\partial x^1}, \ldots, \frac{\partial}{\partial x^n},$$

$f_{ij}(x)$ serve as entries of the matrix A_x for the form (1.70).

We can integrate a differential 2-form over two-dimensional manifolds $D \subset M$. For this, we divide a submanifold into pieces D^α each of which lies in one local coordinate system. Further, we substitute the equation $D_\alpha : x = x_\alpha(u, v)$, $(u, v) \in \Delta_\alpha \subset \mathbb{R}^2$ under the integral sign,

$$\int_{D_\alpha} \omega^{(2)} = \int_{\Delta_\alpha} \sum_{i,j=1}^n f_{ij}(x_\alpha(u, v)) \begin{vmatrix} \dfrac{\partial x_\alpha^i}{\partial u} & \dfrac{\partial x_\alpha^i}{\partial v} \\ \dfrac{\partial x_\alpha^j}{\partial u} & \dfrac{\partial x_\alpha^j}{\partial v} \end{vmatrix} du \, dv,$$

and perform the summation over α. Similarly, we define differential k-forms

$$\omega^{(k)} = \sum f_{i_1 \ldots i_k}(x)\, dx^{i_1} \wedge \ldots \wedge dx^{i_k}.$$

On the set of differential forms, we define the exterior product (forms are multiplied as polynomials, but the permutation of differentials inside each of the monomials leads to the change of the sign of this monomial) and the exterior differentiation

$$d\left[\sum f_{i_1,\ldots,i_k}\, dx^{i_1} \wedge \ldots \wedge dx^{i_k}\right] = \sum df_{i_1,\ldots i_k}(x) \wedge dx^{i_1} \ldots \wedge dx^{i_k}.$$

We define the operation of exterior differentiation such that the Stokes formula

$$\int_{\partial\gamma} \omega = \int_\gamma d\omega$$

holds; here, ω is a differential k-form, γ is an oriented piece of a $(k+1)$-dimensional manifold, and $\partial\gamma$ is its boundary with the induced orientation. The proof of the Stokes formula is in [75], Sec. 2.9.

A differential k-form ω is called closed if $d\omega = 0$ and exact if $\omega = d\theta$, where θ is a certain $(k-1)$-dimensional form. Each exact form is closed [108].

8.2. Poincaré–Cartan Integral Invariant. We here prove a property canonical systems of ordinary differential equations, the existence of an integral invariant, which is needed for the further presentation. The idea of an integral invariant arose in works by Stokes when he studied the stationary turbulent flow of an incompressible fluid. We present the Stokes arguments following the presentation in [3], Sec. 44.

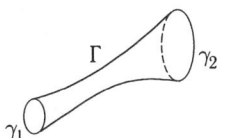

Fig. 1.8

Let v be the velocity vector field of the fluid, and let $\operatorname{curl} v$ be the vorticity field. Integral curves of the vorticity vector field are called the *vortex lines*. We consider a closed one-dimensional contour γ_1 and draw a vortex line through each of its points. We obtain a vortex tube. We consider the contour γ_2 that encloses the same vortex tube (see Fig. 1.8). The lateral surface of this tube with boundary $\gamma_1 - \gamma_2$ is denoted by Γ.

The *circulation of the field v* over the contour γ is

$$\int_\gamma v\, dl,$$

where the form $v\, dl = v_1 dx^1 + v_2 dx^2 + v_3 dx^3$.

Lemma 1.7 (Stokes). *The circulation of the field v is the same over γ_1 and γ_2, i.e.,*

$$\int_{\gamma_1} v\,dl = \int_{\gamma_2} v\,dl.$$

Proof. Applying the Stokes formula to Γ, we have

$$\int_{\gamma_1} v\,dl - \int_{\gamma_2} v\,dl = \int_{\Gamma} (\operatorname{curl} v)_n\,dS = 0.$$

The latter relation holds because $\operatorname{curl} v$ is tangent to Γ everywhere, and its flow through Γ vanishes.

We can express the assertion of the Stokes lemma as follows: the form $v\,dl$ is an integral invariant of the vorticity field.

A. Poincaré showed that for the mapping

$$\mathbb{R}^{2n} \to \mathbb{R}^{2n}, \qquad (x_0, p_0) \mapsto x(t; x_0, p_0), p(t; x_0, p_0)$$

along trajectories of the canonical system

$$\dot{x} = H_p, \qquad \dot{p} = -H_x,$$

the differential form $p\,dx$ is an integral invariant. E. Cartan further generalized this idea. In the subsequent presentation, the role of the circulation is played by the integral of an arbitrary 1-form $\omega^{(1)}$ over a closed contour, and the role of the vorticity played by the bilinear form $d\omega^{(1)}$. The rank of a bilinear skew-symmetrical form is an even number. A differential 2-form $\omega^{(2)}$ is called *nondegenerate* if its rank is maximal at each point. In the odd-dimensional space $\mathbb{R}^{2k+1}$, nondegenerate 2-forms are of rank $2k$. In this case, the matrix of the mapping A_x, which is defined by the form $\omega^{(2)}$, has a one-dimensional kernel. The vector field directed along this kernel is denoted by $\zeta(x)$.

If the two-dimensional plane σ spanned by the vectors ξ and η contains $\zeta(x)$, then $\omega_x^{(2)}\{\xi, \eta\} = 0$. Indeed, one of the vectors, ξ for example, is a linear combination of ζ and η. Therefore,

$$\omega_x^{(2)}\{\xi, \eta\} = \omega_x^{(2)}\{\alpha\zeta + \beta\eta, \eta\} = \alpha\langle A_x\zeta, \eta\rangle + \beta\omega_x^{(2)}\{\eta, \eta\}.$$

The first term in the right-hand side vanishes because ζ belongs to the kernel of A, and the second one vanishes because $\omega_x^{(2)}$ is skew symmetrical.

Now let $\omega^{(1)}$ be a 1-form in $\mathbb{R}^{2k+1}$ such that its differential $\omega^{(2)} = d\omega^{(1)}$ is nondegenerate. Integral lines of the vector field $\zeta(x)$ are called *characteristics of the form* $\omega^{(1)}$. For a one-dimensional contour γ_1, we consider the tube composed of characteristics passing through γ_1 and a contour γ_2 that encloses this tube (see Fig. 1.8). Applying the Stokes formula to the two-dimensional manifolds with the boundary $\gamma_1 - \gamma_2$, we obtain

$$\int_{\gamma_1} \omega^{(1)} - \int_{\gamma_2} \omega^{(1)} = \int_{\Gamma} d\omega^{(1)} = 0.$$

Indeed, the form $d\omega_x^{(1)}$ on the two-dimensional small area σ_x that is tangent to Γ at the point x vanishes, because σ_x contains the vector ζ_x. We can therefore say that the form $\omega^{(1)}$ is an integral invariant for the field of its characteristic directions.

We can now pass to canonical systems. We consider a Hamiltonian $H(t, x, p)$ defined on the space of variables (t, x, p), which is odd-dimensional ($t \in \mathbb{R}, (x, p) \in \mathbb{R}^{2n}$), and the Poincaré–Cartan differential form $\omega^{(1)} = p\,dx - H\,dt$. We apply the above construction to this form. The matrix $A(t, x, p)$ of the form $\omega^{(2)} = d\omega^{(1)}$ has the following appearance:

$$
\begin{pmatrix}
0_n & I_n & \begin{matrix} -\dfrac{\partial H}{\partial x^1} \\ \cdots \\ -\dfrac{\partial H}{\partial x^n} \end{matrix} \\[2em]
-I_n & 0_n & \begin{matrix} -\dfrac{\partial H}{\partial p_1} \\ \cdots \\ -\dfrac{\partial H}{\partial p_n} \end{matrix} \\[2em]
\begin{matrix} \dfrac{\partial H}{\partial x^1} & \cdots & \dfrac{\partial H}{\partial x^n} \end{matrix} & \begin{matrix} \dfrac{\partial H}{\partial p_1} & \cdots & \dfrac{\partial H}{\partial p_n} \end{matrix} & 0
\end{pmatrix}
$$

For all (t, x, p), this matrix is of maximal rank $2n$ because its left upper corner is a nonsingular $2n \times 2n$ matrix. The vector $\zeta(x)$ defined above for the matrix A has a nonzero last component, and we can normalize it by the condition $\zeta_{2n+1} = 1$. We then obtain

$$-\zeta_{n+1} - \frac{\partial H}{\partial x^1} = 0, \ldots, -\zeta_{2n} - \frac{\partial H}{\partial x^n} = 0,$$

$$\zeta_1 - \frac{\partial H}{\partial p_1} = 0, \ldots, \zeta_n - \frac{\partial H}{\partial p_n} = 0, \qquad \zeta_{2n+1} = 1,$$

i.e., the equations of characteristics of the Poincaré–Cartan form

$$\frac{dp}{ds} = -\frac{\partial H}{\partial x}, \qquad \frac{dx}{ds} = -\frac{\partial H}{\partial p}, \qquad \frac{dt}{ds} = 1 \qquad (1.71)$$

yield the canonical system with the Hamiltonian H. Therefore, we have proved the following theorem.

Theorem 1.12 (on the integral invariant). *The differential form*

$$p \, dx - H \, dt$$

is an integral invariant of canonical system (1.71).

This means that the integral of the form $p \, dx - H \, dt$ over any closed contour is preserved along the tube of trajectories of system (1.71).

8.3. Legendre Manifolds. In this subsection, we present the construction that serves as a base for proving the theorem on sufficient conditions for a strong minimum that is given below. Let the Poincaré–Cartan form $p \, dx - H \, dt = \omega^{(1)}$, $H \in C^2(D)$, where $D \subset \mathbb{R}^{2n+1}$ is an open set, be given.

Definition 1.11. A submanifold $M \subset D$ is called an *integral manifold* if the form ω vanishes on any collections of vectors tangent to M, that is,

$$\omega|_M = 0.$$

Definition 1.12. A manifold $\mathfrak{M} \subset D$, $\dim \mathfrak{M} = n + 1$, is said to be *Legendre* with respect to the form $\omega^{(1)}$ if $\mathfrak{M}$ is an integral manifold for the form $d\omega^{(1)}$.

At each point x of a manifold $N \subset \mathbb{R}^n$, $\dim N = k$, we consider the set of vectors $p \in \mathbb{R}^n$ such that for each vector $\xi \in T_x N$, we have

$$\langle p, \xi \rangle = 0. \tag{1.72}$$

The set of elements $\{x, p\}$ for which p satisfies condition (1.72) is called the *normal bundle* (see [53], Chap. 2) of the manifold N. The normal bundle of an arbitrary manifold $N \subset \mathbb{R}^n$ is obviously an integral manifold for the form $\omega^{(1)}$ for any choice of the function H. The dimension of the normal bundle is equal to n, because the condition $\langle p, \xi \rangle = 0$ gives k independent equations if N itself is defined by a set of $n-k$ independent equations.

Example 1.16. For a manifold $N \subset \mathbb{R}^{n+1}$, $\dim N = k$, M denotes the set of elements $\{t, x, p\}$ such that for any $(t, x) \in N$ and $(\tau, \xi) \in T_{tx} N$,

$$\langle p, \xi \rangle - H(t, x, p)\tau = 0. \tag{1.73}$$

Then M is an integral manifold for the form $p \, dx - H \, dt$.

Definition 1.13. The manifold M thus constructed is called the *lift of the manifold* N to the space $\{t, x, p\}$.

We consider the following classical problem of the calculus of variations with a movable left endpoint and a fixed right endpoint.

Problem 3. Let

$$J = \int_{t_0}^{t_1} f(t, x(t), \dot{x}(t))dt \to \inf$$

under the conditions $x(t_1) = x_1$ and

$$\Phi(t_0, x(t_0)) = 0, \quad \Phi : U \to \mathbb{R}^{n+1-k}, \quad \Phi \in C^1(U), \tag{1.74}$$

where $U \subset \mathbb{R}^1 \times \mathbb{R}^n$ is a connected domain whose projection on the axis t contains the segment $[t_0, t_1]$.

If the mapping $d\Phi : T_{t x}U \to \mathbb{R}^{n+1-k}$ is surjective, then condition (1.74) defines a certain k-dimensional smooth manifold N in U. We consider the manifold M constructed in Example 1.16 according to the manifold N and the function $H(t, x, p)$. We can easily recognize that condition (1.73) is transversality condition (1.38).

Definition 1.14. We say that the manifold M *does not contain characteristic directions* at the point (t, x, p) if the vector $\zeta = (1, H_p, -H_x)$ is not tangent to the manifold M, i.e., $\zeta \notin TM$.

Proposition 1.2. *The condition*

$$\mathrm{rk}(\xi_i - H_p(t, x, p)\tau_i) = k, \tag{1.75}$$

where $\{\tau_i, \xi_i\}$, $i = 1, \ldots, k$, is a basis of $T_{t x}N$, is a sufficient condition for M not to contain characteristic directions at the point (t, x, p).

Proof. It suffices to show that the vector $(1, H_p)$ is not tangent to the manifold N, i.e., is not a linear combination of the vectors (τ_i, ξ_i). We suppose the contrary, that is,

$$1 = \sum_{i=1}^{k} \lambda_i \tau_i, \qquad H_p = \sum_{i=1}^{k} \lambda_i \xi_i.$$

Then

$$\sum \lambda_i \xi_i - H_p \sum \lambda_i \tau_i = H_p - H_p = 0;$$

this contradicts (1.75).

Corollary 1.1. *The above proof implies that under condition (1.75), extremals of problem (1.52) satisfying the transversality condition approach the manifold N at a nonzero angle.*

Under condition (1.75), the set of values of p satisfying Eq. (1.73) at the point (t, x) forms a smooth manifold. This manifold is denoted by Q_{tx}.

Basic Construction. Let M be the lift of the manifold N defined by (1.73). Through each point (t_0, x_0, p_0) of the manifold M, we draw an extremal, i.e., a solution to the Hamiltonian system

$$\dot{x} = H_p(t, x, p), \qquad \dot{p} = -H_x(t, x, p) \tag{1.76}$$

defined for $t \geq t_0$. If condition (1.75) holds, we obtain a smooth $(n+1)$-dimensional manifold, denoted by $\mathfrak{M}$, as a result.

Proposition 1.3. *If condition (1.75) holds, then $\mathfrak{M}$ is a Legendre manifold with respect to the Poincaré–Cartan form.*

Proof. We consider an arbitrary smooth closed contour $\gamma \subset \mathfrak{M}$ and draw the tube of trajectories Γ of system (1.76). Solutions to Eqs. (1.76) depend smoothly on the initial data and intersect M at a nonzero angle. Therefore, $\gamma_1 = \Gamma \cap M$ is a smooth contour. By the theorem on the integral invariant,

$$\int_\gamma \omega^{(1)} = \int_{\gamma_1} \omega^{(1)}.$$

But M is an integral manifold for the form $\omega^{(1)}$. Therefore,

$$\int_\gamma \omega^{(1)} = 0 \tag{1.77}$$

for any closed curve $\gamma \subset \mathfrak{M}$. We fix a point $(t_0, x_0, p_0) \in \mathfrak{M}$ and consider

$$\int_{t_0, x_0, p_0}^{t, x, p} \omega^{(1)}$$

as a function of the variable upper limit $(t, x, p) \in \mathfrak{M}$ (by (1.77), the integral does not depend on the path if this path lies in $\mathfrak{M}$). Then the differential of this function defined on $\mathfrak{M}$ is equal to $\omega^{(1)}\big|_{\mathfrak{M}}$. We have thus proved the exactness of the form $\omega^{(1)}\big|_{\mathfrak{M}}$. Its closedness is a consequence of its exactness.

Corollary 1.2. *Let the manifold M be connected, and let two curves l_1 and l_2, where $l_1, l_2 \subset \mathfrak{M}$, connect the point $(t, x, p) \in \mathfrak{M}$ with the manifold M. Then*

$$\int_{l_1} \omega^{(1)} = \int_{l_2} \omega^{(1)}.$$

Proof. Connecting the ends of the curves l_1 and l_2 by a path in M, we obtain a closed contour; by (1.77), the integral over this contour vanishes. But M is an integral manifold for the form $\omega^{(1)}$; therefore, the integral over any curve lying in M vanishes. Consequently,

$$\int_{l_1} \omega^{(1)} = \int_{l_2} \omega^{(1)},$$

as was asserted.

Everywhere in what follows, we understand N, M, Q, and $\mathfrak{M}$ to be the result of applying the basic construction to Problem 3.

§9. Fields of Extremals

9.1. Hilbert Invariant Integral. In the space $\{t, x, p\}$, we consider a domain Δ that along with any point (t_1, x_1, p_1) contains all points of the solution to canonical system (1.76) with the initial values (t_1, x_1, p_1), i.e., $(\tau, x(\tau; t_1, x_1, p_1), p(\tau; t_1, x_1, p_1))$ for $t_0 < \tau < t_1$. Let $D = (\mathfrak{M} \setminus M) \cap \Delta$. We consider the projection operator $\pi : (t, x, p) \mapsto (t, x)$ from $\mathbb{R}^{2n+1}$ to $\mathbb{R}^{n+1}$.

Definition 1.15. Let the projection operator π restricted to the set D be a one-to-one mapping. Then we say that a *field of extremals* $\mathfrak{P}$ with the initial manifold N is given in the domain $G = \pi D$.

If a field is defined in G, then we can set one and only one point $p(t, x)$ in correspondence to each point $(t, x) \in G$ such that $(t, x, p(t, x)) \in D$. In other words, $p = p(t, x)$ is the equation that defines the manifold $\mathfrak{M}$ over the domain D.

In this case, the projections of extremals lying in $\mathfrak{M}$ to the domain G do not intersect each other, because the extremals cannot intersect in $\mathfrak{M}$ (by the uniqueness theorem for system (1.76)), and the mapping $\pi|_{\mathfrak{M}}$ is one-to-one. Therefore, the function

$$g(t, x) = H_p(t, x, p(t, x)) \tag{1.78}$$

is uniquely defined at each point $(t, x) \in G$. It is called the *geodesic inclination of the field* $\mathfrak{P}$. The equation $p = p(t, x)$ allows us to rewrite all the formulas that refer to the set D in the form of formulas on G. Thus, for example, the equations of extremals in G have the form $\dot{x} = g(t, x)$; the restriction of the Poincaré–Cartan form to D can be written as the form

$$p(t, x)dx - H(t, x, p(t, x))dt \tag{1.79}$$

on G. Because $\omega^{(1)}|_{\mathfrak{M}}$ is exact, form (1.79) is also exact.

Remark. The exactness of form (1.79) means that it is the differential of a certain function $S(t, x)$ defined on G, that is,

$$S_x(t, x) = p(t, x), \qquad S_t(t, x) = -H(t, x, p(t, x)).$$

The function $S(t, x)$ is thus a solution to Hamilton–Jacobi equation (1.43). Moreover, $S|_N = $ const (the integral of form (1.79) over any curve $l \subset N$ vanishes).

Therefore, the integral of form (1.79) does not depend on the path γ in G. We rewrite this integral in terms of the integrand f as

$$
\begin{aligned}
I(\gamma) &= \int_{\gamma} p(t,x)dx - H(t,x,p(t,x))dt \\
&= \int_{\gamma} f_{\dot{x}}(t,x,g(t,x))dx \\
&\quad - [-f(t,x,g(t,x)) + f_{\dot{x}}(t,x,g(t,x))g(t,x)]dt. \qquad (1.80)
\end{aligned}
$$

Integral (1.80) is called the *Hilbert invariant integral* for the field $\mathfrak{P}$. It does not depend on the path in the domain G. On the extremals of the field, i.e., on the solutions to the system $\dot{x} = g(t,x)$, integral (1.80) coincides with the initial functional, that is,

$$
\int_{\gamma} f_{\dot{x}}dx - [-f + f_{\dot{x}}g]\,dt = \int_{\gamma} f(t,x,g(t,x))\,dt.
$$

Weierstrass function (1.2) for the field $\mathfrak{P}$ is also defined on G:

$$
\begin{aligned}
\mathcal{E}_{\mathfrak{P}}(t,x,\xi) &= \mathcal{E}(t,x,g(t,x),\xi) \\
&= f(t,x,\xi) - f(t,x,g(t,x)) - p(t,x)(\xi - g(t,x)). \qquad (1.81)
\end{aligned}
$$

Theorem 1.13. Sufficient conditions for a strong minimum in terms of the field of extremals. *Let a field of extremals $\mathfrak{P}$ with the initial manifold N be defined in a domain G, and let Weierstrass function (1.81) be nonnegative for all $(t,x) \in G$ and for any ξ. Let $\hat{x}(\cdot)$ be a certain extremal of the field $\mathfrak{P}$. Then for any curve $x(\cdot)$ such that $x(t_1) = \hat{x}(t_1)$, $x(t_0) \in N$, $x(t) \in G$ for all $t \in [t_0, t_1]$, the inequality $J(\hat{x}(\cdot)) \leq J(x(\cdot))$ holds.*

Proof. By Corollary 1.2, $I(\hat{x}(\cdot)) = I(x(\cdot))$. But $\hat{x}(\cdot)$ is an extremal; therefore, $I(\hat{x}(\cdot)) = J(\hat{x}(\cdot))$. Consequently,

$$
\begin{aligned}
J(x(\cdot)) - J(\hat{x}(\cdot)) &= J(x(\cdot)) - I(\hat{x}(\cdot))J(x(\cdot)) - I(x(\cdot)) \\
&= \int_{t_0}^{t_1} f(t,x(t),\dot{x}(t))dt \\
&\quad - \int_{t_0}^{t_1} \{f_{\dot{x}}(t,x(t),g(t,x(t)))\dot{x}(t)[-f(t,x(t),g(t,x(t))) \\
&\quad + f_{\dot{x}}(t,x(t),g(t,x(t)))g(t,x(t))]\}dt \\
&= \int_{t_0}^{t_1} \mathcal{E}_{\mathfrak{P}}(t,x(t),\dot{x}(t))dt \geq 0.
\end{aligned}
$$

Therefore, to confirm that the strong minimum is attained at the extremal $\hat{x}(\cdot)$, we must embed this extremal in a field.

9.2. Embedding an Extremal in a Field and Focal Points. Let $\hat{x}(\cdot)$ be an extremal of Problem 3 that satisfies transversality conditions (1.38). This means that $(\hat{t}_0, \hat{x}_0, \hat{p}_0) \in M$, where M is the lift of the manifold N defined by conditions (1.74). Let condition (1.75) hold for M, i.e., M does not contain characteristic directions. To construct a field of extremals that contains $\hat{x}(\cdot)$, we apply the basic construction to a certain neighborhood of the point $(\hat{t}_0, \hat{x}_0, \hat{p}_0)$ in the manifold M, where it is possible to choose one coordinate system. We obtain a neighborhood $\mathcal{U}$ of the trajectory $(t, \hat{x}(t), \hat{p}(t))$ in the manifold $\mathfrak{M}$. Let $\alpha = (\alpha_1, \ldots, \alpha_k)$ be local coordinates in N, $t = t_0(\alpha)$, $x = x_0(\alpha)$, and, moreover, $t_0(0) = \hat{t}_0$, $x_0(0) = \hat{x}_0$. The manifold Q_{tx} where the point $(t, x) \in N$ corresponds to the parameter α is denoted by Q_α. Let $\beta = (\beta_1, \ldots, \beta_{n-k})$ be local coordinates in Q_α: $p = p_0(\alpha, \beta)$ and, moreover, $p(0, 0) = \hat{p}_0$. Then (α, β) are local coordinates in M, and (t, α, β) are local coordinates in $\mathfrak{M}$: $x = x(t, \alpha, \beta)$, $p = p(t, \alpha, \beta)$, where $x(t, \alpha, \beta), p(t, \alpha, \beta)$ is a solution to system (1.76) with the initial conditions $(t_0(\alpha), x_0(\alpha), p_0(\alpha, \beta))$. We consider the Jacobian of the mapping $\pi|_u$ at points of the extremal $(t, \hat{x}(t), \hat{p}(t))$ (as usual, the matrix with entries $\partial x^i / \partial \alpha_j$, $i = 1, \ldots, n$, $j = 1, \ldots, k$, is denoted by $\partial x / \partial \alpha$; $\partial \hat{x} / \partial \alpha$, $\hat{H}$, etc., denote the functions obtained under the substitution $\alpha = 0$, $\beta = 0$), that is, we consider

$$\mathfrak{J}(t) = \det \begin{pmatrix} 1 & 0 & 0 \\ \dot{\hat{x}}(t) & \dfrac{\partial \hat{x}}{\partial \alpha}(t) & \dfrac{\partial \hat{x}}{\partial \beta}(t) \end{pmatrix} = \det \left(\dfrac{\partial \hat{x}}{\partial \alpha}, \dfrac{\partial \hat{x}}{\partial \beta} \right). \qquad (1.82)$$

Definition 1.16. A point $\tau > t_0$ is called a *focal point* of the manifold N on the trajectory $\hat{x}(\cdot)$ if $\mathfrak{J}(\tau) = 0$.

Remark. This definition does not depend on the choice of local coordinates, because π and $\mathfrak{M}$ are defined invariantly and the condition $\mathfrak{J}(\tau) = 0$ is equivalent to the condition $\operatorname{Ker} d\pi|_u \neq 0$.

Proposition 1.4. Let $\hat{x}(t)$, $\hat{t}_0 \leq t \leq \hat{t}_1$, be an extremal satisfying the transversality condition, the strengthened Legendre condition, and condition (1.75). Let $\hat{x}(\cdot)$ contain no focal points of the manifold N. Then there exists a neighborhood V of the set $\{t, \hat{x}(t)\}$, $\hat{t}_0 < t \leq \hat{t}_1$, on which the field $\mathfrak{P}$ with the initial manifold N is defined.

To prove this, it suffices to note that at each point $(t, \hat{x}(t))$, $\hat{t}_0 < t \leq \hat{t}_1$, all conditions of the implicit function theorem hold for the mapping $\pi|_u$; hence, each such point has a neighborhood $V_t \subset \mathbb{R}^{n+1}$ in which the mapping $\pi|_u$ is one-to-one. Now $V = \bigcup\limits_{(t_0, t_1]} V_t$ serves as the desired neighborhood.

Theorem 1.14. Sufficient conditions for a strong minimum in terms of one extremal. *Let $\hat{x}(t)$ be a solution to the Euler equation for Problem 3 that satisfies the transversality conditions, strengthened Legendre condition, and condition (1.75). Moreover, let the function f for a fixed (t, x)*

be convex in $\dot{x}$, $\hat{f}(t_0) \neq 0$, and let the semiopen interval $(t_0, t_1]$ contain no focal points of the manifold N. Then the strong minimum is attained at $\hat{x}(\cdot)$.

Proof. Although Proposition 1.4 allows us to embed the extremal $\hat{x}(t)$, $t_0 < t \leq t_1$, in a field of extremals, the point $(\hat{t}_0, \hat{x}_0)$ lies on the boundary of the domain G, which is covered by the field, and the trajectories $x(\cdot)$ lying in an arbitrary small C-neighborhood of $\hat{x}(\cdot)$ could leave the domain G for t close to $\hat{t}_0$. To prove Theorem 1.14, we must therefore construct another field for which the point $(\hat{t}_0, \hat{x}_0)$ lies inside the domain G. To construct this field, we need to "shift" the manifold N to the left from the point $\hat{t}_0$. However, the technical realization of this idea turns out to be sufficiently complicated and occupies practically the remainder of the proof.

We now pass to the construction of this field. We consider the k-parameter family $\mathcal{K}$ of extremals $x(t; \alpha, 0)$, $|\alpha| < \epsilon$, consisting of extremals constructed in Proposition 1.4 for $\beta = 0$. For definiteness, we assume that $\hat{f}(t_0) > 0$. (For $\hat{f}(t_0) < 0$, we must take $\sigma < 0$ and preserve all other arguments). We fix a number $\sigma > 0$ and prolong each of the extremals of the family $\mathcal{K}$ to the left beyond the point of intersection with the manifold N (i.e., for $t < t_0(\alpha)$) up to an instant of time $t_\sigma(\alpha)$ for which

$$\int_{t_\sigma(\alpha)}^{t_0(\alpha)} f(t, x(t; \alpha, 0), \dot{x}(t; \alpha, 0))dt = \sigma. \tag{1.83}$$

The function $t_\sigma(\alpha)$ that is found from Eq. (1.83) turns out to be a smooth function of σ and α for sufficiently small σ and α by the implicit function theorem. Indeed, the derivative of the left-hand side of Eq. (1.83) with respect to $t_\sigma(\alpha)$ at the point $\alpha = 0$, $\sigma = 0$ is equal to $-\hat{f}(t_0)$; it is different from zero by the condition of Theorem 6. For a sufficiently small $\sigma > 0$, the left endpoints of the extremals $(t_\sigma(\alpha), x_\sigma(\alpha))$ (where $x_\sigma(\alpha) = x(t_\sigma(\alpha), \alpha, 0)$) form a smooth manifold N_σ.

Indeed, by Corollary 1.1, we have

$$\mathrm{rk}\left(\dot{x}(t; \alpha, 0) \frac{\partial x}{\partial \alpha}(t; \alpha, 0) \right)\bigg|_{t=\hat{t}_0, \alpha=0} = k + 1. \tag{1.84}$$

We consider α as coordinates on N_σ. To verify that N_σ is a smooth manifold, it suffices to find $\partial x_\sigma / \partial \alpha$ and prove that they are linearly independent. We have the matrix

$$\frac{\partial x_\sigma}{\partial \alpha} = \left(\dot{x}(t; \alpha, 0)|_{t=t_\sigma(\alpha)}, \quad \frac{dt_\sigma}{d\alpha} + \frac{\partial x}{\partial \alpha}(t; \alpha, 0)\bigg|_{t=t_\sigma(\alpha)} \right),$$

whose rank is equal to k for $\alpha = 0$, $\sigma = 0$ by (1.84); hence, it is equal to k for all sufficiently small α, σ.

We apply the basic construction to N_σ. We obtain a manifold $\mathfrak{M}_\sigma$ that contains the extended graph of the extremal $\hat{x}(\cdot)$ (including the point $(\hat{t}_0, \hat{x}_0)$). In what follows, without special stipulations, we mark all the objects of the basic construction that refer to the manifold $\mathfrak{M}_\sigma$ with the subscript σ.

Proposition 1.5. *There exists a number* $\theta > 0$ *independent of* σ *such that*

$$\mathfrak{J}_\sigma(t) \neq 0, \quad \hat{t}_\sigma < t < \hat{t}_\sigma + \theta. \tag{1.85}$$

In other words, the initial part of the extremal $\hat{x}(\cdot)$, *whose length is independent of* σ, *contains no focal points.*

Proof. For brevity in what follows, instead of the expression

$$\frac{\partial x}{\partial \alpha}(t; t_0, x_0(\alpha), p_0(\alpha, \beta))\bigg|_{t_0=t_0(\alpha)},$$

we merely write $\partial x/\partial \alpha$. We separately find the derivative with respect to t_0. Differentiating the identities $x(t_0; t_0, x_0, p_0) \equiv x_0$ with respect to t_0, we obtain $0 = \dot{x}(t_0; t_0, x_0, p_0) + (\partial x/\partial t_0)(t_0; t_0, x_0, p_0)$, i.e., $\partial x/\partial t_0 = -\dot{x}$. In this notation, the Jacobian $\mathfrak{J}_\sigma$ becomes

$$\mathfrak{J}_\sigma(t) = \det\left(\frac{\partial \hat{x}_\sigma}{\partial \alpha} - \dot{\hat{x}}\frac{\partial \hat{t}_\sigma}{\partial \alpha}, \quad \frac{\partial \hat{x}_\sigma}{\partial \beta}\right). \tag{1.86}$$

We have $\mathfrak{J}_\sigma(\hat{t}_\sigma) = 0$, because the last $n-k$ columns vanish for $t = \hat{t}_\sigma$. We must prove that $\mathfrak{J}_\sigma(t) \neq 0$ for $t > \hat{t}_\sigma$. To prove this, we compute the derivatives of these columns. The derivatives with respect to the initial data satisfy the variational equation for system (1.76). We have

$$\frac{d}{dt}\left(\frac{\partial \hat{x}}{\partial \beta}\right) = \widehat{H}_{px}\left(\frac{\partial \hat{x}}{\partial \beta}\right) + \widehat{H}_{pp}\left(\frac{\partial \hat{p}}{\partial \beta}\right).$$

Expanding by the Taylor formula and taking $(\partial \hat{x}_\sigma/\partial \beta)(\hat{t}_\sigma) = 0$ into account, we obtain

$$\frac{\partial \hat{x}}{\partial \beta}(t) = \widehat{H}_{pp}(\hat{t}_\sigma)\frac{\partial \hat{p}}{\partial \beta}(\hat{t}_\sigma)(t - \hat{t}_\sigma) + (t - \hat{t}_\sigma)^2 R(t, \sigma); \tag{1.87}$$

moreover, there exist $C > 0$ and a neighborhood of the point $(\hat{t}_0, 0)$ in which the entries of the matrix R satisfy the inequality

$$|R_{ij}(t, \sigma)| < C. \tag{1.88}$$

Substituting (1.87) in (1.86) and taking $\dot{\hat{x}} = \widehat{H}_p$ into account, we obtain

$$\mathfrak{J}_\sigma(t) = (t - \hat{t}_\sigma)^{n-k} \det\left(\frac{\partial \hat{x}_\sigma}{\partial \alpha} - \widehat{H}_p\frac{\partial \hat{t}_\sigma}{\partial \alpha}, \quad \widehat{H}_{pp}\frac{\partial \hat{p}_\sigma}{\partial \beta} + (t - \hat{t}_\sigma)R\right). \tag{1.89}$$

Lemma 1.8. *The matrix*

$$\left(\frac{\partial \hat{x}_0}{\partial \alpha} - \widehat{H}_p\frac{d\hat{t}_0}{d\alpha}, \quad \widehat{H}_{pp}\frac{\partial \hat{p}_0}{\partial \beta}\right) \tag{1.90}$$

is nonsingular for $t = \hat{t}_0$.

Proof. We assume that there exist $\lambda \in (\mathbb{R}^{n-k})^*$ and $\mu \in (\mathbb{R}^k)^*$ such that

$$\mu\left(\frac{\partial \hat{x}_0}{\partial \alpha} - \widehat{H}_p \frac{\partial \hat{t}_0}{\partial \alpha}\right) + \lambda \widehat{H}_{pp} \frac{\partial \hat{p}_0}{\partial \beta} = 0. \tag{1.91}$$

Taking the inner product of (1.91) and the vector $\lambda(\partial \hat{p}_0/\partial \beta)$, we obtain

$$\mu\left\langle \frac{\partial \hat{x}_0}{\partial \alpha} - \widehat{H}_p \frac{\partial \hat{t}_0}{\partial \alpha}, \lambda \frac{\partial \hat{p}_0}{\partial \beta} \right\rangle + \left\langle \lambda \widehat{H}_{pp} \frac{\partial \hat{p}_0}{\partial \beta}, \lambda \frac{\partial \hat{p}_0}{\partial \beta} \right\rangle = 0.$$

Because the vectors $\partial \hat{p}_0/\partial \beta$ are tangent to the manifold Q_0, the first summand in this sum vanishes; therefore,

$$\left\langle \lambda \widehat{H}_{pp} \frac{\partial \hat{p}_0}{\partial \beta}, \lambda \frac{\partial \hat{p}_0}{\partial \beta} \right\rangle = 0.$$

The matrix $\widehat{H}_{pp}$ is positive definite because the strengthened Legendre condition holds (see (1.20)); hence,

$$\lambda \frac{\partial \hat{p}_0}{\partial \beta} = 0.$$

Because $\partial \hat{p}_0/\partial \beta$ are linearly independent, this implies $\lambda = 0$. Equation (1.91) becomes

$$\mu\left(\frac{\partial \hat{x}_0}{\partial \alpha} - \widehat{H}_p \frac{d\hat{t}_0}{d\alpha}\right) = 0.$$

By condition (1.75), the vectors $d\hat{x}_0/d\alpha_i - \widehat{H}_p(d\hat{t}_0/d\alpha_i)$ are linearly independent; therefore, $\mu = 0$. Therefore, the columns of matrix (1.90) are linearly independent.

The determinant of matrix (1.90) is denoted by q. For definiteness, we assume that $q > 0$. Then, by the smoothness of the functions $x_\sigma(t, \alpha)$ and $p_\sigma(t, \alpha)$, there exist $\sigma_0 > 0$ and $\theta > 0$ such that for all $\hat{t}_0 - \theta < t < \hat{t}_0 + \theta$ and $\sigma < \sigma_0$,

$$\det\left(\frac{\partial \hat{x}_\sigma}{\partial \alpha}(t) - \widehat{H}_p(t)\frac{d\hat{t}_\sigma}{d\alpha}(t), \quad \widehat{H}_{pp}(t)\frac{\partial \hat{p}}{\partial \beta}(t)\right) > \frac{q}{2}. \tag{1.92}$$

The matrices under the determinant sign in (1.89) and (1.92) differ from one another by the summand $(t - \hat{t}_\sigma) \cdot R(t, \sigma)$, where R satisfies inequalities (1.88), i.e., is uniformly bounded in σ and t. This easily implies Proposition 1.5.

We return to the proof of Theorem 1.14. Choosing $\hat{t}_0 - \hat{t}_\sigma < \theta/2$, we obtain

$$\mathfrak{J}_\sigma(t) \neq 0 \quad \text{for } \hat{t}_\sigma < t < \hat{t}_0 + \frac{\theta}{2}. \tag{1.93}$$

The Jacobian $\mathfrak{J}(t) \neq 0$ on the closed interval $\hat{t}_0 + \theta/2 \leq t \leq \hat{t}_1$, and $\mathfrak{J}_\sigma(t)$ converges uniformly to $\mathfrak{J}(t)$ as $\sigma \to 0$. Consequently, for all sufficiently small $\sigma > 0$, we have

$$\mathfrak{J}_\sigma(t) \neq 0 \quad \text{for } \hat{t}_0 + \frac{\theta}{2} \leq t \leq \hat{t}_1. \tag{1.94}$$

Relations (1.93) and (1.94) allow us to construct a field of extremals with the initial manifold N_σ in a neighborhood of the extremal $\hat{x}(\cdot)$. Equation (1.83) means that the initial manifold N lies on the level surface of the function S_σ corresponding to the field $\mathfrak{P}_\sigma$ (see the remark on p. 55), i.e., the value of the initial functional on the part of the extremal lying between N and N_σ is the same for all extremals of the field $\mathfrak{P}_\sigma$. We now show that $\hat{x}(\cdot)$ realizes the strong minimum. Indeed, the trajectory $x(\cdot)$ lying in a sufficiently small neighborhood of the extremal $\hat{x}(\cdot)$ is in the domain covered by the field $\mathfrak{P}_\sigma$. If such a trajectory goes to N and gives the functional J a value less than $J(\hat{x}(\cdot))$, then, prolonging it to N_σ along the extremal of the field $\mathfrak{P}_\sigma$, we obtain a trajectory that lies in the domain covered by the field $\mathfrak{P}_\sigma$ and gives the functional J a value less than that on the corresponding extremal of the field. This contradicts Theorem 1.13.

Chapter 2
Riccati Equation in the Classical Calculus of Variations

§1. Riccati Equation as a Sufficient Condition for Positivity of the Second Variation

We consider the functional

$$J = \int_{t_0}^{t_1} f(t, x(t), \dot{x}(t))\, dt, \tag{2.1}$$

where $x \in \mathbb{R}^n$ and $f: \mathbb{R}^1 \times \mathbb{R}^n \times \mathbb{R}^n \to \mathbb{R}^1$ is a given smooth function of arguments belonging to a domain $\Omega \subset \mathbb{R}^1 \times \mathbb{R}^n \times \mathbb{R}^n$, which is the domain in our subsequent considerations. It is required to find a function $x(\cdot): \mathbb{R}^1 \to \mathbb{R}^n$ of class C^1 that satisfies the fixed boundary conditions

$$x(t_0) = a, \qquad x(t_1) = b \tag{2.2}$$

and is such that functional (2.1) attains the C^1-minimum value at it. To find minimum points of the functional J, we must equate the first variation, i.e., the first derivative of this functional with respect to $x(\cdot)$, to zero,

$$\delta J_{x(\cdot)} h(\cdot) = \int_{t_0}^{t_1} \left[\frac{\partial f}{\partial x}(t, x(t), \dot{x}(t)) h(t) + \frac{\partial f}{\partial \dot{x}}(t, x(t), \dot{x}(t))\, \dot{h}(t) \right] dt = 0,$$

which leads to the Euler equation

$$\frac{\partial f}{\partial x} - \frac{d}{dt} \frac{\partial f}{\partial \dot{x}} = 0 \tag{2.3}$$

for the unknown function $x(t)$.

We assume that there exists a solution to Eq. (2.3) with boundary condition (2.2). This solution is denoted by $\hat{x}(t)$. If $\hat{x}(\cdot)$ in fact realizes the minimum of functional (2.1), then the second variation of the considered functional at the point $\hat{x}(\cdot)$ is a nonnegative quadratic form on the space $C_0^1([t_0, t_1])$, i.e., on the space of functions $h(\cdot)$ having continuous derivatives on $[t_0, t_1]$ and satisfying the boundary conditions

$$h(t_0) = 0, \qquad h(t_1) = 0. \tag{2.4}$$

We recall that the second variation of the initial functional is defined by

$$\delta^2 J_{\hat{x}(\cdot)}(h(\cdot)) = \int_{t_0}^{t_1} \Phi(t, h, \dot{h}) \, dt$$

$$= \frac{1}{2} \int_{t_0}^{t_1} [\langle A(t)\dot{h}, \dot{h} \rangle + 2\langle C(t)\dot{h}, h \rangle + \langle B(t)h, h \rangle] \, dt, \quad (2.5)$$

where $A(t)$, $B(t)$, and $C(t)$ are $n \times n$ matrices with the respective entries

$$\left(\widehat{\frac{\partial^2 f}{\partial \dot{x}_i \partial \dot{x}_j}} \right), \qquad \left(\widehat{\frac{\partial^2 f}{\partial x_i \partial x_j}} \right), \qquad \text{and} \qquad \left(\widehat{\frac{\partial^2 f}{\partial x_i \partial \dot{x}_j}} \right).$$

The angle brackets in (2.5) denote the inner product. We note that the matrices A and B are symmetrical. We assume that the strengthened Legendre condition holds, i.e., the matrix A is positive definite.

As was shown in Chap. 1, the strong positivity of the second variation is a sufficient condition for a minimum. The Riccati equation is connected with the study of the quadratic form $\delta^2 J$. In particular, one of the main uses of the Riccati equation in the classical calculus of variations is for the proof of the positive definiteness of the second variation.

As in Chap. 1, we transform (2.5) by adding the total derivative of a certain quadratic form in h with variable coefficients to the integrand,

$$\frac{d}{dt} \langle W(t)h, h \rangle, \qquad W^\top = W. \quad (2.6)$$

We choose the coefficients for this form below. The addition of (2.6) does not change integral (2.5) itself, because of boundary conditions (2.4).

The addition of the total derivative (2.6) to the integrand of (2.5) transforms (2.5) into the form

$$\int_{t_0}^{t_1} [\langle A(t)\dot{h}, \dot{h} \rangle + 2\langle C(t)\dot{h}, h \rangle + \langle B(t)h, h \rangle + \langle \dot{W}h, h \rangle + 2\langle Wh, \dot{h} \rangle] \, dt. \quad (2.7)$$

We choose the coefficients for the quadratic form W such that the integrand of (2.7) is a full scalar square of a certain vector. We obtain the equation

$$\dot{W} = (C + W)A^{-1}(C^\top + W) - B, \quad (2.8)$$

which is called the (matrix) Riccati differential equation.

Lemma 2.1. If a symmetrical matrix $W(t_0)$ is the initial condition for the Riccati equation, then a solution to it is symmetrical for all t.

Proof. We consider the equation

$$\dot{W}^\top = (C + W^\top)A^{-1}(C^\top + W^\top) - B,$$

which is obtained by transposition of Eq. (2.8). Then if $W(t)$ is a solution to (2.8), $W^\top(t)$ is also a solution to it. These solutions have the same initial conditions; therefore, by the theorem on the uniqueness of a solution, $W(t) = W^\top(t)$ for all t.

We summarize the above arguments in the form of the following theorem.

Theorem 2.1. *If there exists a symmetrical solution $W(t)$ to Eq. (2.8) that is defined on the whole closed interval $[t_0, t_1]$, then functional (2.5) is positive definite on the space $C_0^1([t_0, t_1])$.*

§2. Riccati Equation for a Problem with Differential Constraints

The second approach to the study of the Riccati equation is related to the Hamilton–Jacobi equation for functional (2.5). We recall that the Hamilton–Jacobi equation is a partial differential equation in the unknown function $S(\tau, h)$. The function $S(\tau, h)$ is defined as the minimum value of functional (2.5) (in which it is necessary to set $t_1 = \tau$) attained on smooth functions $h(t)$ with the boundary conditions $h(t_0) = 0$ and $h(\tau) = h$. As shown in Chap. 1, the function $S(\tau, h)$ is a solution to the equation

$$\frac{\partial S}{\partial \tau} = -H\left(\tau, h, \frac{\partial S}{\partial h}\right), \tag{2.9}$$

where $H(\tau, h, p)$ is the Hamiltonian of functional (2.5), i.e., the Legendre transform of the integrand of (2.5) with respect to the variables $\dot{h}$.

In the case considered, the Legendre transform is as follows. We introduce the momentum

$$p = \frac{\partial \Phi}{\partial \dot{h}} = A\dot{h} + C^\top h. \tag{2.10}$$

Because of the strengthened Legendre condition, the determinant of the matrix A does not vanish. We can therefore express $\dot{h}$ in formula (2.10) as a function of h and p:

$$\dot{h} = A^{-1}p - A^{-1}C^\top h.$$

We then find that the Hamiltonian is determined as

$$H(\tau, h, p) = -\Phi + \langle p, \dot{h}\rangle$$
$$= -\frac{1}{2}\left[\langle A(t)\dot{h}, \dot{h}\rangle + 2\langle C(t)\dot{h}, h\rangle + \langle B(t)h, h\rangle\right] + \langle p, \dot{h}\rangle.$$

We substitute $A^{-1}p - A^{-1}C^\top h$ for $\dot{h}$ to obtain the Hamiltonian of (2.5)

$$\begin{aligned}
H(\tau, h, p) &= -\frac{1}{2}\big[\langle p - C^\top h, A^{-1}p - A^{-1}C^\top h\rangle \\
&\quad + 2\langle C(A^{-1}p - A^{-1}C^\top)h, h\rangle + \langle Bh, h\rangle \\
&\quad - 2\langle p, A^{-1}p - A^{-1}C^\top h\rangle\big] \\
&= \frac{1}{2}\big[\langle A^{-1}p, p\rangle - \langle Bh, h\rangle \\
&\quad + \langle CA^{-1}C^\top h, h\rangle - 2\langle p, A^{-1}C^\top h\rangle\big]. \tag{2.11}
\end{aligned}$$

Hamiltonian (2.11) depends quadratically on the variables p and h; therefore, we can seek solutions to the Hamilton–Jacobi equation that look like the quadratic form $S = -\langle W(t)h, h\rangle/2$, where the matrix W is symmetrical. Because $\partial S/\partial h = -Wh$, we have

$$-\frac{\partial S}{\partial \tau} = \frac{1}{2}\langle \dot{W}h, h\rangle$$

$$= \frac{1}{2}\langle A^{-1}Wh, Wh\rangle + \langle A^{-1}C^{\mathsf{T}}h, Wh\rangle + \frac{1}{2}\langle(CA^{-1}C^{\mathsf{T}} - B)h, h\rangle.$$

Symmetrizing the matrix of the quadratic form in the right-hand side, we obtain Riccati equation (2.8) for the matrix W.

2.1. Problem with Differential Constraints. Using this approach, we can also obtain the Riccati equation for more general variational problems, in particular, for problems with differential constraints. This problem is called the *Lagrange problem* in the calculus of variations. In the simplest form, it is formulated as follows. It is required to minimize the integral functional

$$J = \int_{t_0}^{t_1} f(t, x(t), \dot{x}(t))\, dt$$

on curves $x(t) \in C^1$ satisfying the boundary conditions

$$x(t_0) = a, \qquad x(t_1) = b$$

and the set of ordinary differential equations

$$\Phi(t, x(t), \dot{x}(t)) = 0, \tag{2.12}$$

where $x \in \mathbb{R}^n$ and $f\colon R^1 \times \mathbb{R}^n \times \mathbb{R}^n \to \mathbb{R}^1$ and $\Phi\colon \mathbb{R}^1 \times \mathbb{R}^n \times \mathbb{R}^n \to \mathbb{R}^m$ are smooth functions of arguments belonging to some domain $\Omega \subset \mathbb{R}^1 \times \mathbb{R}^n \times \mathbb{R}^n$. The Lagrange problem is nondegenerate if $m < n$. Otherwise, we minimize over a discrete (or even empty) set of trajectories.

2.2. Optimal Control Problem. We can transform the Lagrange problem into another form, which is more convenient for our further study. Fixing the values of the variables t and x for the moment, we consider differential constraints (2.12) as equations in $\dot{x}$. By the condition $m < n$, in the n-dimensional space of variables $\dot{x}$, this system defines a certain manifold $\mathfrak{N}$ (($n-m$)-dimensional in the general case). In a neighborhood of a nonsingular point of the manifold $\mathfrak{N}$,[1] we can introduce coordinates u that vary in a domain $U \in \mathbb{R}^{n-m}$ and parameterize the manifold $\mathfrak{N}$ in the neighborhood considered. In these coordinates, system (2.12) is equivalent to the system

---

[1] This means in a neighborhood of a value of $\dot{x}$ at which the Jacobi matrix $\partial\Phi/\partial\dot{x}$ has the maximal rank, which is equal to m.

$$\dot{x} = \phi(t, x, u), \tag{2.13}$$

where $u \in U$. Substituting (2.13) in functional (2.1), we obtain

$$J = \int_{t_0}^{t_1} F(t, x(t), u(t)) \, dt. \tag{2.14}$$

The optimal control problem is formulated as follows. Let $u(t)$ be an arbitrary measurable function taking its values in a set U, and let $x(t)$ be the absolutely continuous function that is the solution to Eq. (2.13) with $u(t)$ chosen. It is required to minimize functional (2.14) on solutions to the system of differential equations (2.13) with the boundary conditions

$$x(t_0) = a, \qquad x(t_1) = b.$$

The variable u is called the *control*. This term, whose name originated in applied extremal problems, is justified in this case because we can choose the function $u(t)$ arbitrarily. After such a choice, the solution to control system (2.13) is uniquely defined for a fixed initial value $x(t_0)$.

The case studied in optimal control theory differs from the classical statement of the Lagrange problem in that the domain U where the control assumes its value can be arbitrary in optimal control theory. More often, it is closed, and the values of the optimal control typically lie on the boundary of the set U. In the classical Lagrange problem, these values belong to an open set.

2.3. Linear-Quadratic Problem. We can use methods of optimal control theory to solve classical problems in the calculus of variations. For example, we consider the classical Lagrange problem from this standpoint.

The nonnegativity of the second variation of functional (2.1) on solutions to the variational equation for differerential constraint equations (2.12) is a necessary optimality condition for the Lagrange problem. As in the classical calculus of variations, this necessary condition is related to the following *associated* linear-quadratic problem. It is required to minimize the functional

$$\delta^2 J_{\hat{x}(\cdot)}(h(\cdot)) = \int_{t_0}^{t_1} \Phi(t, h, \dot{h}) \, dt$$

$$= \frac{1}{2} \int_{t_0}^{t_1} [\langle A(t)\dot{h}, \dot{h} \rangle + 2\langle C(t)\dot{h}, h \rangle + \langle B(t)h, h \rangle] \, dt$$

on the set of solutions to the linear system of differential equations

$$L(t)\dot{h}(t) + M(t)h(t) = 0 \tag{2.15}$$

with the boundary conditions

$$h(t_0) = h(t_1) = 0.$$

Here, $A(t)$, $B(t)$, and $C(t)$ denote the $n \times n$ matrices with the respective entries

$$\left(\widehat{\frac{\partial^2 f}{\partial \dot{x}_i \partial \dot{x}_j}}\right), \qquad \left(\widehat{\frac{\partial^2 f}{\partial x_i \partial x_j}}\right), \qquad \text{and} \qquad \left(\widehat{\frac{\partial^2 f}{\partial x_i \partial \dot{x}_j}}\right);$$

$L(t)$ and $M(t)$ denote the matrices of partial derivatives $\partial \widehat{\Phi}/\partial \dot{x}$ and $\partial \widehat{\Phi}/\partial x$. We can reformulate the necessary condition for the minimum in the Lagrange problem as the assertion that the minimum in the associated problem equals zero.

Like the Lagrange problem, the associated problem admits the reformulation in terms of optimal control theory. We assume that one of the minors of order m of the matrix $L(t)$ does not vanish for $t \in [t_0, t_1]$. Without loss of generality, we can assume that this minor corresponds to the first m components of the vector $\dot{h}$. We can then resolve system (2.15) with respect to the variables $\dot{h}$. The last $n-m$ components of the vector $\dot{h}$ are denoted by u:

$$\dot{h}_{m+1} = u_1, \quad \ldots, \quad \dot{h}_n = u_{n-m}. \tag{2.16}$$

Then system (2.15) becomes

$$\dot{h} = a(t)h + b(t)u, \tag{2.17}$$

where a and b are matrices of the respective sizes $n \times n$ and $n \times (n-m)$. We note that in accordance with notation (2.16), the last $n-m$ equations of system (2.17) have the form

$$\dot{h}_i = u_{i-m}, \quad i = m+1, \ldots, n.$$

Substituting (2.17) in functional (2.5), we obtain

$$I(h) = \frac{1}{2} \int_{t_0}^{t_1} [\langle P(t)u, u \rangle + 2\langle Q(t)h, u \rangle + \langle R(t)h, h \rangle] \, dt, \tag{2.18}$$

where $P(t)$ is an $(n-m) \times (n-m)$ matrix, $R(t)$ is an $n \times n$ matrix, and $Q(t)$ is an $(n-m) \times n$ matrix. We note that in (2.18), the angle brackets denote the inner product in $\mathbb{R}^{n-m}$ for the first two summands and the last angle brackets denote the inner product in $\mathbb{R}^n$.

The linear-quadratic optimal control problem consists in minimizing functional (2.18) with the boundary conditions

$$h(t_0) = h_0, \qquad h(t_1) = h_1 \tag{2.19}$$

on the set of solutions to system (2.17). The associated minimization problem of the second variations has zero boundary conditions: $h_0 = h_1 = 0$. We note that no constraints are imposed on the value of the control function $u(t)$ in this problem.

To derive the Riccati equation for the linear-quadratic optimal control problem, we need an analogue of the Hamilton–Jacobi equation, which is called the Bellman equation in optimal control theory (more precisely, it could be called the Hamilton–Jacobi equation in the Isaacs–Bellman form).

2.4. Bellman Equation. The state of the system (2.17), (2.18) at an instant of time t is characterized by the *phase vector* h and the value of the functional

$$I_{t_0}^t(u(\cdot), h(\cdot)) = \frac{1}{2} \int_{t_0}^t (\langle Pu, u \rangle + 2\langle Qh, u \rangle + \langle Rh, h \rangle) \, dt$$

being minimized that is realized at the instant of time t. As in the deduction of the Hamilton–Jacobi equation, we consider a family of optimal problems corresponding to the set of various initial states $(t, h, I_{t_0}^t)$ of the control system (2.17), (2.18). We let $S(t, h)$ denote the minimum value of the functional

$$I_t^{t_1} = \frac{1}{2} \int_t^{t_1} (\langle Pu, u \rangle + 2\langle Qh, u \rangle + \langle Rh, h \rangle) \, dt$$

that can be obtained on the set of solutions to system (2.17) with the boundary conditions $h(t) = h$ and $h(t_1) = h_1$. Obviously, the function $S(t, h)$ does not depend on $I_{t_0}^t$.

The introduction of the function S implicitly presupposes that a solution to the linear-quadratic optimal control problem exists for any initial values (t, h). The proof of the existence of a solution to the linear-quadratic problem can be found, e.g., in [63].

We assume that the function $S(t, h)$ is smooth. We let $\hat{u}(\cdot)$ and $\hat{h}(\cdot)$ denote the optimal control and optimal trajectory that correspond to the initial state $(t_0, h_0, 0)$ of the control system.

Lemma 2.2. *The following identity holds:*

$$S(t_0, h_0) = S(t, \hat{h}(t)) + I_{t_0}^t(\hat{u}, \hat{h}). \tag{2.20}$$

Proof. We show that the part of the optimal trajectory $\hat{h}(\cdot)$ that starts at the point $\hat{h}(t)$ is an optimal trajectory with respect to a new initial state $(t, \hat{h}(t), I_{t_0}^t)$ that is realized at the instant of time t. Indeed, if the value of $S(t, \hat{h}(t))$ is less than $I_t^{t_1}(\hat{u}, \hat{h})$ and is attained at a certain trajectory $\tilde{h}(\cdot)$ defined on the interval (t, t_1), then the concatenation trajectory

$$h = \hat{h} \qquad \text{on the interval } (t_0, t),$$
$$h = \tilde{h} \qquad \text{on the interval } (t, t_1)$$

gives functional (2.18) a value less than $S(t_0, h_0)$, which contradicts the definition of the function S.

Differentiating Eq. (2.20) with respect to t, we obtain

$$\frac{\partial S}{\partial t} + \frac{\partial S}{\partial h}(a\hat{h} + b\hat{u}) + \frac{1}{2}(\langle P\hat{u}, \hat{u} \rangle + 2\langle Q\hat{h}, \hat{u} \rangle + \langle R\hat{h}, \hat{h} \rangle) = 0. \tag{2.21}$$

Now let an arbitrary constant control v be chosen on the time interval $(t, t+\delta)$ in the system that is in the state (t, h). Let $\tilde{h}(\cdot)$ denote a solution to system

(2.17) in which $u = v$ with the initial conditions $(t, \hat{h}(t))$ is substituted. Then at the instant of time $t + \delta$, the system passes to the state $(t + \delta, \tilde{h}(t + \delta))$, $I_t^{t+\delta}(v, \tilde{h}(\cdot)))$, where

$$\tilde{h}(t + \delta) = \hat{h}(t) + (a(t)\hat{h}(t) + b(t)v)\delta + o(\delta). \qquad (2.22)$$

Starting from the instant of time $t + \delta$, we use the optimal control corresponding to the obtained initial state. Then the functional takes the value

$$I_t^{t+\delta}(v, \tilde{h}) + S(t + \delta, \tilde{h}(t + \delta)),$$

which should be not less than the minimum value of the functional, that is,

$$I_t^{t+\delta}(v, \tilde{h}) + S(t + \delta, \tilde{h}(t + \delta)) \geq S(t, \hat{h}(t)). \qquad (2.23)$$

Dividing inequality (2.23) by δ and passing to the limit as $\delta \to +0$, we obtain

$$\frac{\partial S}{\partial t} + \frac{\partial S}{\partial h}(a\hat{h} + b\hat{v}) + \frac{1}{2}(\langle Pv, v \rangle + 2\langle Q\hat{h}, v \rangle + \langle R\hat{h}, \hat{h} \rangle) \geq 0. \qquad (2.24)$$

We can join relation (2.20) and (2.24) by writing them in the form of one formula:

$$\min_u \left[\frac{\partial S}{\partial t} + \frac{\partial S}{\partial h}(ah + bu) + \frac{1}{2}(\langle Pu, u \rangle + 2\langle Qh, u \rangle + \langle Rh, h \rangle) \right] = 0. \qquad (2.25)$$

Equation (2.25) is called the *Bellman equation*. The boundary condition for the Bellman equation has the form

$$S(t_1, h) = 0. \qquad (2.26)$$

In the case where the matrix $P(t)$ is positive definite for all t (the strengthened Legendre condition holds), the minimum in Eq. (2.25) is attained at the unique point

$$u = \hat{u}\left(t, h, \frac{\partial S}{\partial h}\right). \qquad (2.27)$$

Substituting this minimizing value u in (2.25), we obtain a first-order partial differential equation.

By the Cauchy–Kowalevskaya theorem, a solution to Eq. (2.25) with initial condition (2.26) is defined in a certain neighborhood of the plane $t = t_1$ in the space of variables (t, h).

Theorem 2.2 (sufficient optimality condition). *Let the matrix $P(t)$ be positive definite. Let $S(t, h)$ be a smooth solution to Eq. (2.25) defined on a certain open set Ω that contains the plane $t = t_1$ and satisfies boundary conditions (2.26). Also, for each initial point $(t_0, h_0) \in \Omega$, let the trajectories of system (2.17) with the control $u = \hat{u}(t, h)$ obtained by the function $S(t, h)$ using (2.27) be defined and lie in the domain Ω on the whole time interval (t_0, t_1). Then $S(t, h)$ is the minimum value of functional (2.18) and the function $\hat{u}(t, h)$ is an optimal control for all points of the domain Ω.*

Proof. For any initial point $(t_0, h_0) \in \Omega$, we consider an arbitrary control $\tilde{u}(\cdot)$. Let $\tilde{h}(\cdot)$ be the corresponding control for system (2.17) with the initial conditions (t_0, h_0). By Bellman equation (2.25), the function

$$S(t, h(t)) + I_{t_0}^t(\tilde{u}(\cdot), \tilde{h}(\cdot))$$

is a nondecreasing function of the variable t. Consequently,

$$S(t_1, h(t_1)) + I_{t_0}^{t_1}(\tilde{u}(\cdot), \tilde{h}(\cdot)) \geq S(t_0, h_0). \tag{2.28}$$

The right-hand side of inequality (2.28) is equal to $I_{t_0}^{t_1}(\hat{u}(\cdot), \hat{h}(\cdot))$; therefore, the boundary condition $S(t_1, h) = 0$ leads to the inequality

$$I_{t_0}^{t_1}(\tilde{u}(\cdot), \tilde{h}(\cdot)) \geq I_{t_0}^{t_1}(\hat{u}(\cdot), \hat{h}(\cdot)).$$

The control defined by formula (2.27) linearly depends on h and $\partial S/\partial h$. Substituting this minimizing value u in (2.25), we obtain a first-order partial differential equation whose right-hand side depends quadratically on h and $\partial S/\partial h$. We can therefore seek its solution as a quadratic form $S = \langle W(t)h, h \rangle / 2$. As a result, as for the ordinary Hamilton–Jacobi equation, we obtain a differential equation with a quadratic right-hand side (the Riccati equation) for the matrix $W(t)$. By Theorem 1, the extendibility of the solution to this equation with the boundary condition $W(t_1) = 0$ to the whole interval $[t_0, t_1]$ gives a solution to the linear-quadratic problem.

If the linear-quadratic problem is obtained as the second variation of a certain functional, then the boundary conditions in this problem are zero. Therefore, the functional I on the set of solutions to linear system (2.17) with zero boundary conditions becomes homogeneous. Therefore, if the functional I assumes a negative value on a certain solution $\tilde{x}(t)$, then $\inf I = -\infty$ on the set $\lambda \tilde{h}(t)$, $\lambda \in \mathbb{R}_+^1$. Consequently, we have the alternatives: either there exists a solution $W(t)$ to the Riccati equation defined on the whole closed interval $[t_0, t_1]$, in which case the functional I is positive definite, or the solution $W(t)$ to the Riccati equation with the boundary condition $W(t_1) = 0$ on the closed interval $[t_0, t_1]$ goes to infinity, in which case, $\inf I = -\infty$.

§3. Riccati Equation and the Grassmann Manifold

The third approach, which is the most essential in what follows, is related to an important geometric object, called the *Grassmann manifold*. We recall that a *k-dimensional topological manifold* M is a topological space in which each point x has a neighborhood U_x that is homeomorphic to an open set $V_x \subset \mathbb{R}^k$. The set U_x and the corresponding homeomorphism $\phi: U_x \to V_x$ is called a *chart* on the manifold M. The coordinates of the points $\phi(y)$ for $y \in U_x$ are called *local coordinates* in this chart. If two charts (U_i, ϕ_i) and

(U_j, ϕ_j) are given at a point x, then we can consider *transition mappings* $g_{ij} = \phi_i(\phi_j)^{-1} \colon \mathbb{R}^k \to \mathbb{R}^k$ defined on the set $\phi_j(U_i \cap U_j)$.

In a sense, the mappings ϕ_i and ϕ_j are invisible in this construction: only transition mappings g_{ij} are given by analytic formulas. To obtain g_{ij} from certain mappings ϕ_i, we must require fulfillment of the natural "chain" condition: the relation $g_{ij}g_{jk} = g_{ik}$ holds for all points $y \in U_i \cap U_j \cap U_k$.

To define a smooth manifold, we must require that all the functions g_{ij} have one and the same class of smoothness, i.e., satisfy one of the following possible conditions:

1. they are continuously differentiable k times, i.e., are of class C^k;
2. they are infinitely differentiable, i.e., are of class C^∞;
3. they are analytic, i.e., are of class C^ω;
4. they are algebraic, i.e., are of class C^α.

A set of charts that cover M together with the corresponding transition functions is called an *atlas*. Two atlases are called *equivalent* if their union is also an atlas. The equivalence class of atlases assigns the structure of a smooth manifold on M. The class of smoothness of the transition mappings is called the *class of smoothness* of the given manifold.

3.1. Grassmann Manifold.

The *Grassmann manifold* $G_n(\mathbb{R}^{2n})$ is the set whose points are n-dimensional linear subspaces $\mathbb{R}^{2n}$.

To define the structure of a manifold on $G_n(\mathbb{R}^{2n})$, we consider the chart U constructed in the following way. In the $2n$-dimensional space of variables (h, p), $h = (h_1, \ldots, h_n)$, $p = (p_1, \ldots, p_n)$, we choose two n-dimensional planes $H_0 = \{(h, 0)\}$ and $H_\infty = \{(0, p)\}$, which are respectively called the *horizontal* plane and the *vertical* plane.

We consider two projections $\pi_0 \colon (h, p) \mapsto (h, 0)$ and $\pi_\infty \colon (h, p) \mapsto (0, p)$. The first of these projections is the projection of the space (h, p) onto H_0 parallel to H_∞, and the second is the projection onto H_∞ parallel to H_0.

The domain of the chart U is the set of n-dimensional subspaces $\mathbb{R}^{2n}$ that are transversal to the vertical plane H_∞ (i.e., intersecting with H_∞ at only one point, the origin). The projection π_0 of any subspace $W \in U$ onto the plane H_0 parallel to H_∞ is one-to-one. Indeed, if a certain nonzero vector is projected into zero, then it belongs to H_∞; this contradicts the transversality of W and H_∞. In what follows, by π_0 we mean the restriction of π_0 to the considered subspace W.

We fix bases in the planes H_0 and H_∞ and consider the matrix Z, which assigns the mapping $\pi_\infty \circ (\pi_0)^{-1}$. Entries of the matrix Z are local coordinates of the plane W in the chart considered. (In what follows, we use one and the same symbol W for a subspace as well as for its local coordinates.) We can write the vectors of the plane W in the form (h, Wh). We define other charts similarly; only the pair of transversal planes H_0 and H_∞ is changed.

To find transition functions, we need the following construction. We consider two charts U and $\tilde{U}$ on the Grassmann manifold. Let a plane W belong to $U \cap \widetilde{U}$. We find coordinates of the plane W in the chart $\widetilde{U}$. For this, we consider the transformation $\mathfrak{A} \colon \mathbb{R}^{2n} \to \mathbb{R}^{2n}$, which transforms a pair of planes H_0, H_∞ to a pair of planes $\widetilde{H}_0$, $\widetilde{H}_\infty$. Let

$$\begin{pmatrix} \mathfrak{A}_{11} & \mathfrak{A}_{12} \\ \mathfrak{A}_{21} & \mathfrak{A}_{22} \end{pmatrix}$$

be the block $(n \times n)$ decomposition of the matrix $\mathfrak{A}$.

Because $W \in \widetilde{U}$, it is transversal to $\widetilde{H}_\infty$. We find the coordinates $\widetilde{W}$ of the plane W in the chart $\widetilde{U}$ as follows. We have

$$(\tilde{h}, \tilde{p}) = ((\mathfrak{A}_{11} + \mathfrak{A}_{12}W)h, (\mathfrak{A}_{21} + \mathfrak{A}_{22}W)h).$$

The matrix $\mathfrak{A}_{11} + \mathfrak{A}_{12}W$ is invertible because if there exists a vector $h^* \in \mathrm{Ker}(\mathfrak{A}_{11} + \mathfrak{A}_{12}W)$, $h^* \neq 0$, then the point (h^*, Wh^*) of the plane W has the coordinates $(0, p^*)$ in the chart $\widetilde{U}$, i.e., the plane W has a nontrivial intersection with the plane H_∞. Consequently, $h = (\mathfrak{A}_{11} + \mathfrak{A}_{12}W)^{-1}\tilde{h}$, and we can write the point $(\tilde{h}, \tilde{p})$ in the form $(\tilde{h}, (\mathfrak{A}_{21} + \mathfrak{A}_{22}W)(\mathfrak{A}_{11} + \mathfrak{A}_{12}W)^{-1}\tilde{h})$.

Therefore, the coordinates of the point $W \in G_n(\mathbb{R}^{2n})$ in the chart $\widetilde{U}$ is the matrix

$$\widetilde{W} = (\mathfrak{A}_{21} + \mathfrak{A}_{22}W)(\mathfrak{A}_{11} + \mathfrak{A}_{12}W)^{-1}. \tag{2.29}$$

Transformation (2.29) is called the *generalized linear-fractional transformation*.

Therefore, generalized linear-fractional transformations are transition mappings from the chart U to the chart $\widetilde{U}$. Because these mappings are analytic, the structure of an analytic manifold is introduced on the Grassmann manifold.

3.2. Riccati Equation as a Flow on the Grassmann Manifold.

Let a linear system of ordinary differential equations with smooth coefficients be given in the space $\mathbb{R}^{2n}$. It is known that solutions of such a system are infinitely continued in time [78]. This means that for each finite instant of time t_1 and for each initial state $(t_0, x_0) \in \mathbb{R}^1 \times \mathbb{R}^{2n}$, the value of the solution $x(t_1)$ to the Cauchy problem is defined at the instant of time t_1. Fixing instants of times t_0 and t_1 and changing $x_0 \in \mathbb{R}^{2n}$, we obtain the mapping $\Gamma(t_0, t_1) \colon \mathbb{R}^{2n} \to \mathbb{R}^{2n}$, which assigns $x_1 \in \mathbb{R}^{2n}$, the final point of the trajectory at the instant of time t_1, to the initial point $x_0 \in \mathbb{R}^{2n}$ at the instant of time t_0. By the linearity of the system considered, $\Gamma(t_0, t_1)$ is a nondegenerate linear operator.

The set of operators $\Gamma(t_0, t_1)$ has the group property

$$\Gamma(t_1, t_2) \circ \Gamma(t_0, t_1) = \Gamma(t_0, t_2).$$

We say that this family forms a *nonstationary flow* Γ on $\mathbb{R}^{2n}$. By the linearity and nondegeneracy, the operators $\Gamma(t_0, t_1)$ transform n-dimensional subspaces into n-dimensional subspaces. Thus, the flow Γ induces the flow $\widetilde{\Gamma}$ on the Grassmann manifold $G_n(\mathbb{R}^{2n})$. We show that the vector field that defines the flow $\tilde{\Gamma}$, is given by the field of right-hand sides of Riccati equations. In other words, if we suppose that planes W are transferred by the action of the flow described by the linear canonical system with Hamiltonian (2.11), then the Riccati equation arises as an equation for local coordinates W on the Grassmann manifold.

The canonical system of equations with Hamiltonian (2.11) has the form

$$\begin{aligned} \dot{h} &= -A^{-1}C^{\top}h + A^{-1}p, \\ \dot{p} &= (B - CA^{-1}C^{\top})h + CA^{-1}p. \end{aligned} \tag{2.30}$$

The mapping $\Gamma(t_0, t_1)$ describing the shift along trajectories of system (2.30) at time $[t_0, t_1]$ is given by the fundamental matrix of solutions to system (2.30).

Theorem 2.3. *If*

$$\Gamma(t_0, t) = \begin{pmatrix} \Gamma_{11} & \Gamma_{12} \\ \Gamma_{21} & \Gamma_{22} \end{pmatrix}$$

is the fundamental matrix of solutions to system (2.30), *then*

$$W = (\Gamma_{21} + \Gamma_{22}W_0)(\Gamma_{11} + \Gamma_{12}W_0)^{-1}$$

is a solution to Riccati equation (2.8).

Proof. The fact that Γ is the fundamental solution means that $\dot{\Gamma} = \mathfrak{A}\Gamma$, where $\mathfrak{A}$ is the matrix of coefficients of system (2.30). Consequently, we obtain

$$\begin{aligned} \dot{\Gamma}_{11} &= -A^{-1}C^{\top}\Gamma_{11} + A^{-1}\Gamma_{21}, \\ \dot{\Gamma}_{12} &= -A^{-1}C^{\top}\Gamma_{12} + A^{-1}\Gamma_{22}, \\ \dot{\Gamma}_{21} &= (B - CA^{-1}C^{\top})\Gamma_{11} + CA^{-1}\Gamma_{21}, \\ \dot{\Gamma}_{22} &= (B - CA^{-1}C^{\top})\Gamma_{12} + CA^{-1}\Gamma_{22}. \end{aligned}$$

Therefore, differentiating the formula

$$W = (\Gamma_{21} + \Gamma_{22}W_0)(\Gamma_{11} + \Gamma_{12}W_0)^{-1},$$

we obtain

$$\begin{aligned} \dot{W} &= [(B - CA^{-1}C^{\top})\Gamma_{11} + CA^{-1}\Gamma_{21} + (B - CA^{-1}C^{\top})\Gamma_{12}W_0 \\ &\quad + CA^{-1}\Gamma_{22}W_0](\Gamma_{11} + \Gamma_{12}W_0)^{-1} \\ &\quad - (\Gamma_{21} + \Gamma_{22}W_0)(\Gamma_{11} + \Gamma_{12}W_0)^{-1}[-A^{-1}C^{\top}\Gamma_{11} + A^{-1}\Gamma_{21} \\ &\quad - A^{-1}C^{\top}\Gamma_{12}W_0 + A^{-1}\Gamma_{22}W_0](\Gamma_{11} + \Gamma_{12}W_0)^{-1} \\ &= (B - CA^{-1}C^{\top}) + CA^{-1}W + WA^{-1}C^{\top} - WA^{-1}W. \end{aligned}$$

The change $W \mapsto -W$ transforms the obtained equation into the equation

$$\dot{W} = (C + W)A^{-1}(C^{\top} + W) - B.$$

§4. Grassmann Manifolds of Lower Dimension

We can consider Grassmann manifolds over various numerical fields. More-over, there exist many "similar" geometric objects that have a close geometric nature. Before giving specific examples of Grassmann manifolds, we present certain modifications of the initial definition.

Definition 2.1. A linear subspace of the space $\mathbb{R}^{2n}$ is called *oriented* if a family of bases is chosen in it such that the passage from one basis to another is realized by a matrix with a positive determinant.

It is sometimes important to distinguish manifolds of oriented and nonori-ented linear subspaces. In the first case, two geometrically coinciding sub-spaces with opposite orientations correspond to two distinct points of the Grassmann manifold; in the second case, two such subspaces are considered as one and the same point of the Grassmann manifold. If it is important to stress the fact that we deal with the first case, we use the notation $G_m^+(\mathbb{R}^n)$. A point on the manifold $G_m(\mathbb{R}^n)$, the initial plane W without accounting for its orientation, corresponds to each point $W \in G_m^+(\mathbb{R}^n)$. This mapping is denoted by $p\colon G_m^+(\mathbb{R}^n) \to G_m(\mathbb{R}^n)$.

In contrast, because there are two opposite orientations on a given plane W, two distinct points of the manifold $G_m^+(\mathbb{R}^n)$ correspond to each point of $G_m(\mathbb{R}^n)$. In other words, the inverse mapping p^{-1} is two-valued. Two inverse images of the point $W \in G_m(\mathbb{R}^n)$ are always far from one another in the sense of the topology of the manifold $G_m^+(\mathbb{R}^n)$; therefore, each inverse image has a neighborhood in which the mapping p is one-to-one. In this case, we say that the mapping p is a *double covering* of the manifold $G_m(\mathbb{R}^n)$.

The *complex Grassmann manifold* $G_n(\mathbb{C}^{2n})$ is the set whose points are complex n-dimensional linear subspaces of the space $\mathbb{C}^{2n}$. We introduce the structure of an analytic manifold on $G_n(\mathbb{C}^{2n})$ in exactly the same way as on $G_n(\mathbb{R}^{2n})$.

It is useful to geometrically imagine the Grassmann manifolds for small values of n. We present several simple examples.

Example 2.1. Let $n = 1$. Then $G_1(\mathbb{R}^2)$ is the set of lines in the plane that pass through the origin. By definition, this is the projective line $\mathbb{R}\mathbb{P}^1$.

Example 2.2. The manifold $G_1^+(\mathbb{R}^2)$ of oriented lines on the plane that pass through the origin is the circle S^1.

Example 2.3. The manifold $G_1(\mathbb{C}^2)$ of complex lines in $\mathbb{C}^2$ that pass through the origin is the complex projective space $\mathbb{C}\mathbb{P}(1)$.

The simplest nontrivial example arises for $n = 2$.

Theorem 2.4. *The manifold $G_2^+(\mathbb{R}^4)$ is homeomorphic to the direct prod-uct $S^2 \times S^2$.*

It is most convenient to prove this theorem in terms of quaternions, which, incidentally, are repeatedly needed in what follows. We therefore use this occasion for a small digression.

4.1. Quaternions. On October 16, 1843, W. R. Hamilton found fundamental formulas for multiplying quaternions. Hamilton himself described this event in a letter to his son [42]: "As I was going to the meeting of the Irish Royal Academy accompanied by your mother, in spite of her discussion with me, my subconscious ideas were so clear that there finally arose a result whose importance I immediately felt. I kept imagining an electric circuit was closed and a spark flared up... I could not suppress the impulse, not in essence philosophical, to write the basic formulas on the stone of Brigham Bridge which we passed by. Of course, they were rubbed out as a superscription long ago. However, a firmer trace remained in books...."[2]

We recall the definition and main properties of quaternions. Quaternions are constructed as the associative algebra over the field $\mathbb{R}$ with the generators 1, i, j, and k and the relations

$$i^2 = j^2 = k^2 = -1,$$
$$ij = -ji = k, \qquad jk = -kj = i, \qquad ki = -ik = j. \tag{2.31}$$

The set of all quaternions is denoted by $\mathbb{H}$.

We therefore can write any quaternion in the form

$$x = x_0 + x_1 i + x_2 j + x_3 k, \tag{2.32}$$

where x_0, x_1, x_2, and x_3 are real numbers. The quaternion $\bar{x} = x_0 - x_1 i - x_2 j - x_3 k$ is called *conjugate* to x.

It follows from (2.31) that

$$x\bar{x} = \bar{x}x = x_0^2 + x_1^2 + x_2^2 + x_3^2.$$

This value is called the *squared module* and is denoted by $|x|^2$. Consequently, for any nonzero x, there exists an inverse element defined by $x^{-1} = \bar{x}/|x|^2$. Similar to complex numbers, the value x_0 is called the *real part* of x and is denoted by $\mathfrak{R}\,x$; the difference $x - x_0$ is called the *imaginary part* of x and is denoted by $\mathfrak{I}\,x$. It is easy to verify that the module of a product of quaternions is equal to the product of the modules; therefore, the set of quaternions whose module equals 1 is a multiplicative group. We can set one and only one point (a_0, a_1, a_2, a_3) of the Euclidean space $\mathbb{R}^4$ with a fixed basis in correspondence to each quaternion $a_0 + a_1 i + a_2 j + a_3 k$. This correspondence is called the *natural identification*. In what follows, we constantly use it.

Proof of Theorem 2.4. We consider $\mathbb{R}^4$ as the set of quaternions $a_0 + a_1 i + a_2 j + a_3 k$. Let an element $W \in G_2^+(\mathbb{R}^4)$ be defined by its orthonormal basis,

---

[2] [Translation of the Russian translation.]

i.e., by the pair of mutually orthogonal unit vectors x and y. Considering these vectors as quaternions, we form two more quaternions

$$u = xy^{-1}, \qquad v = x^{-1}y. \qquad (2.33)$$

Lemma 2.3. *The quaternions u and v are purely imaginary. Their module is equal to 1.*

Proof. The module of the quaternions u and v is equal to 1 as the product of unit quaternions. We prove that they are purely imaginary. We consider the quaternion u (for v, the proof is similar). Let

$$x = a_0 + a_1 i + a_2 j + a_3 k,$$
$$y = b_0 + b_1 i + b_2 j + b_3 k.$$

Because $y^{-1} = \bar{y}$, we have $\Re u = a_0 b_0 + a_1 b_1 + a_2 b_2 + a_3 b_3$. Because x and y are orthogonal vectors, we have $\langle x, y \rangle = 0$. Therefore, the quaternion u is purely imaginary.

Lemma 2.4. *The quaternions u and v do not depend on the choice of an orthonormal basis in the oriented plane W.*

Proof. Let the matrix

$$\begin{pmatrix} \alpha & \beta \\ \gamma & \delta \end{pmatrix}$$

be orthogonal and *unimodular* (i.e., its determinant is equal to $+1$). The passage to a new basis in the plane (x, y) is described by the equations

$$\tilde{x} = \alpha x + \beta y,$$
$$\tilde{y} = \gamma x + \delta y.$$

From the orthogonality condition of base vectors and the unimodularity of the transition matrix, we obtain

$$\begin{aligned}
\tilde{x}\tilde{y}^{-1} &= (\alpha x + \beta y)\overline{(\gamma x + \delta y)} \\
&= (\alpha x + \beta y)(\gamma \bar{x} + \delta \bar{y}) \\
&= \alpha \gamma x \bar{x} + \alpha \delta x \bar{y} + \beta \gamma y \bar{x} + \beta \delta y \bar{y} \\
&= (\alpha \gamma + \beta \delta) + (\alpha \delta - \beta \gamma)u = u.
\end{aligned}$$

In this calculation, we use the fact that $x\bar{x} = 1$, $y\bar{y} = 1$, $x\bar{y} = xy^{-1} = u$, and $y\bar{x} = yx^{-1} = \bar{u} = -u$, and also that $\alpha \gamma + \beta \delta = 0$ and $\alpha \delta - \beta \gamma = 1$.

We have thus obtained the mapping $\lambda \colon G_2^+(\mathbb{R}^4) \to S^2 \times S^2$ under which W is mapped to $(u, v) \in S^2 \times S^2$.

Lemma 2.5. *The mapping $\lambda \colon G_2^+(\mathbb{R}^4) \to S^2 \times S^2$ is continuous.*

The proof of this lemma is obvious, because the quaternions u and v vary continuously when x and y vary continuously.

To prove that the mapping λ is a homeomorphism, we construct the inverse mapping. We consider a point $(u, v) \in S^2 \times S^2$. We set the quaternion $uq - qv$ in correspondence to each quaternion q (u and v are the pair of purely imaginary unit quaternions). We have thus constructed the mapping

$$\mathfrak{B}_{u,v} \colon \mathbb{H} \to \mathbb{H}, \qquad \mathfrak{B}_{u,v}(q) = uq - qv. \tag{2.34}$$

This transformation is linear.

Lemma 2.6. *Under the identification of $\mathbb{H}$ with $\mathbb{R}^4$, the multiplication matrix (to the left as well as to the right) by a purely imaginary quaternion is skew symmetrical.*

Proof. It is easy to verify that multiplication by any of the purely imaginary base quaternions i, j, and k has a skew-symmetrical matrix. Because any purely imaginary quaternion is represented in the form of a linear combination of i, j, and k, its matrix is also skew symmetrical.

Lemma 2.7. *The matrix of the mapping*

$$\Phi_{u,v} \colon \mathbb{H} \to \mathbb{H}, \qquad \Phi_{u,v}(q) = uqv,$$

where u and v are purely imaginary unit quaternions, is orthogonal and symmetrical.

Proof. The orthogonality of $\Phi_{u,v}$ is implied by the fact that $|uqv| = |q|$.

We prove the symmetry. By Lemma 2.6, multiplication by a purely imaginary quaternion is represented by a skew-symmetrical matrix. Let multiplication to the left by u be represented by the matrix Ψ_u, and let multiplication to the right by v be represented by the matrix Ξ_v. Then

$$\Phi_{u,v}^\top = (\Psi_u \Xi_v)^\top = \Xi_v^\top \Psi_u^\top = \Xi_v \Psi_u.$$

Because the multiplication in $\mathbb{H}$ is associative, we have $\Xi_v \Psi_u = \Psi_u \Xi_v = \Phi_{u,v}$.

Corollary 2.1. *Eigenvalues of the matrix $\Phi_{u,v}$ are real, and their module equals 1.*

Lemma 2.8. *The trace of the mapping $\Phi_{u,v}$ equals 0.*

Proof. Let $\Phi_{u,v}(q)_\alpha$, where $\alpha = 1, i, j, k$, denote the α-component of the quaternion $\Phi_{u,v}(q)$. Let $u = a_1 i + a_2 j + a_3 k$ and $v = b_1 i + b_2 j + b_3 k$. It is easy to verify that

$$\Phi_{u,v}(1)_1 = -(a_1 b_1 + a_2 b_2 + a_3 b_3). \tag{2.35}$$

Because

$$\Phi_{u,v}(i) = (a_1 i + a_2 j + a_3 k) i (b_1 i + b_2 j + b_3 k),$$

we have

$$\Phi_{u,v}(i)_i = -a_1b_1 + a_2b_2 + a_3b_3. \tag{2.36}$$

Similarly,

$$\Phi_{u,v}(j)_j = a_1b_1 - a_2b_2 + a_3b_3, \tag{2.37}$$

$$\Phi_{u,v}(k)_k = a_1b_1 + a_2b_2 - a_3b_3. \tag{2.38}$$

Adding relations (2.35)–(2.38), we obtain

$$\operatorname{Tr}\Phi_{u,v} = \Phi_{u,v}(1)_1 + \Phi_{u,v}(i)_i + \Phi_{u,v}(j)_j + \Phi_{u,v}(k)_k = 0.$$

Corollary 2.2. *The eigenvalues $+1$ and -1 of the matrix $\Phi_{u,v}$ are of multiplicity 2.*

The proof is implied by Corollary 2.1 and Lemma 2.8.

Lemma 2.9. *The image of the mapping $\mathfrak{B}_{u,v}$, $\operatorname{Im}\mathfrak{B}_{u,v}$, is a two-dimensional linear subspace that corresponds to the eigenvalue $+1$ of the matrix $\Phi_{u,v}$.*

Proof. We note that $u^2 = -u\bar{u} = -1$ and $v^2 = -v\bar{v} = -1$. Because

$$\Phi_{u,v}(uq - qv) = u(uq - qv)v = -qv + uq,$$

any vector lying in the image of $\mathfrak{B}_{u,v}$ is an eigenvector of the mapping $\Phi_{u,v}$ with the eigenvalue $+1$.

The matrix of the mapping $\mathfrak{B}_{u,v}$ is skew symmetrical and is hence of an even rank. This rank cannot be equal to 0 (because the matrix $\mathfrak{B}_{u,v} \neq 0$) and is not equal to 4 (because the image does not contains vectors corresponding to the eigenvalue -1). Consequently, $\operatorname{rk}\mathfrak{B}_{u,v} = 2$ and $\operatorname{Im}\mathfrak{B}_{u,v}$ is two-dimensional.

Let $\mu: S^2 \times S^2 \to G_2(\mathbb{R}^4)$ denote the mapping defined by $\mu(u,v) = \operatorname{Im}\mathfrak{B}_{u,v}$. It is obvious from formula (2.34) defining the operator $\mathfrak{B}_{u,v}$ that the mapping μ is continuous. For each point $(u,v) \in S^2 \times S^2$, this mapping defines a two-dimensional plane. To choose the orientation of this plane, we need an additional construction.

In what follows, we mark the symbols corresponding to points of the manifold $G_2^+(\mathbb{R}^4)$ with the superscript $+$. For example, W^+ denotes the plane $W \in G_2(\mathbb{R}^4)$ if the orientation is given on it and it is considered as a point of the manifold $G_2^+(\mathbb{R}^4)$.

Lemma 2.10. *Let $W \in G_2(\mathbb{R}^4)$ be a (nonoriented) plane with an orthonormal basis x, y, and let $\lambda(x,y) = (u,v)$. Then $\mu(u,v) = W$.*

Proof. The mapping λ transforms the plane W to a pair of purely imaginary quaternions u and v:

$$u = xy^{-1}, \qquad v = x^{-1}y, \qquad (u,v) \in S^2 \times S^2.$$

The mapping μ transforms this pair in the image of the mapping $\mathfrak{B}_{u,v}$, the two-dimensional plane consisting of the eigenvectors of $\Phi_{u,v}$ with the eigenvalue $+1$.

At the same time, the vectors x and y are eigenvectors of $\Phi_{u,v}$ with the eigenvalue $+1$. Indeed, the fact that $uxv = x$ and $uyv = y$ is a simple consequence of definition (2.33) of the quaternions u and v. Consequently, the vectors x and y lie in the plane W.

We have thus constructed the mapping μ, which acts from $S^2 \times S^2$ into $G_2(\mathbb{R}^4)$. Using this mapping, we construct the mapping $\mu^+: S^2 \times S^2 \to G_2^+(\mathbb{R}^4)$. For this, we must introduce the orientation in the plane $\operatorname{Im}\mathfrak{B}_{u,v}$.

Let $W_0^+ \in G_2^+(\mathbb{R}^4)$ be a two-dimensional plane with an orienting basis x_0, y_0. The image of this plane under the mapping λ is denoted by (u_0, v_0). The orientation of the plane $\operatorname{Im}\mathfrak{B}_{u_0,v_0}$ is defined such that it coincides with the orientation of the plane W_0^+ that is generated by the basis x_0, y_0. This means that at a given point (u_0, v_0), we choose one of the inverse images of the point $\operatorname{Im}\mathfrak{B}_{u,v}$ with respect to the double covering $p: G_2^+(\mathbb{R}^4) \to G_2(\mathbb{R}^4)$. This inverse image is denoted by $\operatorname{Im}\mathfrak{B}_{u,v}^+$.

Lemma 2.11. *The choice* $\operatorname{Im}\mathfrak{B}_{u,v}^+$ *of the inverse image of one point can be extended up to a continuous mapping of a certain neighborhood* $U \subset S^2 \times S^2$ *in* $G_2^+(\mathbb{R}^4)$.

Proof. The mapping p is a covering, and a neighborhood of the point $\operatorname{Im}\mathfrak{B}_{u,v}$ on the manifold $G_2(\mathbb{R}^4)$ is hence homeomorphic to a neighborhood of the point $\operatorname{Im}\mathfrak{B}_{u,v}^+$ (with orientation introduced in the corresponding plane) on the manifold $G_2^+(\mathbb{R}^4)$. Consequently, there exists a neighborhood U of the point $(u_0, v_0) \in S^2 \times S^2$ such that the mapping $\mu^+: U \to G_2^+(\mathbb{R}^4)$ with the values at the point (u_0, v_0) constructed above is defined in it.

Corollary 2.3. *In the neighborhood* U, *the mapping* μ^+ *is inverse to the mapping* λ.

Lemma 2.12. *The mapping* μ^+ *admits a continuous extension to the whole manifold* $S^2 \times S^2$. *The obtained extension of the mapping* μ^+ *is inverse to the mapping* λ.

To prove the first part of Lemma 2.12, we must briefly digress into the field of topology.

4.2. Homotopic Paths. Let M be a smooth manifold.

Definition 2.2. A *closed path* is a continuous mapping $\gamma: S^1 \to M$ of the circle S^1 into M.

In other words, a closed path is a periodic function $\gamma(s)$ of the scalar argument s that takes its values in M. A path is called *identical* if this function is constant.

Definition 2.3. Two paths γ_1 and γ_2 are called *homotopic* if there exists a continuous mapping $f \colon S^1 \times [0,1] \to M$ such that

$$f(s,0) = \gamma_1(s), \qquad f(s,1) = \gamma_2(s).$$

In other words, a homotopy is a process of continuous deformation in which one path passes into the other.

Definition 2.4. A manifold M is called *simply connected* if any closed path in M is homotopic to the identical one.

Example 2.4. The interior of a disk is a simply connected manifold. A homotopy is realized by a homothetic contraction of the disk to its center.

Example 2.5. The circle is not simply connected, because a closed path that performs a complete turn along the circle cannot be contracted to a point by its continuous varying on the circle.

Example 2.6. The sphere S^2 is simply connected. To prove this, it suffices to consider the stereographic projection of the sphere from a point not belonging to a given path onto the plane and use Example 2.1.

Example 2.7. The direct product of two spheres $S^2 \times S^2$ is simply connected. We can perform a homotopy along each of the factors separately.

Proof of Lemma 2.12. We consider an arbitrary continuous curve $x \colon [0,1] \to S^2 \times S^2$. By Lemma 2.11, the mapping μ^+ is extended up to a continuous mapping of the closed interval $[0,1]$ into the manifold $G_2^+(\mathbb{R}^4)$.

The mapping μ^+ is not continuous iff there exists two distinct curves that go to one and the same point and lead to opposite orientations of the plane $\mathrm{Im}\,\mathfrak{B}_{uv}$. In this case, these curves form a closed path γ such that when varying along it, the orientation of the plane $\mathrm{Im}\,\mathfrak{B}_{uv}$ is changed to the opposite one.

By Example 3.4, the path γ is homotopic to the identical one. Because the orientation varies continuously in the process of homotopy, we can obtain an arbitrarily small closed path on which the orientation varies continuously and assumes opposite values; this contradicts Lemma 2.11. The first part of the lemma is proved.

To prove the second part of Lemma 2.12, it suffices to observe the following. First, by Lemma 2.10, the mapping $p \circ \mu^+$ coincides with the mapping $p \circ \lambda^{-1}$. By the first part of Lemma 2.12, the orientation of the planes obtained as a result of these mappings also coincide. Therefore, $\mu^+ = \lambda^{-1}$.

Remark. The first part of the statement of Lemma 2.12 is a consequence of the general assertion stating that a bundle is orientable if its basis is simply connected ([91], Chap. 5, Sec. 6).

Lemma 2.12 completes the proof of Theorem 1.

Exercise. Prove that the manifold $G_2(\mathbb{R}^4)$ is diffeomorphic to the manifold $S^2 \times S^2$, in which the points (u, v) and $(-u, -v)$ are identified, i.e., it is diffeomorphic to the quotient manifold of the direct product of two two-dimensional spheres by the action of the group $\mathbb{Z}_2$: $(u, v) \mapsto (-u, -v)$.

We give one more interpretation of $G_2(\mathbb{R}^4)$, which is important in what follows. In $\mathbb{R}^4$, we consider the three-dimensional plane $x_1 = 1$. The projective compactification of this plane (that is, the addition to the three-dimensional plane of improper elements each of which corresponds to the family of parallel lines of the initial plane) is the three-dimensional projective space $\mathbb{RP}^3$. Each two-dimensional space of $\mathbb{R}^4$ intersects $\mathbb{RP}^3$ along a projective line (proper or improper). Therefore, $G_2(\mathbb{R}^4)$ is interpreted as the set of all lines in $\mathbb{RP}^3$. Therefore, e.g., any ruled surface (a hyperboloid, etc.) is a curve in $G_2(\mathbb{R}^4)$.

Theorem 2.3 shows an important role that Grassmann manifolds play in the study of the Riccati equation. To describe the geometry of these manifolds themselves, we need some initial facts in the theory of Lie groups and algebras. The following chapter is devoted to the presentation of the foundation of this theory.

Chapter 3
Lie Groups and Lie Algebras

§1. Lie Groups: Definition and Examples

A *Lie group* is a smooth manifold on which a group structure is defined. Moreover, the group operations (multiplication and inversion) are required to be given by smooth functions in local coordinates.

In this definition, smoothness can be understood as smoothness of the order $k \geq 1$, as infinite differentiability, or as analyticity. All three variants lead to equivalent theories. There are essential distinctions only between the analytic and the algebraic case. To be more specific, we consider the case C^∞ in what follows.

We present the simplest examples.

The space $\mathbb{R}^n$, considered as an additive group, is the simplest example of a Lie group.

The quotient group $\mathbb{R}/\mathbb{Z}$ (or, what is the same, the multiplicative group of complex numbers whose module is equal to 1), considered as a manifold, is the circle S^1.

The direct products of groups $\mathbb{R}/\mathbb{Z} \times \ldots \times \mathbb{R}/\mathbb{Z}$, considered as manifolds, are multidimensional tori $S^1 \times \ldots \times S^1$.

The group of quaternions whose module is equal to 1, considered as a manifold, is the three-dimensional sphere S^3. This group is denoted by Spin(3).

In what follows, we consider *linear* or *matrix* groups.[1] To prove that the groups presented below are smooth manifolds, we need the following theorem.

Theorem 3.1. *Let $F: \mathbb{R}^n \to \mathbb{R}^m$ be a mapping of finite-dimensional linear spaces, and let $n \geq m$. Let $F(x_0) = y_0$, and let the mapping F be continuously differentiable at the point x_0. Moreover, let the differential of this mapping (the linear mapping of $\mathbb{R}^n$ into $\mathbb{R}^m$ given by the Jacobi matrix $F_*(x_0)$) be surjective (i.e., the Jacobi matrix has the maximum rank). Then*

1. *in a neighborhood of the point x_0, the full inverse image of the point y_0, i.e., $M = \{x \in \mathbb{R}^n \mid F(x) = y_0\}$, is a smooth manifold and*
2. *the tangent plane to the manifold M at the point x_0 is obtained by a parallel translation of the subspace $\mathrm{Ker}(F_*(x_0))$ to the point x_0.*

Exercise. Prove this theorem using the implicit function theorem.

Example 3.1. We prove that $\mathrm{GL}(n, \mathbb{R})$, the group of nonsingular matrices of order n is a smooth manifold. Let $\mathfrak{M}_n(\mathbb{R})$ be the linear space of all $n \times n$ matrices with real entries. We consider the mapping $\mathfrak{W}: \mathfrak{M}_n(\mathbb{R}) \to \mathbb{R}^{n^2}$ that

---

[1] The term "linear group" is used as an abbreviation of the term "groups of linear operators."

sets a vector of the space $\mathbb{R}^{n^2}$ whose coordinates are the entries x_{ij} of the matrix $X \in \mathfrak{M}_n(\mathbb{R})$ taken in a certain fixed order (for example, we first write the first row, then the second, and so on) in correspondence to each matrix. The mapping $\mathfrak{W}$ is one-to-one. We consider the closed subset D of the space $\mathbb{R}^{n^2}$ defined by $D = \{x \in \mathbb{R}^{n^2} \mid \det \mathfrak{W}^{-1}(x) = 0\}$. Then $\mathfrak{W}(\mathrm{GL}(n)) = \mathbb{R}^{n^2} \setminus D$ is an open set in the space $\mathbb{R}^{n^2}$ and is therefore a smooth manifold. To prove that this group is a Lie group, we must verify the smoothness of the group operations. But coordinates of the product of elements of this group are written in the form of polynomials in coordinates of summands. The coordinates of the inverse element of the matrix A are written as the ratio of the same polynomials; moreover, the polynomial in the denominator is the determinant of the matrix A, which is assumed to be different from zero.

In further examples, the verification of the smoothness of the group operations is based on the same argument. To prove that the matrix groups considered below are Lie groups, it suffices to prove that they are smooth manifolds.

Example 3.2. We consider $\mathrm{SL}(n, \mathbb{R})$, the group of $n \times n$ matrices with the determinant 1. Using the same construction as in Example 3.1, we conclude that $\mathrm{SL}(n, \mathbb{R})$ is identified with the set $\mathfrak{W}(\mathrm{SL}(n, \mathbb{R})) = \{x \in \mathbb{R}^{n^2} \mid \det \mathfrak{W}^{-1}(x) = 1\}$. To prove that $\mathrm{SL}(n, \mathbb{R})$ is a smooth manifold, we apply Theorem 3.1 considering the mapping $\det : \mathfrak{M}_n(\mathbb{R}) \to \mathbb{R}^1$ as F. The partial derivative of the mapping $\det$ in x_{ij} is equal to A_{ij}, the algebraic complement to the entry x_{ij}. Therefore, the differential of the mapping F at the point X_0 applied to the vector $h \in \mathfrak{M}_n(\mathbb{R})$ is equal to

$$F_*(X_0)h = \sum_{ij} A_{ij}(X_0)h_{ij}. \tag{3.1}$$

Because $\det X_0 = 1$, we have $\mathrm{rk}\, X_0 = n$. Consequently, at least one of the minors $A_{ij}(X_0)$ is different from zero; therefore, the mapping $F_*(X_0)$ is surjective. It follows from Theorem 3.1 that the group $\mathrm{SL}(n, \mathbb{R})$ is a smooth manifold.

Example 3.3. We consider $\mathrm{O}(n)$, the group of orthogonal matrices of order n, i.e., matrices that leave the quadratic form $\sum_{i=1}^n x_i^2$, which assigns the Euclidean structure on the space $\mathbb{R}^n$, invariant. The image of this form under the transformation with the matrix S has the matrix $SIS^\top$. Therefore, orthogonal matrices are defined by the relation

$$SS^\top - I = 0 \tag{3.2}$$

in the space $\mathbb{R}^{n^2}$.

To apply Theorem 3.1, we consider the mapping $F(S) = SS^\top - I$, which maps from the space $\mathbb{R}^{n^2}$ into the space $\mathrm{Sym}(n)$ of symmetrical $n \times n$ matrices.

Let H be an element of the space tangent to $\mathfrak{M}_n(\mathbb{R})$. Because the space $\mathfrak{M}_n(\mathbb{R})$ is linear, the space tangent to it is identified with it. Then

$$F_*(S)H = SH^\top + HS^\top. \tag{3.3}$$

In the right-hand side of (3.3), we have an expression that is obtained by symmetrization[2] of the matrix $HS^\top$. The symmetrization of the set $\mathfrak{M}_n(\mathbb{R})$ generates the whole space $\mathrm{Sym}(n)$. Because $\det S \neq 0$, the set $HS^\top$ (for all possible H) coincides with the whole space $\mathfrak{M}_n(\mathbb{R})$. Consequently, the values of the right-hand side of (3.3) range over the whole set $\mathrm{Sym}(n)$. That is, $\mathrm{Im}\, F_* = \mathrm{Sym}(n)$. By Theorem 3.1, the set $\{X \mid F(X) = 0\}$ is a smooth manifold.

Remark. The group $\mathrm{O}(n)$ consists of two connected components: the component consisting of orthogonal matrices with the determinant $+1$, which is denoted by $\mathrm{SO}(n)$, and the component consisting of orthogonal matrices with the determinant -1. The set $\mathrm{SO}(n)$ is also a manifold because it is a connected component of the manifold $\mathrm{O}(n)$.

Example 3.4. We consider $\mathrm{U}(n)$, the group of complex $n \times n$ matrices acting on the space $\mathbb{C}^n$ and leaving the Hermitian form $\sum_{i=1}^n x_i \bar{x}_i$, which assigns the structure of a Hermitian space on $\mathbb{C}^n$, invariant. The image of this form under the mapping with the matrix S has the matrix $S I \bar{S}^\top$. Therefore, unitary matrices are defined by the relation

$$S\bar{S}^\top - I = 0 \tag{3.4}$$

in the space $\mathbb{C}^{n^2}$. The fact that Eq. (3.4) defines a smooth manifold in the space $\mathbb{C}^{n^2}$ is proved in the same way as in Example 3.3.

A slightly more difficult exercise is to prove that the set in the space $\mathbb{C}^{n^2}$ that corresponds to the group $\mathrm{SU}(n)$ (of unitary matrices with the determinant 1) is a smooth manifold.

Exercises.

1. Prove that $\mathrm{U}(n)$ and $\mathrm{SU}(N)$ are Lie groups.

2. Are the groups $\mathrm{O}(n)$, $\mathrm{SO}(n)$, $\mathrm{U}(n)$, and $\mathrm{SU}(n)$ simply connected topological spaces?

3. Prove that all matrices belonging to the group $\mathrm{SO}(n)$ are points of the sphere of radius $\sqrt{n}$ in the space $\mathbb{R}^{n^2}$.

The following example is the most important for what follows.

Example 3.5. We consider $\mathrm{Sp}(n, \mathbb{R})$, the group of symplectic transformations of the space $\mathbb{R}^{2n}$ that leave the skew-symmetrical nondegenerate bilinear

---

[2] We recall that the *symmetrization* of a matrix A is the expression $A + A^\top$; the *skew-symmetrization* (alternation) is the expression $A - A^\top$.

form $J(x, y) = \sum_{i=1}^{n}(x_i y_{i+n} - y_i x_{i+n})$ invariant. The matrix of this form looks like

$$J = \begin{pmatrix} 0 & I_n \\ -I_n & 0 \end{pmatrix}. \tag{3.5}$$

We recall that any nondegenerate skew-symmetrical bilinear form defines the *symplectic structure* on the space $\mathbb{R}^{2n}$.

Because the skew-symmetrical $m \times m$ matrix is necessarily degenerate for an odd m, the symplectic structure exists only on even-dimensional spaces. We can reduce any nondegenerate skew-symmetrical bilinear form on $\mathbb{R}^{2n}$ to the form with matrix (3.5).

The set of linear transformations of the space $\mathbb{R}^{2n}$ that preserve the symplectic structure (i.e., leave the skew-symmetrical form J invariant) is called the *symplectic group* and is denoted by $\mathrm{Sp}(n, \mathbb{R})$. It is easy to see that this set is in fact a group: the product of transformations preserving J also preserves J; the inverse mapping is defined because a degenerate transformation cannot preserve the nondegenerate form; obviously, the inverse mapping also preserves the form J. The equation that defines a symplectic group has the form

$$SJS^\top - J = 0. \tag{3.6}$$

We verify that Eq. (3.6) assigns a smooth manifold. We consider the mapping $F = SJS^\top - J$ and find the differential F. We have

$$F_*(S)H = HJS^\top + SJH^\top. \tag{3.7}$$

It is easy to see that the mapping F transforms the linear space of matrices $\mathfrak{M}_n(\mathbb{R})$ into the linear space of skew-symmetrical matrices. The skew-symmetrical part of the matrix $HJS^\top$ stands in the right-hand side of Eq. (3.7). The skew-symmetrization operation maps the set of all matrices onto the space of skew-symmetrical matrices, and multiplication by the nonsingular matrix $JS^\top$ maps it onto the whole $\mathfrak{M}_n(\mathbb{R})$. Therefore, the mapping F_* is surjective, and the set of all symplectic matrices hence forms a smooth manifold.

Example 3.6. The group $\mathrm{Sp}(n)$ is the unitary restriction of $\mathrm{Sp}(n, \mathbb{R})$. It consists of matrices lying in the intersection $\mathrm{Sp}(n, \mathbb{R}) \cap \mathrm{U}(n)$.

Example 3.7. The group $\mathrm{O}(m, n)$ consists of transformations that leave the quadratic form $\sum_{i=1}^{m} x_i^2 - \sum_{j=m+1}^{m+n} x_j^2$ invariant.

The Lorentz group $\mathrm{O}(1, 1)$ that preserves the Minkowski metric $\Omega = x_1^2 - x_2^2$ on the two-dimensional plane is the simplest example of groups of this kind. Let K denote the matrix of the form Ω:

$$K = \begin{pmatrix} 1 & 0 \\ 0 & -1 \end{pmatrix}.$$

The equation characterizing the group $O(1,1)$ has the form $F(S) = SKS^\top - K = 0$. The determinant of any matrix S that satisfies this equation is equal to ± 1. The value of the differential F_* at the vector H is equal to $HKS^\top + SKH^\top$. The surjectivity is verified in the same way as in the preceding examples. It is easy to show that the connected component of the identity of this group (which is denoted by $SO(1,1)$) consists of matrices of the form

$$\begin{pmatrix} \cosh t & \sinh t \\ \sinh t & \cosh t \end{pmatrix}. \tag{3.8}$$

The transformation with matrix (3.8) of the two-dimensional plane with Cartesian coordinates (x, y) leaves the hyperbola $x^2 - y^2 = 1$ invariant; it is therefore called the *hyperbolic turn*.

§2. Lie Algebras

In what follows, allowing for a certain roughness of speech, we do not make a distinction between a Lie group and the manifold corresponding to it.

Among all smooth manifolds, Lie groups occupy a special position. Their geometric structure is almost entirely (up to factorization by a discrete subgroup) reflected in the algebraic structure of its tangent plane at the point e, the identity of the group G. Before describing this structure (which is called a *Lie algebra*), we compute tangent spaces at the point e for the linear groups considered in Sec. 3.1.

Example 3.8. The manifold $GL(n)$ is an open set in the space $\mathfrak{M}_n(\mathbb{R})$; therefore, the tangent space at the identity is identified with the whole space $\mathfrak{M}_n(\mathbb{R})$.

Example 3.9. The manifold $SL(n)$ is given by the equation $\det X = 1$. Because the algebraic complements A_{ij} are equal to δ_{ij} for the identity matrix I, the kernel of the mapping $F_*(I)$, where $F(X) = \det X$, consists of trace-free matrices, which forms the space tangent to $SL(n)$ at the point e by Theorem 3.1.

Example 3.10. The manifold $O(n)$ is given by the equation $XX^\top = I$. The kernel of the mapping $F_*(I)$, where $F(X) = XX^\top$, consists of matrices that satisfy the equation $H + H^\top = 0$, i.e., skew-symmetrical matrices.

Example 3.11. The manifold $U(n)$ is given by the equation $X\overline{X}^\top = I$. The kernel of the corresponding mapping consists of complex matrices that satisfy the equation $H + \overline{H}^\top = 0$; such matrices are called *skew-Hermitian*.

Example 3.12. The manifold $Sp(n, \mathbb{R})$ is given by the equation $XJX^\top = J$. The kernel of the corresponding mapping consists of matrices that satisfy the equation $HJ + JH^\top = 0$, i.e., those matrices H for which the matrix HJ is symmetrical.

To describe an additional structure on the space tangent to a Lie group, we need the concept of left-invariant vector fields.

In what follows, the space of smooth functions on a manifold $\mathfrak{A}$ is denoted by $C^\infty(\mathfrak{A})$.

2.1. Vector Fields on a Manifold. As usual, by a *vector tangent* to a manifold $\mathfrak{A}$, we mean a vector tangent to a smooth curve lying on $\mathfrak{A}$ [1]. However, it is often useful to identify a tangent vector with the operator of differentiation in the direction of this vector.

Definition 3.1. A *vector field* X on a manifold $\mathfrak{X}$ is a first-order linear differential operator that acts on functions belonging to the space $C^\infty(\mathfrak{X})$. That is, X is a linear mapping $X\colon C^\infty(\mathfrak{X}) \to C^\infty(\mathfrak{X})$ having the property

$$X(fg) = X(f)g + X(g)f,$$

which is called the *Leibnitz rule*.

Remark. This definition is equivalent to the standard definition (see [72]) using a local coordinate system because in local coordinates, the coefficients of a linear differential operator are transformed as coordinates of a (contravariant) vector.

Let $\phi\colon \mathfrak{X} \to \mathfrak{Y}$ be a smooth mapping of manifolds.

Definition 3.2. Two functions $g \in C^\infty(\mathfrak{X})$ and $h \in C^\infty(\mathfrak{Y})$ are called ϕ-compatible if

$$g(x) = h(\phi(x)),$$

i.e., on all inverse images of the point $\phi(x)$ under the mapping ϕ, $x \in \mathfrak{X}$, the function g is equal to the value of the function h on the image of $\phi(x)$.

Two vector fields X and Y are called ϕ-compatible ($Y = \phi_* X$) if their values of ϕ-compatible functions coincide, i.e.,

$$X(g(x)) = Y(h(y)).$$

Remark. In fact, the definition of $\phi_* X$ is only another statement of the usual definition of the differential of a mapping ϕ because if X is a differential operator

$$\sum X^i(x)\frac{\partial}{\partial x^i}$$

corresponding to a vector with coordinates $X^i(x)$, then

$$\phi_* X(h(y)) = X(h(\phi(x))) = \sum X^i(x)\frac{\partial h}{\partial y^\alpha}\frac{\partial \phi^\alpha}{\partial x^i}.$$

Therefore, the coefficients of the differential operator $\phi_* X$ are

$$\frac{\partial \phi^\alpha}{x^i} X^i = (JX)^\alpha,$$

where J is the Jacobian. This is the usual definition of the differential.

Vector fields form a linear space. At the same time, the superposition (product) of two operators does not belong to this space because the superposition of two first-order differential operators is a second-order differential operator. However, using two vector fields X and Y on $\mathfrak{X}$, we can compose the expression $[X, Y] = XY - YX$, which is a first-order differential operator. Indeed,

$$[X,Y]f(x) = X^i \frac{\partial}{\partial x^i} \left[Y^j \frac{\partial f}{\partial x^j} \right] - Y^i \frac{\partial}{\partial x^i} \left[X^j \frac{\partial f}{\partial x^j} \right]$$

$$= \left(X^i \frac{\partial Y^j}{\partial x^i} - Y^i \frac{\partial X^j}{\partial x^i} \right) \frac{\partial f}{\partial x^j} \tag{3.9}$$

because the second-order terms are mutually annihilated by the theorem that second derivatives are independent of the order of differentiation.

The operation thus introduced is called the *commutator* of two vector fields. The operation of commutation is functorial in the following sense.

Lemma 3.1. *Let $\phi \colon \mathfrak{X} \to \mathfrak{U}$ be a smooth mapping between smooth manifolds, and let X and Y be two vector fields on $\mathfrak{X}$. If the vector fields U and V are ϕ-compatible with X and Y, i.e.,*

$$U = \phi_* X, \qquad V = \phi_* Y,$$

then the corresponding commutators are compatible,

$$[U, V] = \phi_*[X, Y].$$

Proof. Let a function $h(y)$ be compatible with a function $g(x)$. Then $UVh(y)$ is compatible with $XYg(x)$. Indeed, by definition, the function $Vh(y)$ is compatible with the function $Yg(x)$, and $UVh(y)$ is hence compatible with $XYg(x)$. The same is true for the functions $VUh(y)$ and $YXg(x)$. Therefore, the differences of these functions, i.e., $[U, V]h(y)$ and $[X, Y]g(x)$ are compatible.

Now let a Lie group G be given. For simplicity, we consider only linear groups. On G, we can consider transformations called *left translations*. Namely, if $a \in G$, then $\Lambda_a \colon G \to G$ is defined as premultiplication by a, i.e., $\Lambda_a g = ag$.

Definition 3.3. A vector field X on a Lie group G is called *left invariant* if $(\Lambda_a)_* X = X$ for all $a \in G$.

Lemma 3.2. *For each tangent vector $X(e)$ at a point e, there exists a unique left-invariant vector field with the value $X(e)$ at the point e.*

Proof. If the vector $X(e)$ is given, we can define the field $\widetilde{X}$ as $\widetilde{X}(g) = (\Lambda_g)_* X(e) = gX(e)$. This field is left invariant. Indeed, $(\Lambda_a)_* (\Lambda_g)_* X = agX(e) = \widetilde{X}(ag)$. Moreover, two left-invariant fields with the same value $X(e)$ obviously coincide because $(\Lambda_g)_* X(e) = gX(e)$ for any left-invariant field.

Example 3.13. A field of constant vectors is a left-invariant field on $\mathbb{R}^n$.

On the quotient group $\mathbb{R}^n/\mathbb{Z}^n$ (torus), the image of a constant field under the natural action onto the quotient space is a left-invariant vector field. The commutators of two left-invariant fields vanish in these cases.

Below in Chap. 5, Sec. 6, we deal with left-invariant vector fields on Spin(3) (the three-dimensional sphere).

Remark. For linear Lie groups, the differential of the left translation $(\Lambda_a)_*$ is also an operator of multiplication by the matrix a because the operator of multiplication by a is linear in the ambient space of all matrices and the differential of a linear operator coincides with this operator itself.

By the above remark, for a linear group G, the value of a left-invariant vector field at any point $g \in G$ is obtained from a fixed matrix A lying in the space tangent to the manifold G at the point e via premultiplication by the matrix g.

We use the mapping $\mathfrak{W}: G \to \mathbb{R}^{n^2}$ in Example 3.1. As coordinates of the point $\mathfrak{W}(g)$ in the space $\mathbb{R}^{n^2}$, we can take the components g_{ij} of the matrix g. Let gA and gB be left-invariant vector fields constructed according to two matrices A and B. Then the operator of differentiation corresponding to gA is $(gA)_{ij}\partial/\partial g_{ij}$. We compute the commutator of the left-invariant vector fields gA and gB. We have

$$\left[(gA)_{kr}\frac{\partial(gB)_{ij}}{\partial g_{kr}} - (gB)_{kr}\frac{\partial(gA)_{ij}}{\partial g_{kr}}\right]\frac{\partial}{\partial g_{ij}} = \left[(gA)_{ir}B_{rj} - (gB)_{ir}A_{rj}\right]\frac{\partial}{\partial g_{ij}}$$

$$= (g(AB - BA))_{ij}\frac{\partial}{\partial g_{ij}}.$$

We have proved the following theorem.

Theorem 3.2. *The commutator of the left-invariant vector fields corresponding to the matrices A and B is the left-invariant vector field corresponding to the commutator $[A, B]$ of these matrices.*

2.2. Lie Algebras. By Lemma 3.2, the space of left-invariant vector fields is isomorphic to the plane tangent to the group G at the point e and is therefore a linear space whose dimension coincides with the dimension of the group G. In this space, we can introduce one more algebraic operation. Namely, we consider the commutator of left-invariant vector fields.

Lemma 3.3. *If X and Y are left-invariant vector fields, then the field $[X, Y]$ is also left invariant.*

Proof. By Lemma 3.1, we have

$$\Lambda_a[X, Y] = [\Lambda_a X, \Lambda_a Y] = [X, Y].$$

The operation of commutation transforms the linear space of all left-invariant vector fields into an algebra, called the *Lie algebra* $\mathfrak{G}$ of the group G. It is easy to verify that this operation has the properties

1. the mapping $X, Y \mapsto [X, Y]$ is bilinear,
2. $[X, Y] = -[Y, X]$, and
3. $[X, [Y, Z]] + [Y, [Z, X]] + [Z, [X, Y]] = 0$ (the Jacobi identity).

Often, properties 1–3 are taken as axioms, and abstract Lie algebras that satisfy this set of axioms are considered (see [95]).

Exercise. Prove that isomorphic Lie groups have isomorphic Lie algebras.

Remark. The Lie algebra characterizes the Lie group "almost uniquely." This means that Lie groups with isomorphic Lie algebras locally coincide, i.e., one group covers the other (see [18, 108]).

§3. Lie Groups of Lower Dimension

Because $SO(2)$ and $U(1)$ are identified with complex numbers whose module equals 1, they are isomorphic. Considered as manifolds, they are the circle S^1.

3.1. Topological Structure of the Groups $SO(3)$ and $Spin(3)$. To describe the topological structure of the group $SO(3)$, we consider the group of quaternions whose module is equal to 1. This group is the three-dimensional sphere S^3. The operator of premultiplication by the quaternion q is denoted by L_q, and the operator of postmultiplication by q is denoted by R_q. We note that these two operators commute by the associativity of the field of quaternions.

Theorem 3.3. *The topological space $SO(3)$ is homeomorphic to the real projective space $\mathbb{RP}^3$.*

Proof. We consider the mapping $L_q \circ R_q^{-1} : \mathbb{R}^4 \to \mathbb{R}^4$ defined by the formula

$$L_q \circ R_q^{-1} : X \mapsto q X q^{-1}.$$

Because $|qXq^{-1}| = |X|$, the transformation $L_q \circ R_q^{-1}$ is an orthogonal transformation. Moreover, the real line remains invariant; therefore, $L_q \circ R_q^{-1}$ transforms the purely imaginary space into itself. Because $L_1 \circ R_1 = \mathrm{id}$ and the sphere is a connected topological space, the determinant (continuously depending on q) of the mapping $L_q \circ R_q^{-1}$ is equal to $\det L_q (\det R_q)^{-1} = 1$ for all q.

Consequently, $L_q \circ R_q^{-1}$ induces a rotation of the space $\mathbb{R}^3$, which is denoted by ϕ_q. The transformation $\phi_q : S^3 \to SO(3)$ is a group homomorphism. Indeed,

$$\phi_{q_1}\phi_{q_2} = \phi_{q_1 q_2}.$$

We show that ϕ_q is an epimorphism. To do this, we first show that there exists an element of S^3 that transforms the quaternion k to any purely imaginary quaternion. We consider an element S^3 of the form $l = (\cos\alpha)i + (\sin\alpha)j$. Then $lk = -kl$ and $l^{-1} = -l$. We find the image of the quaternion k under the mapping $L_q \circ R_q^{-1}$, where $q = \cos\beta + l\sin\beta$. We have

$$
\begin{aligned}
(\cos\beta + l\sin\beta)k(\cos\beta - l\sin\beta) &= (\cos\beta + l\sin\beta)^2 k \\
&= (\cos 2\beta + l\sin 2\beta)k \\
&= (\cos 2\beta)k - (\cos\alpha\sin 2\beta)j + (\sin\alpha\sin 2\beta)i.
\end{aligned}
$$

Because α and β, which can be considered as spherical coordinates on the sphere S^2 in the space of purely imaginary quaternions whose module is equal to 1, are arbitrary, any vector of this sphere can serve as the image of the quaternion k.

Further, for $\alpha = 0$, the quaternion i passes to i. Any rotation around the axis i can be obtained for a certain value of β.

We have thus proved that the homomorphism $\phi_q: S^3 \to SO(3)$ is surjective. Its kernel is formed by those quaternions $q \in S^3$ for which the relation $x = qxq^{-1}$ holds for all purely imaginary $x \in S^2$. This implies $xq = qx$.

Therefore, the quaternion q should commute with all purely imaginary quaternions. But only the quaternion $a + bi$ commutes with the quaternion i, and only the quaternion $c + dj$ commutes with the quaternion j. Consequently, the kernel of the homomorphism ϕ_q consists of the real quaternion q. Because $|q| = 1$, we have $\operatorname{Ker}\phi_q = \pm 1$. (We note that the mapping ϕ_q does not depend on the sign of q.) Therefore, diametrically opposite points of the sphere are identified under the homomorphism ϕ_q.

In particular, Theorem 3.1 implies the nonobvious fact that the closed path

$$
g(t) = \begin{pmatrix} \cos 2\pi t & -\sin 2\pi t & 0 \\ \sin 2\pi t & \cos 2\pi t & 0 \\ 0 & 0 & 1 \end{pmatrix}, \quad 0 \le t \le 1, \tag{3.10}
$$

which is the image of the curve $q(t) = \cos\pi t + k\sin\pi t$, $0 \le t \le 1$, connecting the points 1 and -1 in S^3, cannot be contracted to a point in the space $SO(3)$. However, the same path passed twice, i.e., for $0 \le t \le 2$, can be contracted to a point because it is the image of a closed path in S^3 in this case.

Remark. The closed path (3.10) is a generator of the fundamental group $\pi_1(SO(3)) = \pi_1(\mathbb{RP}^3) = \mathbb{Z}_2$.

The group S^3 is simply connected ($\pi_1(S^3) = 0$); therefore, it is a universal covering of the group $SO(3)$.

3.2. Topological Structure of the Group SL(2, ℝ). We now consider the group SL(2, ℝ). The entries of a matrix $A \in SL(2, \mathbb{R})$ are denoted by a_{ij}, $i, j = 1, 2$. The equation defining SL(2, ℝ) in these coordinates has the form

$$a_{11}a_{22} - a_{12}a_{21} = 1. \tag{3.11}$$

We first describe the manifold defined by the equation

$$a_{11}a_{22} - a_{12}a_{21} = 0. \tag{3.12}$$

Because Eq. (3.12) is homogeneous, it describes a cone in $\mathbb{R}^4$. To find a directrix of this cone, we consider its intersection T with the sphere S^3. If we turn the coordinate axes by the angle $\pi/4$ in the planes $\{a_{11}, a_{22}\}$ and $\{a_{12}, a_{21}\}$, then the quadratic form $a_{11}a_{22} - a_{12}a_{21}$ is reduced to the canonical form $x^2 + y^2 - z^2 - w^2 = 0$, and the intersection of the cone with the sphere $x^2 + y^2 + z^2 + w^2 = 1$ is given by the set of equations

$$x^2 + y^2 = 1/2,$$

$$z^2 + w^2 = 1/2.$$

Consequently, T is the direct product of two circles, i.e., the two-dimensional torus, which divides the sphere S^3 into two congruent parts $\{x^2 + y^2 > z^2 + w^2\}$ and $\{x^2 + y^2 < z^2 + w^2\}$. The torus T serves as a directrix of the cone (3.12).

Equation (3.11) implies that the manifold SL(2, ℝ) is a three-dimensional hyperboloid with asymptotic cone (3.12).

Remark. We note that the torus constructed does not look like the torus that we usually imagine as embedded in the three-dimensional space. The torus T is the so-called flat torus whose curvature vanishes at all points. It is a homogeneous space, which has the same structure at all its points.

3.3. Topological Structure of the Groups Sp(1, ℝ), U(1), and SU(2). The group Sp(1, ℝ) coincides with SL(2, ℝ). Indeed,

$$\begin{pmatrix} x & y \\ z & w \end{pmatrix} \begin{pmatrix} 0 & 1 \\ -1 & 0 \end{pmatrix} \begin{pmatrix} x & z \\ y & w \end{pmatrix} = \begin{pmatrix} 0 & xw - yz \\ yz - xw & 0 \end{pmatrix}.$$

This expression is equal to J iff $xw - yz = 1$.

As previously mentioned, the group U(1) is identified with the set of complex numbers whose module equals 1 and hence with the unit circle S^1.

We now describe the group SU(2).

Theorem 3.4. *The group SU(2) is isomorphic to the group of quaternions whose module equals 1. As a manifold, it is the sphere S^3.*

Proof. We can write each quaternion in a unique way as

$$x = x_1 + x_2 i + x_3 j + x_4 k = z_1 + z_2 j,$$

where $z_1 = x_1 + x_2 i$ and $z_2 = x_3 + x_4 i$. That is, $(z_1, z_2) \in \mathbb{C}^2$; therefore, the set of quaternion is interpreted as the two-dimensional complex space $\mathbb{C}^2$.

The linear operator corresponding to postmultiplication by a fixed quaternion $y = y_1 + y_2 i + y_3 j + y_4 k = w_1 + w_2 j$ yields the element $xy = (z_1 w_1 - z_2 \overline{w}_2) + (z_1 w_2 + z_2 \overline{w}_1) j$ when acting on x. We can write this element as the product

$$xy = (z_1, z_2) \begin{pmatrix} w_1 & w_2 \\ -\overline{w}_2 & \overline{w}_1 \end{pmatrix}.$$

By the associativity of the quaternion division ring, the matrix product corresponds to sequentially postmultiplying by fixed quaternions. Therefore, we have obtained the representation of the quaternion algebra by complex 2×2 matrices. If w is a quaternion whose module equals 1, then for any quaternion q, we have $|q| = |qw|$. Therefore, the corresponding matrices belong to SU(2).

Exercises.

1. Prove that the matrices

$$\begin{pmatrix} i & 0 \\ 0 & -i \end{pmatrix}, \qquad \begin{pmatrix} 0 & 1 \\ -1 & 0 \end{pmatrix}, \qquad \begin{pmatrix} 0 & i \\ i & 0 \end{pmatrix}.$$

correspond to multiplication by the quaternions i, j, and k. Using this fact, prove that these matrices form a basis of the Lie algebra of the group SU(2).

2. By a direct calculation, verify that for $|w| = 1$, the matrix $\begin{pmatrix} w_1 & w_2 \\ -\overline{w}_2 & \overline{w}_1 \end{pmatrix}$ is unitary and unimodular. Any matrix of this form is the image of a certain quaternion whose module equals 1.

3. Prove that the Lie algebra of the group SU(2) consists of trace-free skew-Hermitian matrices of size 2×2. By a direct computation, prove that the Lie algebras $\mathfrak{S}U(2)$ and $\mathfrak{S}O(3)$ are isomorphic in contrast to the corresponding groups SU(2) and SO(3).

§4. Adjoint Representation and Killing Form

4.1. Adjoint Representation. Let G be a linear Lie group. We can consider each element $g \in G$ as a linear operator acting on an n-dimensional (real or complex) space. Further let $\mathfrak{S}$ be the Lie algebra of the group G. For each $s \in G$, the mapping

$$A_s : G \to G, \qquad g \mapsto sgs^{-1}, \tag{3.13}$$

is an automorphism of the group G. In particular, $A_s(e) = e$.

The differential of the mapping A_s is the linear transformation that transforms the plane tangent to the group G at the point e (i.e., the algebra $\mathfrak{G}$) into itself. We denote this mapping by $\operatorname{Ad} s$.

In a natural way, we can extend the transformation A_s to the set of all matrices $\operatorname{GL}(n)$ by the same formula (3.13). We obtain the linear mapping of the space of matrices $\operatorname{GL}(n)$ into itself. Consequently, its differential coincides with this mapping. The restriction of A_s to the group G is no longer linear because G is not a linear variety. However, we can consider the Lie algebra $\mathfrak{G}$ as a linear subspace in $\operatorname{GL}(n)$. Therefore, the differential of the restriction of A_s to the group G is as written before in the form of premultiplication of a matrix g by the matrix s and postmultiplicatuion by the matrix s^{-1}:

$$(A_s)_*(g) = sgs^{-1}. \tag{3.14}$$

The obtained transformation is denoted by $\operatorname{Ad} s \colon \mathfrak{G} \to \mathfrak{G}$. Obviously, the relation

$$\operatorname{Ad} s_1 \circ \operatorname{Ad} s_2 = \operatorname{Ad}(s_1 \circ s_2)$$

holds for arbitrary elements $s_1, s_2 \in G$, i.e., the mapping $\operatorname{Ad} s$ considered as a function of s defines a representation of the group G in the group of linear transformations of the space $\mathfrak{G}$. In other words, we have constructed the homomorphism

$$\operatorname{Ad} \colon G \to \operatorname{GL}(\mathfrak{G}).$$

This representation is called the *adjoint representation of the Lie group G*.

We find the action of the differential of this mapping at the point e on a vector $h \in \mathfrak{G}$. Differentiating formula (3.14), we obtain

$$\left.\frac{d}{ds}\right|_{s=e} ((\operatorname{Ad} s)g)h = (hgs^{-1} - sgs^{-1}hs^{-1})|_{s=e}$$

$$= hg - gh = [h, g]. \tag{3.15}$$

In a natural way, formula (3.15) leads us to the consideration of the *adjoint representation of a Lie algebra*.

Let $\mathfrak{G}$ be the Lie algebra of the group G. With each $a \in \mathfrak{G}$, we can associate the linear transformation that is the commutation of any element $g \in \mathfrak{G}$ with the element a. Namely,

$$\operatorname{ad} a \colon \mathfrak{G} \to \mathfrak{G}, \qquad (\operatorname{ad} a)g = [a, g].$$

In this case, the sum of operators corresponds to the sum of elements of the algebra $\mathfrak{G}$, and the matrix commutator

$$(\operatorname{ad} a)(\operatorname{ad} b) - (\operatorname{ad} b)(\operatorname{ad} a) \colon g \mapsto [a, [b, g]] - [b, [a, g]]$$

corresponds to the commutator $[a, b]$ of two elements a and b.

Indeed, $(\text{ad}\,[a,b])g = [[a,b],g]$. By the Jacobi identity, we have $[[a,b]g] = [a,[b,g]] - [b,[a,g]]$. Therefore, we have constructed the representation of the Lie algebra $\mathfrak{G}$ in the algebra of linear operators over $\mathfrak{G}$. This representation is called the adjoint representation of a Lie algebra.

By (3.15), the differential of the adjoint representation $\text{Ad}\,G$ of the group G is the adjoint representation $\text{ad}\,\mathfrak{G}$ of the Lie algebra $\mathfrak{G}$.

4.2. Killing Form. The adjoint representation of a Lie algebra allows us to define an invariant bilinear symmetrical form on $\mathfrak{G}$.

Definition 3.4. The *Killing form* is the trace of the product of the operators $\text{ad}\,g_1$ and $\text{ad}\,g_2$:

$$B(g_1, g_2) = \text{Tr}\,[(\text{ad}\,g_1) \circ (\text{ad}\,g_2)], \quad g_1, g_2 \in \mathfrak{G}.$$

The Killing form is bilinear and symmetrical. The latter is implied by the fact that the trace of the product of two matrices does not depend on the order of the factors.

Exercise. Compute the Killing form for the algebras $SL(2, \mathbb{R})$ and $SO(3)$.

Theorem 3.5. *The Killing form is invariant under automorphisms of a Lie algebra $\mathfrak{G}$.*

Proof. Let $\sigma \colon \mathfrak{G} \to \mathfrak{G}$ be an automorphism of the algebra $\mathfrak{G}$. Then the formula $\text{ad}\,(\sigma g) = \sigma(\text{ad}\,g)\sigma^{-1}$ implies

$$\begin{aligned} B(\sigma g_1, \sigma g_2) &= \text{Tr}((\text{ad}\,\sigma g_1)(\text{ad}\,\sigma g_2)) \\ &= \text{Tr}(\sigma\,\text{ad}\,g_1\,\text{ad}\,g_2\sigma^{-1}) = B(g_1, g_2). \end{aligned}$$

Theorem 3.6. *For any elements $X, Y, Z \in \mathfrak{G}$, the following relation holds:*

$$B([X,Y], Z) = B([Y,Z], X) = B([Z,X], Y). \tag{3.16}$$

Exercise. Prove this theorem using the fact that the trace of the product of two matrix does not depend on the order of factors, the trace of the product of three (or any other number of) matrices is not changed under a cyclic permutation of the factors, the trace of the difference of matrices equals the difference of their traces, and finally, the mapping ad is a Lie algebra representation.

4.3. Subalgebras and Ideals.

Definition 3.5. A subspace $\mathfrak{H}$ is called a *subalgebra* of a Lie algebra $\mathfrak{G}$ if $\mathfrak{H}$ is a linear subspace, and for any elements a and b belonging to $\mathfrak{H}$, their commutator also belongs $\mathfrak{H}$, i.e., $[a, b] \in \mathfrak{H}$.

We consider a Lie subgroup H of the Lie group G (i.e., a subgroup that is a smooth submanifold of the manifold G).

Theorem 3.7. *The Lie algebra $\mathfrak{H}$ of the subgroup $H \subset G$ is a subalgebra of the algebra $\mathfrak{G}$.*

Proof. On the subgroup H, we consider a vector field X that is invariant with respect to left translation by elements from H. As shown, vectors tangent to H at the identity e of the group G uniquely correspond to such fields. Let a vector X_e correspond to the field X. Using this vector, we construct a left-invariant vector field on G. This field coincides with the field X on the subgroup H. We therefore construct an embedding of the Lie algebra $\mathfrak{H}$ of the group H in the Lie algebra $\mathfrak{G}$ of the group G. Clearly, $\mathfrak{H}$ forms a linear subspace in $\mathfrak{G}$. Moreover, because the commutator of the elements h_1 and h_2 in $\mathfrak{H}$ again belongs to $\mathfrak{H}$, we find that $\mathfrak{H}$ is a subaslgebra of the Lie algebra $\mathfrak{G}$.

Remark. The converse statement also holds: a certain subgroup H of the Lie group G corresponds to each subalgebra $\mathfrak{H} \subset \mathfrak{G}$. (The proof can be found in [18].) However, this subgroup can be nonclosed (in the sense of the topology of the manifold G). An everywhere dense torus winding is an example of such a nonclosed subgroup.

Example 3.14. The one-dimensional subspace $l = t\lambda$ of the Lie algebra $\mathfrak{G}$, where $t \in \mathbb{R}$ and $\lambda \in \mathfrak{G}$, is a subalgebra because the commutator of proportional vectors vanishes. A one-parameter subgroup L of the Lie group G corresponds to this subalgebra. For linear groups, the subgroup L is formed by exponentials of elements of the subspace l, i.e., by elements written in the form

$$\exp(t\lambda) = \sum_{n=0}^{\infty} \frac{(t\lambda)^n}{n!}.$$

This series converges for any $t \in \mathbb{R}$ (see [18]).

Example 3.15. Any of the linear groups considered in Examples 3.1–3.3 and 3.5 is a Lie subgroup of the group $\mathrm{GL}(n, \mathbb{R})$. The corresponding Lie algebras are subalgebras of the algebra $\mathfrak{M}(n, \mathbb{R})$.

Remark. The group $\mathrm{U}(n)$ is a subgroup of the group $\mathrm{GL}(n, \mathbb{C})$. However, because the condition on complex-conjugate matrices enters its definition (and it is therefore not a complex submanifold of $\mathbb{C}^{n^2}$), it is not a complex-analytic Lie subgroup of the group $\mathrm{GL}(n, \mathbb{C})$. (Prove that the dimension of the Lie algebra $\mathrm{U}(\mathfrak{n})$ over the field $\mathbb{R}$ is equal to n^2.) Nevertheless, $\mathrm{U}(n)$ is a Lie subgroup of the group $\mathrm{GL}(2n, \mathbb{R})$ because it is a real-analytic submanifold.

If a closed subgroup H of the Lie group G is given, then we can consider the set of left cosets G/H, i.e., the set of elements of the form $\{aH \mid a \in G\}$, and the set of right cosets $H \backslash G$, i.e., the set of elements of the form $\{Hb \mid b \in G\}$.

Definition 3.6. A subgroup H is called *invariant* (*normal*) if the left coset of any element $a \in G$ coincides with its right coset.

This definition implies that any subgroup aHa^{-1} that is conjugate to an invariant subgroup H coincides with H. In this case, the set of left cosets can be endowed with the group structure by setting $(aH)(bH) = abH$. The obtained group is called the quotient group of the Lie group G by the invariant subgroup H.

Remark. The proof that a quotient group is a smooth manifold is in [18]. We note that the closedness condition of the subgroup H is essential. For example, the quotient group of the torus by its everywhere dense winding is not a manifold.

We now reveal which Lie algebras correspond to invariant subgroups of a Lie group. We consider the Lie algebra $\mathfrak{H}$ of an invariant subgroup $H \subset G$.

Definition 3.7. A subalgebra $\mathfrak{H} \subset \mathfrak{G}$ is called an *ideal* of a Lie algebra $\mathfrak{G}$ if $[h, g] \in \mathfrak{H}$ for any elements $h \in \mathfrak{H}$ and $g \in \mathfrak{G}$.

Proposition 3.1. *Let $\phi: G_1 \to G_2$ be a continuous Lie group homomorphism. Then its kernel*

$$\operatorname{Ker} \phi = \{g \in G_1 \mid \phi(g) = e\}$$

is an invariant closed subgroup of G.

Proof. The closedness of $\operatorname{Ker} \phi$ is implied by the continuity of the mapping ϕ. Because ϕ is a homomorphism, $\operatorname{Ker} \phi$ is a subgroup. If $\phi(g) = e$, then its invariance is implied by the relation $\phi(aga^{-1}) = \phi(a)\phi(g)\phi(a)^{-1} = e$.

Theorem 3.8. *The Lie algebra $\mathfrak{H}$ corresponding to an invariant subgroup $H \subset G$ forms an ideal in $\mathfrak{G}$.*

Proof. We consider the natural projection

$$\phi: G \to G/H.$$

The kernel of this mapping coincides with H. The differential of the projection ϕ at the point e is the mapping

$$\phi_*(e): \mathfrak{G} \mapsto \mathfrak{K},$$

where $\mathfrak{K}$ is the Lie algebra of the group G/H. The kernel of the differential ϕ_* is $\mathfrak{H}$, and its image is $\mathfrak{K}$.

Let $h \in \mathfrak{H}$ and $g \in \mathfrak{G}$. Then, by Lemma 3.1, we have

$$\phi_*(e)[g, h] = [\phi_*(e)(g), \phi_*(e)(h)] = [\phi_*(e)(g), 0] = 0.$$

This means that $[g, h] \in \mathfrak{H}$.

Remark. The converse statement also holds: an invariant subgroup H of the group G corresponds to each ideal $\mathfrak{H} \subset \mathfrak{G}$.

§5. Semisimple Lie Groups

Definition 3.8. A Lie group G is called *simple* if it has no nontrivial invariant subgroups.

The Lie algebra of a simple noncommutative Lie group has no nontrivial ideals. Such Lie algebras are called *simple*. We note that Lie algebras all of whose elements commute with each other (the so-called Abelian Lie algebras), except for the one-dimensional case, are not simple.

Definition 3.9. The set of elements of a Lie algebra that commute with all elements of this algebra is called the *center* of the Lie algebra.

Obviously, the center of a Lie algebra is its ideal, which is contained in the kernel of the Killing form.

Definition 3.10. A Lie algebra is called *semisimple* if its Killing form is nondegenerate. The corresponding Lie group is also called semisimple.

Proposition 3.2. *The kernel of the Killing form B is a Lie algebra ideal.*

Proof. We recall that the *kernel of the Killing form* is an element $X \in \mathfrak{G}$ such that $B(X, Y) = 0$ for any $Y \in \mathfrak{G}$.

Let $Y \in \operatorname{Ker} B$, and let X and Z be arbitrary elements of the algebra. By (3.16), we obtain $B(X, [Y, Z]) = B(Y, [Z, X]) = 0$. Consequently, $[Y, Z] \in \operatorname{Ker} B$; therefore, the kernel of the Killing form is in fact an ideal.

Theorem 3.9. *A semisimple Lie algebra $\mathfrak{G}$ contains no Abelian ideals.*

Proof. Let $\mathfrak{A}$ be an Abelian ideal, and let $X \in \mathfrak{A}$ and $Y \in \mathfrak{G}$. We consider $B(X, Y) = \operatorname{Tr}(\operatorname{ad} X \circ \operatorname{ad} Y)$. The transformation $\sigma = \operatorname{ad} X \circ \operatorname{ad} Y$ transforms the whole Lie algebra to the ideal $\mathfrak{A}$. Indeed, the commutation with the element X transforms any element of the algebra to an element that belongs to $\mathfrak{A}$. The transformation σ transforms the ideal $\mathfrak{A}$ to zero (because the ideal $\mathfrak{A}$ is Abelian). Therefore, $\sigma^2 = 0$ and hence $\operatorname{Tr} \sigma = 0$, i.e., the Killing form is degenerate.

Example 3.16. We note that $U(n) = S^1 \times SU(n)$. The Lie algebra of this group consists of skew-Hermitian matrices. Its center consists of matrices of the form $\exp(i\alpha) I_n$, $\alpha \in \mathbb{R}$. Therefore, $\operatorname{ad} \exp(i\alpha) I_n = 0$, i.e., $B(\exp(i\alpha) I_n, g) = 0$ for all $g \in \mathfrak{G}$. Consequently, the Killing form is degenerate, and the Lie algebra $\mathfrak{U}(n)$ is therefore not semisimple.

Exercises.

1. Prove that the Lie algebra $\mathfrak{SU}(n)$, $n > 1$, is simple.

2. Prove that the Lie algebra of the group $GL(n)$ has a nontrivial center and invariant subgroup $SL(n)$. Consequently, it is not semisimple. At the same

time, the Lie algebra of the group $SL(n)$, i.e., the space of trace-free matrices, is semisimple.

3. Prove that the group $U(n)$ has a nontrivial invariant subgroup $SU(n)$.

Remark. The groups $SO(n)$ for $n > 4$, $SU(n)$ for $n > 1$, and $Sp(n)$ are simple Lie groups (see [79]).

Let $\mathfrak{A}^\perp$ denote the orthogonal complement of a subspace $\mathfrak{A}$ with respect to the Killing form.

Theorem 3.10. *Let the Killing form B of an algebra $\mathfrak{G}$ be nondegenerate, and let $\mathfrak{A}$ be an ideal in $\mathfrak{G}$. Then $\mathfrak{A}$ is a semisimple Lie algebra, $\mathfrak{A}^\perp$ is an ideal, and $\mathfrak{G} = \mathfrak{A} \oplus \mathfrak{A}^\perp$.*

Proof. We prove that the orthogonal complement of the ideal $\mathfrak{A}$ is an ideal. Let $Y \in \mathfrak{A}^\perp$. For any $Z \in \mathfrak{G}$ and $X \in \mathfrak{A}$, by (3.16), we have the relation $B(X, [Y, Z]) = B(Y, [Z, X]) = 0$ because $[Z, X] \in \mathfrak{A}$ and $Y \in \mathfrak{A}^\perp$. This means that $[Y, Z] \in \mathfrak{A}^\perp$ for any $Z \in \mathfrak{G}$.

The nondegeneracy of the form B implies $\dim \mathfrak{A} + \dim \mathfrak{A}^\perp = \dim \mathfrak{G}$. Therefore, to prove that their sum is direct, we only need to verify that $\mathfrak{A} \cap \mathfrak{A}^\perp = 0$.

First, we show that $\mathfrak{A} \cap \mathfrak{A}^\perp$ is an Abelian ideal. Indeed, $\mathfrak{A}$ and $\mathfrak{A}^\perp$ are ideals in the algebra $\mathfrak{G}$. Let $X, Y \in \mathfrak{A} \cap \mathfrak{A}^\perp$ and $Z \in \mathfrak{G}$. Because $[Y, Z] \in \mathfrak{A} \cap \mathfrak{A}^\perp$ in this case, by Theorem 3.6, the relation $B(Z, [X, Y]) = B(X, [Y, Z]) = 0$ holds for any Z. From the nondegeneracy of the form B, we obtain $[X, Y] = 0$. Therefore, $\mathfrak{A} \cap \mathfrak{A}^\perp$ is an Abelian ideal. This ideal is zero by Theorem 3.9.

That the algebra $\mathfrak{A}$ contains no nonzero vector orthogonal to $\mathfrak{A}$ implies that it is semisimple. Otherwise, this vector is orthogonal to $\mathfrak{A}^\perp$ and hence to the whole algebra $\mathfrak{G}$; this contradicts the nondegeneracy of the form B.

Theorem 3.11. *A semisimple Lie algebra $\mathfrak{G}$ admits a unique (up to the order of summands) decomposition into a direct sum of simple ideals.*

Proof. If $\mathfrak{G}$ contains no nontrivial ideals, then it is simple and the theorem is thus proved. Now let $\mathfrak{A}$ be an ideal of $\mathfrak{G}$. By Theorem 3.10, the algebra $\mathfrak{A}$, as well as the algebra $\mathfrak{A}^\perp$, is semisimple. In turn, these algebras can be further decomposed into a direct sum of ideals up to the point where we obtain a decomposition into simple ideals,

$$\mathfrak{G} = \sum_{i=1}^{k} \mathfrak{G}_i.$$

We show that the ideals $\mathfrak{G}_i$ are uniquely defined. Indeed, let $\mathfrak{L}$ be a simple ideal that is different from all the $\mathfrak{G}_i$. Then $\mathfrak{G}_i \cap \mathfrak{L} = 0$ because the intersection of these ideals is an ideal contained in each of them; this contradicts their simplicity. Consequently, any vector $l \in \mathfrak{L}$ is orthogonal (with respect to the Killing form) to any vector from $\mathfrak{G}_i$. Because each vector $g \in \mathfrak{G}$ is represented in the form of a linear combination of vectors from $\mathfrak{G}_i$, the ideal $\mathfrak{L}$ is the center of $\mathfrak{G}$; this contradicts its semisimplicity.

5.1. Compact Lie Algebras.

Definition 3.11. A real Lie algebra $\mathfrak{G}$ is called *compact* if its Killing form B is negative definite.

The term "compact algebra" is used because such Lie algebras correspond to compact (in the sense of the topology of the corresponding manifold) Lie groups. Namely, the following theorem holds.

Theorem 3.12. *Let G be a compact semisimple Lie group. Then the Killing form of its Lie algebra is negative definite.*

To prove this theorem, we need the invariant integration over a Lie group.

A two-sided invariant measure $\mu\,(dx)$, called the *Haar measure* is defined and unique on any compact Lie group. The invariance means that the integral by this measure is not changed under left and right translations on the group and also under the inversion:

$$\int f(ya)\mu(dy) = \int f(ay)\mu(dy) = \int f(y^{-1})\mu(dy)$$

$$= \int f(y)\mu(dy), \quad f \in C^\infty(G), \quad a \in G. \qquad (3.17)$$

By the compactness of G, we can normalize the measure μ by the condition

$$\int \mu(dy) = 1.$$

(For a detailed proof, see [79], Sec. 29.)

Proof of Theorem 3.12. We first prove two lemmas.

Lemma 3.4. *Let $\mathfrak{G}$ be a compact Lie algebra. Then there exists a positive-definite quadratic form that is invariant with respect to $\operatorname{Ad} G$.*

Proof. We consider an arbitrary positive-definite quadratic form (x, x) on the space $\mathfrak{G}$. Using the invariant integration, we transform it into an invariant form. For this, we consider an arbitrary element $y \in G$ and apply the transformation $\operatorname{Ad} y$ to the elements x. We define a new quadratic form $Q(x, x)$ by the formula $Q(x, x) = \int ((\operatorname{Ad} y)x, (\operatorname{Ad} y)x)\,\mu(dy)$. The quadratic form Q is positive definite because the integrand is positive (for all nonzero $x \in \mathfrak{G}$ and $y \in G$). We prove the invariance of the form constructed. Taking into account that Ad is a representation of the group G and using (3.17), we obtain

$$Q(\operatorname{Ad}(z)x, (\operatorname{Ad} z)x) = \int_{y\in G} (\operatorname{Ad}(yz)x, \operatorname{Ad}(yz)x)\,\mu(dy) = Q(x, x). \qquad (3.18)$$

The lemma is proved.

The invariance of the form Q with respect to the adjoint representation $\operatorname{Ad} G$ of the group G implies the following invariance property of the form Q with respect to the adjoint representation $\operatorname{ad} \mathfrak{G}$ of the algebra $\mathfrak{G}$.

Lemma 3.5. *For all* $h \in \mathfrak{G}$, *the following relation holds:*

$$Q(x, (\operatorname{ad} h)y) + Q((\operatorname{ad} h)x, y) = Q(x, [h, y]) + Q([h, x], y) = 0. \qquad (3.19)$$

To prove this lemma, it suffices to differentiate (3.18) with respect to z for $z = e$ and use (3.15).

We return to the proof of the theorem. In the algebra $\mathfrak{G}$, we choose a basis $\{e_i\}$, $i = 1, \ldots, n$, in which Q has the diagonal form

$$Q = \sum_{i=1}^{n} x_i^2.$$

Relation (3.19) means that the operator $x \mapsto [h, x]$ in this basis is given by a skew-symmetrical matrix. We then have

$$B(h, h) = \operatorname{Tr}((\operatorname{ad} h)^2) = \sum_{i=1}^{n} (e_i, (\operatorname{ad} h)^2 e_i)$$

$$= -\sum_{i=1}^{n} ((\operatorname{ad} h)e_i, (\operatorname{ad} h)e_i) \le 0.$$

Because the group G is semisimple, the form B is nondegenerate. Consequently, the Killing form B is negative definite. The theorem is proved.

Corollary 3.1. *The Killing form assigns a two-sided invariant metric on any compact semisimple Lie group.*

Proof. The Killing form of a Lie algebra assigns an inner product in this algebra or, equivalently, in the tangent space at the point e of the group G. The left and right translations by an element $a \in G$ define an inner product in the tangent spaces at all other points of the group G. This inner product does not depend on the method that induced it from the Lie algebra (left or right translation or a combination of translations). This is implied by Theorem 3.5 on the invariance of the Killing form with respect to automorphisms of the Lie algebra.

Therefore, a compact semisimple Lie group forms a Riemannian manifold because the metric on it is nondegenerate and positive definite.

Remark. The metric is nondegenerate on a semisimple noncompact Lie group but it is not positive definite. (Manifolds equipped with such a metric are called pseudo-Riemannian.)

§6. Homogeneous and Symmetrical Spaces

Let G be a semisimple compact Lie group, and let K be its closed subgroup. The Killing form on the corresponding Lie algebra $\mathfrak{G}$ is negative definite. This form defines the structure of a Riemannian manifold with a two-sided invariant (with respect to left and right translations) metric on G. Indeed, the differential of the left translation Λ_a maps the tangent plane at the point e into the tangent plane at the point a. If we first apply the left translation Λ_a and then the right translation $R_{a^{-1}}$, then we obtain a mapping $G \to G$ that maps the point e into e. The differential of this mapping at the point e maps the Lie algebra $\mathfrak{G}$ into itself. We have thus obtained the representation of the group G that coincides with the adjoint representation of G. Because the Killing form is invariant, its image coincides with itself; this ensures the invariance of the metric.

Definition 3.12. A mapping $s\colon \mathfrak{X} \to \mathfrak{X}$ is called an *isometry of a Riemannian* (or *pseudo-Riemannian*) *manifold* $\mathfrak{X}$ if the first quadratic form $g_{ij}(x)$ for any point $x \in \mathfrak{X}$ remains invariant with respect to the differential $s_*(x)$. A mapping s is called a *local isometry* if there exists a neighborhood $U \subset X$ for any $x \in \mathfrak{X}$ such that $s|_U$ is an isometry.

Therefore, the group G is mapped into the isometry group of the Riemannian manifold G (i.e., into the group of diffeomorphisms of G that preserve the metric).

If a closed subgroup K of the Lie group G is not invariant, then we can consider the set of left (or right) cosets G/K. A left coset of the group G is the set of elements of the group G of the form aK, $a \in G$. In contrast to the case where K is an invariant subgroup, this set is no longer a group. Nevertheless, G/K is a smooth manifold ([18], Vol. 1, Chap. 2, Sec. 3). The mapping $\phi\colon G \to G/K$ that sets its left coset gK in correspondence to each element $g \in G$ is called the *natural mapping* (*natural projection*).

The set G/K is called a *homogeneous space* of the group G. The term "homogeneous" is related to the fact that this space has the same geometric structure at all its points: the group G acts on it as a group of transformations $g\colon G/K \to G/K$, $g\colon aK \mapsto (ga)K$; therefore, under the left action of the group G, a left coset again passes to a left coset. This action is transitive because for any elements $a, b \in G$, there exists an element g such that $aK \mapsto (ga)K = bK$. In this case, K is the set of elements of the group G that leave the point $o = \phi(e) \in G/K$, which corresponds to the identity $e \in G$, fixed. Therefore, K is often called a *stationary subgroup*. Because the action of K on G/K transforms tangent directions at a fixed point to other ones, this subgroup is also called an *isotropy subgroup*.

In the Lie algebra $\mathfrak{G}$ of a group G, we consider the orthogonal (in the sense of the Killing form) complement $\mathfrak{P}$ to a subalgebra $\mathfrak{K}$, i.e., $\mathfrak{P} = \mathfrak{K}^\perp$. Each left coset $\mathfrak{G}/\mathfrak{K}$ has its representative in $\mathfrak{P}$; therefore, we can identify $\mathfrak{P}$

with the tangent space G/K at the point $o = \phi(e)$. Let $\mathfrak{P}_g = g\mathfrak{P}$ denote the distribution of subspaces obtained from the subspace $\mathfrak{P}$ by applying left translations by elements of the group G.

Theorem 3.13. *The restriction of the Killing form to $\mathfrak{P}$ and the extension of the obtained quadratic form to all subspaces $\mathfrak{P}_g$ define a Riemannian metric on the manifold G/K that is invariant with respect to the left action of the group G.*

Proof. The left multiplication by elements of the group G allows us to construct a distribution of linear subspaces in the tangent bundle of the manifold G that is constructed as follows. The differential of the left translation by an element g transforms a subspace $\mathfrak{K}$ of the tangent space at the point e to a certain subspace $\mathfrak{K}_g$ of the tangent space at the point g. At each point $g \in G$, the corresponding subspace $\mathfrak{K}_g$ of the tangent space $T_g(G)$ is defined as the left translation by the element g of all vectors belonging to $\mathfrak{K}$. Obviously, $\mathfrak{K}_g$ is tangent to the orbit of the group K that passes through the point g. By construction, this distribution is left invariant.

Similarly, according to a subspace $\mathfrak{P} \subset \mathfrak{G}$, we construct a left-invariant distribution of linear subspaces $\mathfrak{P}_g$, which turns out to be the orthogonal complement to $\mathfrak{K}_g$ by the invariance of the Killing form at each point $g \in G$. We can consider this distribution as the space tangent to the manifold G/K at the point g.

Therefore, the group G is a subgroup of the group of motion of the homogeneous space G/K.

6.1. Symmetrical Spaces.

We can find generators of the group of motions for certain homogeneous spaces. These generators have a clear geometric sense: they are certain reflections. We describe spaces that have a complete set of such reflections.

To introduce this class of spaces, we first study the *exponential mapping*. Let $\mathfrak{X}$ be an arbitrary Riemannian manifold. The equations of geodesics (see (1.16)) are second-order equations. Therefore, a unique geodesic passes through each point $x \in \mathfrak{X}$ and each direction $\lambda p \in T_x\mathfrak{X}$, $\lambda \in \mathbb{R}$. When λ is changed, the parameter on a geodesic is multiplied by λ. This fact gives us a possibility to define the mapping of the tangent plane $T_x\mathfrak{X}$ to the manifold $\mathfrak{X}$ itself. Namely, we draw a geodesic $\gamma_{x,p}(t)$ through x that corresponds to the tangent vector p at the point x, i.e., $\gamma(0) = x, \dot{\gamma}(0) = p$. We define the mapping

$$\text{Exp: } (x, p) \mapsto \gamma_{x,p}(1).$$

We note that a distinct point of one and the same geodesic γ corresponds to proportional vectors.

By the theorem that solutions to a differential equation depend smoothly on the initial data (see [78]), the mapping Exp is a smooth mapping of a neighborhood U of the origin of the tangent plane onto a certain neighborhood of the point of intersection. Moreover, a unique geodesic connecting x with y passes through each point y that is sufficiently close to the point x; therefore, the mapping Exp defines a diffeomorphism of the neighborhood U of the point $0 \in T_x\mathfrak{X}$ onto the neighborhood V of the point $x \in \mathfrak{X}$.

Definition 3.13. The mapping $\sigma_x : V \to \mathfrak{X}$, which transforms the point $\mathrm{Exp}(x,p)$ to the point $\mathrm{Exp}(x,-p)$, is called the *geodesic symmetry* at the point x.

The geodesic symmetry σ_x has the following properties:

1. The mapping σ_x is involutive.
2. The point $y = x$ is a unique fixed point of σ_x in a sufficiently small neighborhood of the point x.

Sometimes we have the following additional property:

3. The mapping σ_x is an isometry.

Definition 3.14. A Riemannian manifold (or a pseudo-Riemannian manifold) is called a *locally symmetrical space at a point x* if, for this point x, there exists a certain mapping that has properties 1–3.

A manifold is called *symmetrical* if it is locally symmetrical at each of its points.

To stress the distinction of a symmetrical space from a space that is locally symmetrical at a certain point, the term *globally asymmetrical space* is sometimes used.

The following manifolds yield the simplest examples of symmetrical spaces:

the n-dimensional Euclidean space $\mathbb{R}^n$;
the torus $\mathbb{R}^n/\mathbb{Z}^n$ endowed with the induced metric;
the standard sphere S^n;
the projective space $\mathbb{RP}^n$.

The ellipsoid

$$\sum_{k=1}^{n} x_i^2/a_i^2 = 1$$

with the metric induced by the ambient Euclidean space $\mathbb{R}^n$ is a locally symmetrical space only at the points of intersections with the principal axes. It is not a globally symmetrical space.

A geodesic symmetry, the reflection at a point x, is an involutive transformation (its square is the identity mapping). A combination of two such isometries (first with respect to the point x and then with respect to the point y) transforms a geodesic passing through these points into itself. This transformation is called a *transvection*. In the Euclidean space, this is a parallel translation; on the sphere, it is a rotation around the axis that is orthogonal to the equator on which the considered geodesic lies.

Therefore, if a manifold is globally symmetrical, then we can construct a subgroup of the isometry group of the manifold $\mathfrak{X}$ by combining geodesic symmetries. In fact, this subgroup coincides with the whole isometry group, which turns out to be a Lie group [17, 44]. Therefore, the transformations σ_y are generators of the isometry group.

Lie groups are the most important examples of symmetrical spaces.

Theorem 3.14. *A linear Lie group G is a symmetrical space.*

Proof. For an arbitrary point $x \in G$, we construct the mapping $\sigma_x : y \mapsto xy^{-1}x$.

We show that the mapping $\sigma_x(y)$ is an involutive isometry with a unique fixed point x, i.e., it has the following properties:

a. the mapping σ_x is an isometry,
b. the mapping σ_x is involutive, and
c. the point $y = x$ is a unique fixed point of σ_x in a sufficiently small neighborhood of the point x.

That σ_x is a combination of three isometries implies property a.

The formula

$$\sigma_x^2(y) = x(x^{-1}yx^{-1})x = y$$

implies property b.

We prove property c. Indeed, the condition $xy^{-1}x = y$ is equivalent to $(xy^{-1})^2 = e$. We show that in a sufficiently small neighborhood of the point e, there exists a unique element a such that $a^2 = e$ and $a = e$.

Lemma 3.6. *Let a self-mapping of the matrix algebra $\mathfrak{M}(n)$ be defined by $F(z) = z^2 - e$. Then $z = e$ is an isolated solution to the equation $F(z) = 0$.*

Proof. For any element h that belongs to the Lie algebra $\mathfrak{M}(n)$, we have $F_*(e)h = (zh + hz)|_{z=e} = 2h$. Consequently, the Jacobian of the mapping F is different from zero, and by the implicit function theorem, the solution $z = e$ to the equation $F(z) = 0$ is isolated.

By this lemma, we have $xy^{-1} = e$, i.e., $x = y$. The theorem is proved.

The following assertion plays an important role in the study of symmetrical spaces.

Proposition 3.3. *Let σ_x be an involutive isometry of the manifold G, and let a point x be an isolated fixed point of σ_x. Let U be a neighborhood of the point x such that a unique geodesic lying in U passes through each two points of this neighborhood. Let U contain no other fixed points of the mapping σ_x.*

Let a point $y \in U$ be such that $\sigma_x(y) \in U$. Through the points y and $\sigma_x(y)$, a geodesic γ is drawn. Then the segment of this geodesic γ that connects the points y and $\sigma_x(y)$ contains the point x.

Proof. The image of γ with respect to the mapping σ_x also passes through the points y and $\sigma_x(y)$. Because a geodesic passes into a geodesic under an isometry and only one geodesic passes through two points of U, we have $\sigma_x(\gamma) = \gamma$. Consequently, σ_x defines a self-mapping of a segment of a geodesic. By the Brauer theorem, it has a fixed point. But σ_x has only one fixed point; hence, $x \in \gamma$.

Theorem 3.15 (on one-parametric subgroups). *Let B be a two-sided invariant Riemannian metric on a compact Lie group G. Then geodesics passing through the point e are one-parameter subgroups. Conversely, each one-parameter subgroup is a geodesic.*

Proof. Obviously, it suffices to consider a part of a geodesic that lies in the neighborhood U of the point e. We consider a one-parameter subgroup $\{Z(t)\}$ in the group G that passes through the point e (in particular, $Z(0) = e$). We set $x = Z(1/2)$ and define the mapping σ_x by $\sigma_x(y) = xy^{-1}x$. The proof of Theorem 3.14 implies that the mapping σ_x is an involutive isometry with respect to the point x. Because $\{Z(t)\}$ is a subgroup, we have $\sigma_x(e) = x^2 = Z(1)$ and $\sigma_x(Z(1)) = e$.

By Proposition 3.3, the point $Z(1/2)$ belongs to the geodesic γ passing through the points e and $Z(1)$. In exactly the same way, for any $r = m/2^n$, $m, n \in \mathbb{N}$, the point $Z(r)$ belongs to the same geodesic γ. Indeed, without loss of generality, we can suppose that m is odd. We first substitute the point $Z((m-1)/2^n)$ for the variable y in the expression $Z(m/2^n)y^{-1}Z(m/2^n)$ and then $Z((m+1)/2^n))$. Because $\{Z(t)\}$ is a subgroup, it follows that these two points pass into one another. This gives us the possibility to apply Proposition 3.2 and to prove using induction that the images of all numbers of the form $m/2^n$ on the one-parameter subgroup $Z(t)$ lie on the geodesic γ. By the continuity of a geodesic as well as a one-parameter subgroup of the group, we have $Z(t) \in \gamma$ for all values of t.

The converse statement is implied by the fact that in each direction at the point e, only one geodesic and only one one-parameter subgroup pass.

Because left and right translations are isometries, all other geodesics are obtained by translations of geodesics passing through the point e.

Symmetrical spaces were introduced and studied by E. Cartan, who also classified them [17, 44, 115]. Moreover, it turns out that all symmetrical spaces are some special homogeneous spaces; we now pass to their description.

Definition 3.15. Let G be a Lie group, and let K be a subgroup of G. The pair (G, K) is called *Riemannian symmetrical* if the group G is connected, the subgroup K is closed, and there exists an involutive automorphism σ of the group G whose set of fixed points coincides with K.

The set of fixed points of an automorphism σ is denoted by G^σ.

We explain the definition. Its sense consists in the fact that σ is a "reflection in the submanifold K." After factorization by a subgroup K, the automorphism σ passes to the "reflection at a point." Indeed, $G^\sigma = K$ implies that we can reduce σ up to a mapping $\tilde{\sigma}: G/K \to G/K$. Namely, we set

$$\tilde{\sigma}(gK) = \sigma(g)K. \tag{3.20}$$

Because $\sigma(gk) = \sigma(g)\sigma(k) = \sigma(g)k$, the correctness of (3.20) is implied by the fact that the mapping σ assumes values lying in one and the same coset $\sigma(g)K$ on all elements of the coset gK.

Remark. Usually, a Riemannian symmetrical pair is a pair (G, K) such that the set G^σ "almost coincides" with K, namely, the component of the identity of the group K is contained in G^σ, and G^σ itself is contained in K, i.e., $K_0 \subset G^\sigma \subset K$, where K_0 is the component of the identity. In this book, we restrict ourselves to the simpler definition above. Moreover, for simplicity, we restrict ourselves to the consideration of semisimple Lie groups.

Let $\phi: G \to G/K$ be the natural projection.

Theorem 3.16. *Let σ be an involutive automorphism of the group G for which a closed subgroup K remains fixed. Then the quotient manifold G/K is a symmetrical space, and the mapping $\tilde{\sigma}$ is its involutive isometry at the point $o = \phi(e)$.*

Proof. The involutivity of $\tilde{\sigma}$ is a direct consequence of the involutivity of σ.

Let o be the image of the identity e under the natural projection of G onto G/K. We show that o is an isolated fixed point of the mapping $\tilde{\sigma}$. Formula (3.20) means that for any element $g_1 \in G$, there exists an element $k(g_1) \in K$ such that $\sigma(g_1) = g_1 k(g_1)$. If $k(g_1) = e$, then by $\sigma(g_1 k) = \sigma(g_1)\sigma(k) = g_1 k$, the whole coset $g_1 K$ consists of fixed elements. Therefore, in a sufficiently small neighborhood of the class K, there cannot be a class that contains an element g_1 such that $k(g_1) = e$, because this would contradict $G^\sigma = K$.

On the other hand, because σ is an automorphism, $\sigma^2(g_1) = \sigma(g_1 k(g_1)) = g_1 k(g_1)^2$. By the involutiveness, $\sigma^2(g_1) = g_1$. Consequently, $k(g_1)^2 = e$. By Lemma 3.6, the point e is an isolated solution to the equation $k^2 = e$.

If there exists a coset that is fixed with respect to $\tilde{\sigma}$ and arbitrarily close to the class K for a certain $k(g_1)$ but never vanishes at e, then this contradicts the continuity of the mapping σ.

To show that $\tilde{\sigma}$ is an isometry, we recall that on a homogeneous space G/K, a Riemannian metric is defined that is invariant with respect to the action of the group G, which is an extension of the restriction of the Killing form (taken with the minus sign) to the subspace $\mathfrak{P}$ (see Sec. 3.6). The automorphism σ leaves the subspace $\mathfrak{P}$ itself and its translations $\mathfrak{P}_g$ invariant, as well as the restriction of the Killing form to them. This means that $\tilde{\sigma}$ is an isometry.

To understand which quotient spaces of a Lie group are symmetrical spaces, we need the infinitesimal characteristics of such quotient spaces, i.e., their characterization in terms of the Lie algebra.

Let (G, K) be a Riemannian pair, G be a semisimple compact Lie group, and σ be an involutive automorphism of the group G whose set of fixed points coincides with K.

The differential of the mapping $\sigma: G \to G$ at the identity is denoted by $s = \sigma_*|_e$. The mapping s is identical on $\mathfrak{K}$, the Lie algebra of the group K. As before, we consider $\mathfrak{P} = \mathfrak{K}^\perp$, the orthogonal complement to $\mathfrak{K}$ with respect to the Killing form. On this subspace, $s(X) = -X$. If π is the differential of the natural projection $\phi: G \to G/K$, then we have $\mathrm{Ker}(\pi) = \mathfrak{K}$. Therefore, π is an isomorphism of $\mathfrak{P}$ onto $T_e G/K$.

This construction leads us to the following definition.

Definition 3.16. An *orthogonal Lie algebra with involution* is a pair $(\mathfrak{G}, s)$ such that

1. $\mathfrak{G}$ is a semisimple compact Lie algebra,
2. s is an involutive automorphism of the algebra $\mathfrak{G}$, and
3. the set of fixed points of the mapping s forms a subalgebra of the algebra $\mathfrak{G}$.

Let K be a closed subgroup of a semisimple compact Lie group G, and let $\mathfrak{K}$ be the corresponding subalgebra of the Lie algebra $\mathfrak{G}$. Let s be an automorphism of $\mathfrak{G}$ satisfying conditions 1–3, i.e., the automorphism that transforms $\mathfrak{G}$ to an orthogonal Lie algebra with involution and with the set of fixed points $\mathfrak{K}$.

We lift s up to an automorphism σ of the group G as follows. Let $U(e) \subset G$ be a neighborhood of the point e on which the mapping Exp is one-to-one. Let $g \in U(e)$, and let $\gamma(t)$ be a geodesic that satisfies the conditions $\gamma(0) = e$ and $\gamma(1) = g$. Let Γ denote the tangent vector to this geodesic at the point e. We act on the vector Γ by the mapping s. We draw a geodesic through $s(\Gamma)$, which is denoted by $s(\gamma)(t)$. We set the point $\sigma(g) = s(\gamma)(1)$ in correspondence to a point g.

Proposition 3.4. *If $(\mathfrak{G}, s)$ is an orthogonal Lie algebra with involution, then (G, K) forms a symmetrical Riemannian pair.*

Proof. If $g \in K$, then the geodesic $\gamma(t)$ belongs to K. Consequently, $\Gamma \in \mathfrak{K}$. By condition 3, we have $s(\Gamma) = \Gamma$ and hence $\sigma(g) = g$. This means that $K \subset G^\sigma$.

Conversely, if $\sigma(g) = g$, then $s(\Gamma) = \Gamma$ by the single-valuedness of the exponential mapping. Consequently, $\Gamma \in \mathfrak{K}$ and therefore $g \in K$.

Remark. An orthogonal Lie algebra with involution uniquely defines a universal covering of the symmetrical space $\tilde{G}/\tilde{K}$. (For the definition of a universal covering and the proof of this assertion, see [18].)

§7. Totally Geodesic Submanifolds

Definition 3.17. A submanifold N of a Riemannian manifold M is said to be *geodesic at a point* $x \in N$ if geodesics of the ambient spaces M passing through x and tangent to N at this point belong to N.

A submanifold $N \subset M$ is said to be *totally geodesic* if it is geodesic at each of its points.

Remark. We give an equivalent definition: a submanifold $N \subset M$ is said to be *totally geodesic* if any geodesic of the manifold N (with respect to the metric induced on N by the metric of the ambient manifold M) is at the same time a geodesic of the ambient manifold M.

Exercise. Prove that these definitions are equivalent.

Example 3.17. Affine subspaces of the Euclidean space $\mathbb{R}^n$ are totally geodesic in $\mathbb{R}^n$ (and only they are).

Example 3.18. For the standard n-dimensional sphere

$$S^n = \left\{ \sum_{i=1}^{n+1} x_i^2 = 1 \right\},$$

sections of S^n by planes (of arbitrary dimension) passing through the origin are totally geodesic submanifolds.

Totally geodesic submanifolds in general Riemannian manifolds are relatively rare phenomenon, which are, as a rule, related to one or another symmetry of the manifold considered. However, homogeneous and symmetrical spaces (whose symmetry is basic in their definition) have a large number of totally geodesic submanifolds. The Lie group theory gives us a technique for an effective description of totally geodesic submanifolds.

7.1. Lie Group Isometries. Let G be a compact semisimple linear Lie group, and let $\mathfrak{G}$ be its Lie algebra. The Killing form on the algebra $\mathfrak{G}$ defines a two-sided invariant Riemannian metric on G. This means that both the left and right translations by elements of the Lie group G are isometries of the manifold G. Consequently, the direct product of the groups $F = G \times G$ is a subgroup of the group of isometries of the manifold G. It is convenient to write the action of an element $a = (a_1, a_2) \in F$ on the manifold G in the form $a \colon g \mapsto a_1 g a_2^{-1}$.

We choose the exponent of degree -1 in this formula so that the action F on the manifold G is a representation of the group $G \times G$ with the usual (componentwise) multiplication law, that is, $ab = (a_1 b_1, a_2 b_2)$. Indeed, the superposition of mappings $a \circ b$ has the form

$$a \circ b \colon g \mapsto a_1 b_1 g \, (b_2)^{-1} (a_2)^{-1} = (a_1 b_1) g \, (a_2 b_2)^{-1},$$

i.e., corresponds to the action on the element g of the product ab.

Lemma 3.7. *The stationary subgroup K of the point $e \in G$ is the diagonal of the group $G \times G$.*

Proof. We have $a_1 e a_2^{-1} = e$ iff $a_1 = a_2$.

Corollary 3.2. *The quotient group F/K is isomorphic to G.*

The Lie algebra of the group $G \times G$ is $\mathfrak{G} \times \mathfrak{G}$; the Lie algebra $\mathfrak{K}$ of the group K is the diagonally embedded subalgebra of the Lie algebra $\mathfrak{G} \times \mathfrak{G}$, i.e., the set of pairs of the form (g, g), where $g \in \mathfrak{G}$. If B is the Killing form on $\mathfrak{G}$, then the Killing form on F is given by the relation

$$\widetilde{B}((g_1, g_2)(h_1, h_2)) = B(g_1, h_1) + B(g_2, h_2).$$

Therefore, the orthogonal complement $\mathfrak{P}$ to the diagonal $\mathfrak{K}$ is the subspace $\mathfrak{P} = (g, -g)$, where $g \in \mathfrak{G}$. Consequently, we have the decomposition

$$\mathfrak{G} \times \mathfrak{G} = \mathfrak{K} \oplus \mathfrak{P}. \tag{3.21}$$

Let $[A, B]$ denote the linear span of the set of elements of the form $[a, b]$, where $a \in A$ and $b \in B$. The following formulas define the structure of the Lie algebra $\mathfrak{G} \times \mathfrak{G}$, which is related to decomposition (3.21):

$$
\begin{aligned}
&[\mathfrak{K}, \mathfrak{K}] \subset \mathfrak{K} &&\text{because } [(g, g), (h, h)] = ([g, h], [g, h]), \\
&[\mathfrak{K}, \mathfrak{P}] \subset \mathfrak{P} &&\text{because } [(g, g), (h, -h)] = ([g, h], -[g, h]), \\
&[\mathfrak{P}, \mathfrak{P}] \subset \mathfrak{K} &&\text{because } [(g, -g), (h, -h)] = ([g, h], [g, h]).
\end{aligned}
\tag{3.22}
$$

Definition 3.18. A subspace $\mathfrak{S}$ of a space $\mathfrak{P}$ is called a *triple Lie system* if for any $s_1, s_2, s_3 \in \mathfrak{S}$, we have the inclusion

$$[s_1, [s_2, s_3]] \in \mathfrak{S}.$$

The image of the subspace $\mathfrak{S}$ under the exponential mapping in F is denoted by $\exp \mathfrak{S}$.

Theorem 3.17. *Let G be a compact semisimple linear Lie group, and let $\mathfrak{S} \subset \mathfrak{P}$ be a triple Lie system. Then the orbit S of a point e under the action of $\exp \mathfrak{S}$ is a totally geodesic manifold in G.*

Proof. We first prove several lemmas.

Lemma 3.8. *The subspace $[\mathfrak{S}, \mathfrak{S}]$ of the Lie algebra $\mathfrak{G} \times \mathfrak{G}$ is a subalgebra of the Lie algebra $\mathfrak{K}$.*

Proof. By (3.22), the space $[\mathfrak{S}, \mathfrak{S}]$ is a subspace of the subalgebra $\mathfrak{K}$. Let $X, Y, U, V \in \mathfrak{S}$. By the Jacobi identity, the commutator of the elements $[X, Y]$ and $[U, V]$ of the subspace $[\mathfrak{S}, \mathfrak{S}]$ is equal to

$$[[X, Y], [U, V]] = -[V, [[X, Y], U]] - [U, [V, [X, Y]]].$$

Because $\mathfrak{S}$ is a triple Lie system, this commutator belongs to $[\mathfrak{S}, \mathfrak{S}]$. The lemma is proved.

Lemma 3.9. *The subspace* $\mathfrak{T} = \mathfrak{S} + [\mathfrak{S}, \mathfrak{S}]$ *is a Lie subalgebra of the algebra* $\mathfrak{G} \times \mathfrak{G}$.

Proof. Let $s_i \in \mathfrak{S}$, $i = 1, \ldots, 6$. Then

$$[s_1 + [s_2, s_3], s_4 + [s_5, s_6]] = [s_1, s_4] + [s_1, [s_5, s_6]]$$
$$+ [[s_2, s_3], s_4] + [[s_2, s_3], [s_5, s_6]].$$

Here, the first summand obviously belongs to $[\mathfrak{S}, \mathfrak{S}]$; the second and third summands belong to $\mathfrak{S}$ because $\mathfrak{S}$ is a triple Lie system; the fourth summand belongs to $[\mathfrak{S}, \mathfrak{S}]$ by Lemma 3.8. The lemma is proved.

We pass now to the proof of the theorem. We consider the exponential mapping of the subalgebra $\mathfrak{T}$ into the group $G \times G$. The image of this mapping

$$\{\exp s \mid s \in \mathfrak{T}\}$$

is denoted by T. The stationary subgroup of the point e with respect to the action of the group T is denoted by R.

Lemma 3.10. *The Lie algebra of the group T/R coincides with* $\mathfrak{S}$.

Proof. Elements of the group T are isometries of the group G that correspond to pairs $(\exp(h_1 t), \exp(h_2 t))$, $(h_1, h_2) \in \mathfrak{S} + [\mathfrak{S}, \mathfrak{S}]$. The stationary subgroup R of the point e is defined by the condition

$$\exp(h_1 t)\, e\, \exp(-h_2 t) = e,$$

which holds iff $h_1 = h_2$. Consequently, the Lie algebra $\mathfrak{R}$ of the group R belongs to $\mathfrak{K}$. In addition, by Lemma 3.8, $[\mathfrak{S}, \mathfrak{S}] \subset \mathfrak{K}$. Consequently, the Lie algebra of the group T/R coincides with $\mathfrak{S}$. The lemma is proved.

Because R is a diagonal subgroup in T, we can always choose a representative of a coset of the space T/R such that its second coordinate (in the group $G \times G$) is equal to 1. For this, it suffices to multiply both coordinates by the element of the group T that is inverse to the second coordinate. This remark and Lemma 3.10 imply that the orbit S of the point e under the action of $\exp \mathfrak{S}$ coincides with the orbit of the point e under the action of the group T/R and that the space tangent to this orbit coincides with the projection of $\mathfrak{S}$ onto the first factor of the direct product $\mathfrak{G} \times \mathfrak{G}$.

By Theorem 3.15, one-parameter subgroups $\exp(tX)$, where $X \in \mathfrak{G}$, are geodesics on the group G. These geodesics are tangent to the orbit S iff $X \in \mathfrak{S}$. But they belong to the orbit S. Consequently, the submanifold S is geodesic at the point e, the group T/R acts transitively on the orbit S and transforms it into itself. Because elements of T/R are isometries, they transform geodesics into geodesics. Consequently, the orbit S is a geodesic submanifold at each of its points, i.e., it is a totally geodesic submanifold; this is what was required. The theorem is proved.

Remark. The converse statement also holds; although we do not need it in what follows, we present it for completeness.

Theorem 3.18. *If S is a totally geodesic submanifold of a group G that passes through the point e, then its tangent space at the point e is a triple Lie system.*

For the proof, see [44], Chap. 4, Sec. 7.

Exercise. Let $\mathbb{R}^s \subset \mathbb{R}^n$. Then k-dimensional subspaces lying in $\mathbb{R}^s$, $n > s > k$, form a submanifold $G_k(\mathbb{R}^s)$ of the manifold $G_k(\mathbb{R}^n)$. Prove that this submanifold is totally geodesic.

7.2. Geodesics in the Quotient Space of Lie Groups.

We recall the main notation and constructions related to the concept of a Riemannian symmetrical pair (G, K, σ). Moreover, we use the general definition of a Riemannian symmetrical pair without the assumption that the group G is semisimple and compact.

Let K be a closed subgroup of a Lie group G, and let σ be an involutive automorphism of the group G whose set of fixed points is K. The mapping ϕ is the natural projection of the group G onto G/K. A geodesic symmetry of the manifold G/K at the point $o = \phi(e)$ is denoted by $\tilde{\sigma}$. Let $\mathfrak{G}$ be the Lie algebra of the group G, and let $\mathfrak{K}$ be the Lie algebra of the group K. The differential of the automorphism σ at the point e is denoted by $s \colon \mathfrak{G} \to \mathfrak{G}$. The automorphism s is involutive and is therefore reduced to the diagonal form and has the eigenvalues ± 1. The eigensubspace corresponding to the eigenvalue -1 is denoted by $\mathfrak{P}$. The subalgebra $\mathfrak{K}$ is an eigensubspace that corresponds to the eigenvalue $+1$. Moreover, $\mathfrak{G} = \mathfrak{K} \oplus \mathfrak{P}$. The space $\mathfrak{P}$ is identified with the space tangent to the quotient group G/K at the point $\phi(e)$. The group G acts on the quotient group G/K as an isometry group.

We note that when considering quotient spaces of semisimple compact Lie groups, we choose the orthogonal complement to the subalgebra $\mathfrak{K}$ with respect to the Killing form in the Lie algebra $\mathfrak{G}$ as the space tangent to G/K. By the nondegeneracy and positive definitness of the Killing form on a semisimple compact group G, both described methods for choosing the representative of the tangent space G/K yield one and the same subspace of the group G because the eigenspace corresponding to the eigenvalue -1 of the automorphism s is the orthogonal complement to $\mathfrak{K}$.

The decomposition of an element $X \in \mathfrak{G}$ into components corresponding to the summands $\mathfrak{K}$ and $\mathfrak{P}$ in the direct sum $\mathfrak{G} = \mathfrak{K} \oplus \mathfrak{P}$ can be given by the explicit formula

$$X = \frac{1}{2}(X + sX) + \frac{1}{2}(X - sX). \tag{3.23}$$

Indeed, it is easy to verify that the summand $X + sX$ in (3.23) is an eigenvector s with the eigenvalue $+1$, i.e., an element of the subspace $\mathfrak{K}$, and the summand $X - sX$ is an eigenvector s with the eigenvalue -1, i.e., an element of the subspace $\mathfrak{P}$.

Theorem 3.19. *Let* $X \in \mathfrak{P}$. *Then a geodesic* $\gamma(t)$ *of the manifold* G/K *passing through the point* $o = \phi(e)$ *and tangent to the vector* X *is given by the formula*

$$\gamma(t) = \phi(\exp(tX)), \quad t \in \mathbb{R}. \tag{3.24}$$

Proof. The geodesic symmetry of the manifold G/K at the point $\gamma(t)$ is denoted by $\tilde{\sigma}_t$. We consider the transvection (see Sec. 6) of the manifold G/K along the geodesic γ:

$$T_{2t} = \tilde{\sigma}_t \circ \tilde{\sigma}_0.$$

For this transvection, the geodesic γ passes to itself, and hence

$$\phi T_t(o) = \gamma(t). \tag{3.25}$$

Lemma 3.11. *The mapping* $t \mapsto T_t$ *is a one-parameter subgroup of the group* G.

Proof. Indeed, the symmetry $\tilde{\sigma}_0$ transforms $\gamma(s)$ to $\gamma(-s)$; the symmetry $\tilde{\sigma}_t$ transforms $\gamma(-s)$ to $\gamma(2t+s)$. Therefore, $T_{2\tau} \circ T_{2t} : \gamma(s) \mapsto \gamma(2\tau+2t+s)$. At the same time, the action $T_{2\tau+2t}$ on $\gamma(s)$ leads us to exactly the same result. Consequently, T_t is a one-parameter subgroup of the isometry group of the manifold G/K. Because the isometry group of the manifold G/K is identified with the left action of the group G on the manifold G/K, the group T_t is a one-parameter subgroup of the group G. The lemma is proved.

Therefore, by Theorem 3.15,

$$T_t = \exp(tZ), \quad Z \in \mathfrak{G}. \tag{3.26}$$

Because a vector Z passes to the vector $-Z$ under the isometry σ of the group G, we can assert that the vector Z belongs to $\mathfrak{P}$. By (3.25) and (3.26), the tangent to the geodesic γ at the point e is equal to $\phi_*(Z)$, and, by the condition of the theorem, it is equal to $\phi_*(X)$. But both vectors X and Z lie in $\mathfrak{P}$; therefore, $X = Z$. The theorem is proved.

The group G acts on G/K as an isometry group; therefore, geodesics passing through the point $\phi(g) \in G/K$ have the form $g\phi(\exp(tX))$. By the same arguments, totally geodesic submanifolds of G/K have the form $g\phi(S)$, where S is a totally geodesic submanifold of the group G passing through the point e.

Chapter 4
Grassmann Manifolds

In Chap. 2, we showed that the Riccati equation defines a flow on the Grassmann manifold $G_n(\mathbb{R}^{2n})$. Here, we study a more general Grassmann manifold, the manifold of k-dimensional subspaces of the n-dimensional Euclidean space $\mathbb{R}^n$, which is denoted by $G_k(\mathbb{R}^n)$.

§1. Three Approaches to the Description of the Grassmann Manifolds

We describe three different approaches to Grassmann manifolds: the first uses local coordinates; the second, group actions; and the third, the Plücker embedding.

1.1. Local Coordinates on the Grassmann Manifold.
On the manifold $G_k(\mathbb{R}^n)$, we construct a certain chart $\mathfrak{A}$. For this, we consider a pair of transversal planes: one of them is a k-dimensional plane V_0, which is said to be *horizontal*, and the other is a $(n-k)$-dimensional plane V_∞, which is said to be *vertical*. We consider the set $U \subset G_k(\mathbb{R}^n)$ consisting of all k-dimensional planes transversal to the plane V_∞.

For any $W \in U$, we consider two operators of projection on V_0 and V_∞ parallel to V_∞ and V_0 respectively. The restrictions of these operators to the plane W are respectively denoted by π_0 and π_∞ .

We note that the operator $\pi_0 \colon W \to V_0$ is invertible because W does not contain vertical vectors. Therefore, we can define an operator $\pi_\infty \circ \pi_0^{-1} \colon V_0 \to V_\infty$. Choosing a basis in V_0 and V_∞, we obtain a rectangular $k \times (n-k)$ matrix $\widetilde{W}$, which corresponds to this operator. The mapping that sets the matrix $\widetilde{W}$ in correspondence to each plane $W \in U$ is denoted by ϕ.

The mapping ϕ is invertible because for each such matrix $\widetilde{W}$ (or, equivalently, for each operator $\widetilde{W} \colon V_0 \to V_\infty$), we can consider the set of points $\mathbb{R}^n$ that are given in the basis composed of the bases of the planes V_0 and V_∞ by the vectors $(x, \widetilde{W}x)$, where $x \in V_0$ and $\widetilde{W}x \in V_\infty$. This set of points forms a k-dimensional plane with coordinates $\widetilde{W}$ in the chart (U, ϕ).

We can construct a chart on the manifold $G_k(\mathbb{R}^n)$ using each pair of transversal planes (V_0, V_∞) by the method described above. To find transition functions from one chart to another, we need the following construction.

We consider two charts U and $\widehat{U}$ on the Grassmann manifold. Let a plane W belong to $U \cap \widehat{U}$. We find the coordinates of the plane W in the chart $\widehat{U}$.

For this, we consider the transformation $\mathfrak{A} \colon \mathbb{R}^n \to \mathbb{R}^n$ that transforms a pair of planes H_0, H_∞ into the pair of planes $\widehat{H}_0$, $\widehat{H}_\infty$. Let

$$\begin{pmatrix} \mathfrak{A}_{11} & \mathfrak{A}_{12} \\ \mathfrak{A}_{21} & \mathfrak{A}_{22} \end{pmatrix}$$

be a block decomposition of the matrix $\mathfrak{A}$ in the basis composed of the bases of the planes V_0 and V_∞.

Because $W \in \widehat{U}$, this plane is transversal to $\widehat{H}_\infty$. The coordinates $\widehat{W}$ of the plane W in the chart $\widehat{U}$ are found as follows. We have

$$(\hat{h}, \hat{p}) = ((\mathfrak{A}_{11} + \mathfrak{A}_{12}W)h, (\mathfrak{A}_{21} + \mathfrak{A}_{22}W)h).$$

The matrix $\mathfrak{A}_{11} + \mathfrak{A}_{12}W$ is invertible because if there exists a vector $h^* \in \mathrm{Ker}(\mathfrak{A}_{11} + \mathfrak{A}_{12}W)$, $h^* \neq 0$, then the point (h^*, Wh^*) of the plane W has the coordinates $(0, p^*)$ in the chart $\widehat{U}$, i.e., it has a nontrivial intersection with H_∞. Consequently,

$$h = (\mathfrak{A}_{11} + \mathfrak{A}_{12}W)^{-1}\hat{h},$$

and the point $(\hat{h}, \hat{p})$ can be written in the form

$$(\hat{h}, (\mathfrak{A}_{21} + \mathfrak{A}_{22}W)(\mathfrak{A}_{11} + \mathfrak{A}_{12}W)^{-1}\hat{h}).$$

Therefore, the matrix

$$\widehat{W} = (\mathfrak{A}_{21} + \mathfrak{A}_{22}W)(\mathfrak{A}_{11} + \mathfrak{A}_{12}W)^{-1} \tag{4.1}$$

is the coordinates of the point $W \in G_k(\mathbb{R}^n)$. Transformation (4.1) is called a *generalized linear-fractional transformation.*

Therefore, generalized linear-fractional transformations are transition functions from the chart U to the chart $\widehat{U}$. Because these functions are analytic, we have introduced the structure of an analytic manifold on the Grassmann manifold.

It is easy to verify that the group $\mathrm{GL}(n)$ is homomorphically mapped into the group of generalized linear-fractional transformations of the space of $k \times (n-k)$ matrices. Under the multiplication of all entries of the matrix

$$\begin{pmatrix} A & B \\ C & D \end{pmatrix}$$

by a nonzero scalar, the matrix

$$(C + DW)(A + BW)^{-1}$$

is not changed; therefore, the action of the group of generalized linear-fractional transformation on the Grassmann manifold can be considered the action of the group $\mathrm{SL}(n)$.

1.2. Invariant Description of Grassmann Manifolds. Let V_0 be a k-dimensional plane in $\mathbb{R}^n$. We consider the orbit of this plane under the action of the group $O(n)$. In this case, one and the same point of the orbit can be obtained under the action of different transformations belonging to the group $O(n)$. The set of these transformations is in a one-to-one correspondence with transformations that leave the plane V_0 fixed. This subgroup of $O(n)$ is called the *stabilizer of the plane* V_0. In the basis composed of the bases of the planes V_0 and V_∞, where V_0 is orthogonal to V_∞, elements of this group have the form

$$\begin{pmatrix} A & 0 \\ 0 & B \end{pmatrix},$$

where A is a $k \times k$ matrix and B is an $(n-k) \times (n-k)$ matrix. Consequently, the stabilizer of the point V_0 is isomorphic to $O(k) \times O(n-k)$.

We have thus proved the following assertion.

Proposition 4.1. *The space $G_k(\mathbb{R}^n)$ is identified with the homogeneous space $O(n)/(O(k) \times O(n-k))$, i.e.,*

$$G_k(\mathbb{R}^n) = O(n)/(O(k) \times O(n-k)). \tag{4.2}$$

Corollary 4.1. *The manifolds $G_k(\mathbb{R}^n)$ and $G_{n-k}(\mathbb{R}^n)$ are diffeomorphic.*

The proof is directly implied by (4.2). The diffeomorphism is carried out by a permutation of factors of the direct product $O(k) \times O(n-k)$, or in geometric terms, by a mapping that sets the orthogonal complement in correspondence to each plane.

1.3. Metric on the Grassmann Manifold. The group $O(n)$ is compact and semisimple. Therefore, the Killing form on the corresponding Lie algebra $\mathfrak{O}(n)$ is negative definite. By Theorem 3.13, the structure of a Riemannian manifold with an invariant (with respect to translations by elements of $O(n)$) metric is defined on the Grassmann manifold. Therefore, $O(n)$ is the isometry group of the Grassmann manifold, and $SL(n)$ is a group of transformations, which does not preserve this metric in general.

The Grassmann manifold of oriented k-dimensional planes is identified with $SO(n)/(SO(k) \times SO(n-k))$; the Grassmann manifold of nonoriented k-dimensional planes is identified with $O(n)/(O(k) \times O(n-k))$.

We similarly define the Grassmann manifolds over the field $\mathbb{C}$ and the division ring $\mathbb{H}$.

1.4. Grassmann Manifolds as Symmetrical Spaces. We have seen that $G_k(\mathbb{R}^n) = O(n)/(O(k) \times O(n-k))$ and $O(n)$ is the group of motions of $G_k(\mathbb{R}^n)$. We consider the point $o = \phi(e) \in G_k(\mathbb{R}^n)$, where ϕ is the natural projection onto the quotient space. A certain k-dimensional plane $\mathcal{X}$ corresponds to the point o. The orthogonal complement of this plane is the $(n-k)$-dimensional plane $\mathcal{Y}$.

We consider the linear change of variables $\mathbb{R}^n$, which is identical on $\mathcal{X}$ and is equal to $(-I_{n-k})$ on $\mathcal{Y}$. Under this change of variables, an element V passes to the conjugate one, i.e., $\sigma : V \mapsto \sigma V \sigma$, where

$$\sigma = \sigma^{-1} = \left(\begin{array}{cc} I & 0 \\ 0 & -I \end{array} \right).$$

In this case, the matrix

$$V = \left(\begin{array}{cc} A & B \\ C & D \end{array} \right) \in \mathrm{O}(n)$$

passes to the matrix

$$\left(\begin{array}{cc} A & -B \\ -C & D \end{array} \right).$$

Fixed points of this action (i.e., those matrices V for which $\sigma V \sigma = V$) are block-diagonal matrices and only they are. The set of these matrices coincides with the stationary subgroup K of the point e.

Therefore, the transformation σ is an involutive isometry of the group $\mathrm{O}(n)$ endowed with an invariant metric; the set of fixed points σ coincides with K. By definition, the pair (G, K) is a Riemannian symmetrical pair, and $G_k(\mathbb{R}^n)$ is hence a symmetrical Riemannian manifold by Theorem 3.16.

1.5. Plücker Embeddings. Let W be a certain k-dimensional plane of the space $\mathbb{R}^n$. In the plane W, we choose an arbitrary basis $e_1, \ldots, e_k$ and compose the $n \times k$ matrix A whose rows are coordinates of the vectors e_i in the space $\mathbb{R}^n$.

Let κ be the set of tuples consisting of k elements of the set of subscripts $\{1, \ldots, n\}$. The set κ consists of $s = C_n^k$ elements. We order the set κ in a certain way. We let a_i denote the minor of the order k of the matrix A whose columns belong to the element of the set κ with the number i. We consider the s-dimensional vector $a = \{a_1, \ldots, a_s\}$ composed of all possible minors of the order k of the matrix A that are located in the chosen order. The vector a is called the *Plücker coordinates* of the plane W or, when it is clear which plane we refer to, the *Plücker vector*.

Proposition 4.2. *The Plücker coordinates define the embedding*

$$\phi \colon G_k(\mathbb{R}^n) \to \mathbb{R}\mathrm{P}^{C_n^k - 1}$$

of the Grassmann manifold into the projective space of dimension $C_n^k - 1$.

Proof. We note that the vector composed of the Plücker coordinates is defined with an accuracy up to multiplication by a scalar factor. Indeed, in the plane W, we choose any other basis $f_1, \ldots, f_k$, whose vectors are linear combinations of the vectors of the old basis. In this case, the transition matrix $\Gamma = (\gamma_i^j)$ is nonsingular. Therefore,

$$f_j = \sum_{i=1}^{k} \gamma_j^i e_i, \quad j = 1, \ldots, k, \quad \det \|\gamma_i^j\| \neq 0.$$

This formula implies that each of the minors a_i, $i = 1, \ldots, C_n^k$, is multiplied by one and the same nonzero factor that is equal to the determinant of the matrix Γ. Consequently, independently of the choice of a basis of the plane W, we set one and only one vector of homogeneous coordinates of the projective space of dimension $C_n^k - 1$ in correspondence. Because this mapping is given by polynomial functions, its smoothness is obvious.

The mapping ϕ is called the *Plücker embedding*. It is not surjective.

We describe the image of the Plücker embedding. Because the manifolds $G_k(\mathbb{R}^n)$ and $G_{n-k}(\mathbb{R}^n)$ are diffeomorphic (Corollary 4.1), we can assume that $2k \leq n$. We consider the rectangular $2k \times n$ matrix

$$B = \begin{pmatrix} A \\ A \end{pmatrix}$$

whose first k and last k rows coincide with the matrix A. Let J be a certain fixed subset of the set of subscripts $\{1, \ldots, n\}$ that is composed of $2k$ elements. For each J, we write a certain equation that points of the Plücker embedding ϕ satisfy.

Let L be a subset of the set of subscripts J consisting of k elements. Let $\overline{L}$ denote the complement of the set L with respect to J. We consider the minor of order $2k$ of the matrix B whose column indices belong to the set J. This minor vanishes because it has equal rows. Using the Laplace formula, we decompose this minor with respect to the first k rows:

$$0 = \sum_{L \subset J} (-1)^{\nu(L)} a_L a_{\overline{L}}, \tag{4.3}$$

where $\nu(L)$ is the parity of the tuple L. We have obtained a quadratic equation with respect to the coordinates a_i of the Plücker vector. The image of the Plücker embedding lies on the intersection of quadrics of form (4.3), which corresponds to all tuples of the subscripts J that consist of $2k$ elements.

There are other quadratic relations for the Plücker coordinates (see the appendix).

Example 4.1. For $k = 1$, the Grassmann manifold $G_1(\mathbb{R}^n)$ is the projective space $\mathbb{RP}^{n-1}$; the Plücker coordinates are homogeneous coordinates of the line, which defines a point of the space $\mathbb{RP}^{n-1}$; Eq. (4.3) is reduced to an identity. The image of the Plücker embedding is the whole $\mathbb{RP}^{n-1}$.

Example 4.2. We consider the manifold $G_2(\mathbb{R}^4)$. In this case, the number s of minors of order 2 in the 2×4 matrix A is equal to $C_4^2 = 6$. Consequently, the image of the Plücker embedding lies in the five-dimensional projective space with homogeneous coordinates $(a_{12}, a_{13}, a_{14}, a_{23}, a_{24}, a_{34})$, where a pair of subscripts defines the column index of the matrix A.

The matrix B is a square 4×4 matrix. It has only one minor of the order 4, i.e., there is only one relation of form (4.3), which has the form

$$a_{12}a_{34} - a_{13}a_{24} + a_{14}a_{23} = 0. \tag{4.4}$$

If we turn the axes by the angle $\pi/4$ in each of the planes (a_{12}, a_{34}), (a_{13}, a_{24}), and (a_{14}, a_{23}), Eq. (4.4) becomes

$$x_1^2 + x_2^2 + x_3^2 = x_4^2 + x_5^2 + x_6^2, \tag{4.5}$$

where x_i are new homogeneous coordinates in the space $\mathbb{RP}(5)$. We can normalize these coordinates by the condition

$$\sum_{i=1}^{6} x_i^2 = 2.$$

Then Eq. (4.5) transforms into the set of equations

$$x_1^2 + x_2^2 + x_3^2 = 1, \qquad x_4^2 + x_5^2 + x_6^2 = 1,$$

which corresponds to the direct product of two two-dimensional spheres; this is in full correspondence with Theorem 2.4.

§2. Lagrange–Grassmann Manifolds

There remains one more manifold of the Grassmann type to consider, a type very important for the theory of extremal problems: the *Lagrange–Grassmann manifold*.

In this section, we consider the linear space $\mathbb{R}^{2n}$ endowed with a symplectic structure, i.e., a nondegenerate bilinear skew-symmetrical form ω. In the space $\mathbb{R}^{2n}$, we can always choose a coordinate system in which the matrix of the form ω looks like

$$J = \begin{pmatrix} 0 & I_n \\ -I_n & 0 \end{pmatrix}.$$

The group $\mathrm{Sp}(n, \mathbb{R})$ (see Example 3.5) is a group of linear transformations that preserve this symplectic structure.

Definition 4.1. A plane $L \subset \mathbb{R}^{2n}$, $\dim L = n$, is said to be *Lagrangian* if $\omega|_L = 0$.

The space L is called *isotropic* if $\omega|_L = 0$. In contrast to Lagrangian planes, isotropic planes can have an arbitrary dimension that does not exceed n.

Definition 4.2. An isotropic subspace $\pi \subset \mathbb{R}^{2n}$ is called a *Lagrangian plane* if for any isotropic subspace $\rho \supset \pi$, we have $\rho = \pi$, i.e., π is the maximal isotropic subspace.

Exercise. Prove that the definitions just given are equivalent.

Example 4.3. Let $p_1, \ldots, p_n, q_1, \ldots, q_n$ be a basis of the space $\mathbb{R}^{2n}$ in which the matrix of ω is equal to J. Let $k \in \mathbb{N}$, $0 \leq k \leq n$. The plane generated by the vectors $p_{i_1}, \ldots, p_{i_k}, q_{j_1}, \ldots, q_{j_{n-k}}$ is Lagrangian whenever none of the subscripts i is equal to one of the subscripts j.

Because a symplectic transformation leaves the form ω invariant, the images of Lagrangian planes under the action of $\mathrm{Sp}(n, \mathbb{R})$ are also Lagrangian planes.

Definition 4.3. The value of the form ω on a pair of vectors $x, y \in \mathbb{R}^{2n}$ is called the *skew-inner product* of these vectors. We say that two vectors $x, y \in \mathbb{R}^{2n}$ are *skew-orthogonal* if $\omega(x, y) = 0$.

Let l be the linear space $\mathbb{R}^{2n}$. The set of vectors that are skew-orthogonal to l is denoted by $l^{\perp}$.

Exercise. Prove that (a) $(l_1 + l_2)^{\perp} = l_1^{\perp} \cap l_2^{\perp}$; (b) $(l_1 \cap l_2)^{\perp} = l_1^{\perp} + l_2^{\perp}$; (c) $\dim l + \dim l^{\perp} = 2n$; and (d) $(l^{\perp})^{\perp} = l$.

2.1. Coordinates on a Lagrange–Grassmann Manifold. We consider the set $\Lambda(\mathbb{R}^{2n})$ of Lagrangian planes in the space $(\mathbb{R}^{2n}, \omega)$.

We choose two n-dimensional Lagrangian planes that are in general position (i.e., they intersect only at the origin). We let V_0 and V_∞ denote them and fix some bases in these planes. Let W be an arbitrary Lagrangian plane that is transversal to V_∞. As in the manifold $G_n(\mathbb{R}^{2n})$, we define coordinates of the plane W as entries of the matrix of the mapping $W = (\pi_\infty \circ \pi_0^{-1})$.

As in Chap. 2, we use one and the same notation W for subspaces as well as for their local coordinates.

We find conditions that the matrix W should satisfy for the plane W to be Lagrangian. We fix the Euclidean structure on $\mathbb{R}^{2n}$ and assume that the planes V_0 and V_∞ are orthogonal. We have

$$\left\langle \begin{pmatrix} x \\ Wx \end{pmatrix}, J \begin{pmatrix} y \\ Wy \end{pmatrix} \right\rangle = 0$$

for any vectors $x, y \in V_0$, i.e.,

$$(x, Wx) \begin{pmatrix} 0 & I_n \\ -I_n & 0 \end{pmatrix} \begin{pmatrix} y \\ Wy \end{pmatrix} = -\langle Wx, y \rangle + \langle x, Wy \rangle = 0.$$

Therefore, the matrix W should be symmetrical. The obtained condition is a necessary and sufficient condition for the plane W to be Lagrangian.

Let

$$\begin{pmatrix} A & B \\ C & D \end{pmatrix} \in \mathrm{Sp}(n, \mathbb{R}).$$

Then the action of this symplectic matrix on a matrix W when passing from one chart to another is the generalized linear-fractional transformation

$$W \mapsto (C + DW)(A + BW)^{-1}. \tag{4.6}$$

Therefore, the set of all Lagrangian planes $\Lambda(\mathbb{R}^{2n})$ in the symplectic space $\mathbb{R}^{2n}$ is a smooth manifold, which is called a *Lagrange–Grassmann manifold*. In a given chart, symmetrical $n \times n$ matrices are coordinates of points of this manifold. In contrast to the dimension of the Grassmann manifold $G_n(\mathbb{R}^{2n})$, which is equal to n^2, the dimension of the Lagrange–Grassmann manifold $\Lambda(\mathbb{R}^{2n})$ is equal to $n(n+1)/2$.

Exercises.

1. Deduce formula (4.6).

2. Verify that a symmetrical matrix W passes to a symmetrical one under transformation (4.6).

3. Consider two distinct pairs of transversal Lagrangian planes. Prove that there exists a transformation $A \in \mathrm{Sp}(n, \mathbb{R})$ that transforms one of these pairs into the other.

2.2. Lagrange–Grassmann Manifold as a Homogeneous Space. It is convenient to use the complex linear space $\mathbb{C}^n$ to study Lagrange–Grassmann manifolds.

We can consider a vector $(x + iy) \in \mathbb{C}^n$ as a vector with the coordinates $(x, y) \in \mathbb{R}^{2n}$ and complex linear transformations of $\mathbb{C}^n$ as transformations of the real space $\mathbb{R}^{2n}$. In this case, the $2n \times 2n$ matrix

$$\begin{pmatrix} A & -B \\ B & A \end{pmatrix},$$

where $A, B \in \mathfrak{M}_n$, corresponds to a complex matrix $A + iB$ because

$$(A + iB)(x + iy) = (Ax - By) + i(Ay + Bx) = \begin{pmatrix} A & -B \\ B & A \end{pmatrix} \begin{pmatrix} x \\ y \end{pmatrix}.$$

If a matrix $A + iB$ is unitary, i.e., $(A + iB)(A^{\top} - iB^{\top}) = I_n$, then we have the relations

$$AA^{\top} + BB^{\top} = I_n, \qquad AB^{\top} = BA^{\top}. \tag{4.7}$$

We recall that *unitary matrices* are matrices that preserve the following positive-definite Hermitian form on $\mathbb{C}^n$:

$$\Omega(z, w) = \sum_{i=1}^{n} z_i \bar{w}_i.$$

Lemma 4.1. *A transformation $X \in \mathrm{SU}(n)$ is unitary if it belongs simultaneously to $\mathrm{SO}(2n)$ and $\mathrm{Sp}(n, \mathbb{R})$.*

Proof. The identification of $\mathbb{C}^n$ with $\mathbb{R}^{2n}$ allows us to consider a vector $z = x + iy \in \mathbb{C}^n$ as a $2n$-dimensional vector with real coordinates (x, y). For brevity, this $2n$-dimensional vector is denoted by z as previously.

We isolate the real and imaginary parts of the form Ω considering its values on the vectors $z = x + iy$ and $w = u + iv$:

$$\Omega(z, w) = \sum_{k=1}^{n}(x_k + iy_k)(u_k - iv_k)$$

$$= \sum_{k=1}^{n}(x_k u_k + y_k v_k) + i \sum_{k=1}^{n}(y_k u_k - x_k v_k)$$

$$= \langle (x, y), (u, v) \rangle + i(x, y)J\begin{pmatrix} u \\ v \end{pmatrix}$$

or

$$\Omega(z, w) = \langle z, w \rangle + i\omega(z, w). \tag{4.8}$$

We have thus found that the real part $\Re\Omega$ of the form Ω assigns the standard Euclidean inner product on the space $\mathbb{C}^n$, considered as $\mathbb{R}^{2n}$, and the imaginary part $\Im\Omega$ defines a symplectic structure on $\mathbb{C}^n$. The invariance of the form Ω with respect to a unitary transformation $\mathfrak{A} \in U(n)$ means that this transformation, considered as a transformation of $\mathbb{R}^{2n}$, preserves $\Re\Omega$ as well as $\Im\Omega$, i.e., $\mathfrak{A}$ is simultaneously orthogonal and symplectic. To prove the converse statement, it only remains to verify that any mapping $\mathfrak{A} \in SO(2n) \cap Sp(n, \mathbb{R})$ is $\mathbb{C}$-linear. Indeed, because $\mathfrak{A} \in Sp(n, \mathbb{R})$, it commutes with the operator of multiplication by i, which corresponds to the matrix J.

An arbitrary complex linear subspace of the space $\mathbb{C}^n$ is not a Lagrangian plane. Moreover, the following assertion holds.

Proposition 4.3. *A Lagrangian subspace L of the space $\mathbb{R}^{2n}$ contains no complex line.*

Proof. We suppose the contrary, i.e., that there exists a nonzero vector $a \in L$ such that $ia \in L$. Then the value $\Omega(ia, a) = i\Omega(a, a)$ is a purely imaginary number that is different from zero. Consequently, by (4.8), we have $\omega(ia, a) \neq 0$; this contradicts the fact that L is isotropic.

Corollary 4.2. *If a plane L^n is Lagrangian, then its complexification coincides with the whole space $\mathbb{R}^{2n}$.*

Proof. We suppose the contrary. If the complexification L does not coincide with $\mathbb{R}^{2n}$, then the intersection $L \cap iL$ is nontrivial because the subspaces L and iL are n-dimensional; therefore, L contains at least one complex line, which contradicts Proposition 4.3.

Theorem 4.1. *The unitary group $U(n)$ acts transitively on the set of Lagrangian planes.*

Proof. We choose an orthonormal (with respect to the Euclidean metric in $\mathbb{R}^{2n}$) basis $p_1, \ldots, p_n$ in a Lagrangian plane L. Because L is Lagrangian, the skew-inner products of vectors of this basis vanish, and the vectors $p_1, \ldots, p_n$ hence form an orthonormal basis of the complex space $\mathbb{C}^n$.

Now let L_1 and L_2 be two arbitrary Lagrangian planes. We choose orthonormal bases in each of these planes. The unitary transformation that transforms one of these bases into the other transforms the Lagrangian plane L_1 into the Lagrangian plane L_2.

Corollary 4.3. *The symplectic group acts transitively on the set of Lagrangian planes.*

The proof is implied by the fact that the symplectic group contains the group of unitary transformations as a subgroup.

We consider the real n-dimensional subspace V of the space $\mathbb{C}^n$ that consists of vectors $z = x + iy$ for which $y = 0$. The plane V is Lagrangian. Indeed, we consider two vectors $z = x + iy$ and $w = u + iv$. Because $z, w \in V$, we have $y = v = 0$. Then

$$\omega(z, w) = \sum_{k=1}^{n}(y_k u_k - x_k v_k) = 0.$$

We now consider the manifold of oriented Lagrangian planes $\Lambda^+(\mathbb{R}^{2n})$. The unitary group acts transitively on the set of Lagrangian planes. The stationary subgroup of a plane V consists of unitary transformations $\mathfrak{A}$ such that they leave V fixed and do not change its orientation. Such transformations have the form

$$\mathfrak{A} = \begin{pmatrix} A & 0 \\ 0 & A \end{pmatrix},$$

where $A \in \mathrm{SO}(n)$.

Because

$$\begin{pmatrix} A & 0 \\ 0 & A \end{pmatrix} \begin{pmatrix} 0 & I \\ -I & 0 \end{pmatrix} \begin{pmatrix} A^\top & 0 \\ 0 & A^\top \end{pmatrix} = \begin{pmatrix} 0 & A \\ -A & 0 \end{pmatrix} \begin{pmatrix} A^\top & 0 \\ 0 & A^\top \end{pmatrix}$$

$$= \begin{pmatrix} 0 & I \\ -I & 0 \end{pmatrix},$$

we have $\mathfrak{A} \in \mathrm{Sp}(n, \mathbb{R})$. Consequently,

$$\Lambda^+(\mathbb{R}^{2n}) = \mathrm{U}(n)/\mathrm{SO}(n).$$

In exactly the same way, we can easily show that the manifold of nonoriented Lagrangian planes is a homogeneous space

$$\Lambda(\mathbb{R}^{2n}) = \mathrm{U}(n)/\mathrm{O}(n).$$

We have thus introduced the structure of a real analytic manifold on the manifold of Lagrangian planes.

The group $U(n)$ is not semisimple. Nevertheless, the Lagrange–Grassmann manifold $\Lambda(n) = U(n)/O(n)$ is a Riemannian manifold. This fact is implied by the compactness condition of the image of the orthogonal subgroup under the adjoint representation of the unitary group, i.e., the subgroup $\mathrm{Ad}_{U(n)} O(n)$.

Indeed, in the case where a compact group (in our case, $\mathrm{Ad}_{U(n)} O(n)$) acts on the Lie algebra $\mathfrak{U}(n)/\mathfrak{O}(n)$, we can use Lemma 3.4 to construct a positive-definite quadratic form B on this space that is invariant with respect to the action of this group. On the manifold $\Lambda(n) = U(n)/O(n)$, we can construct a left-invariant metric using the action of the group $U(n)$. This metric is defined by the form B on the tangent space at the point e and is extended to the whole manifold using left translations. In this way, we transform $\Lambda(n)$ into a Riemannian manifold on which the group $U(n)$ acts as a subgroup of the isometry group.

We can obtain another proof that $\Lambda(n)$ is a Riemannian manifold using Lie algebra theory. Namely, Proposition 3.4 also remains valid under the following more general definition of an orthogonal Lie algebra with involution.

Definition 4.4. An *orthogonal Lie algebra with involution* is a pair $(\mathfrak{G}, s)$ such that

1. $\mathfrak{G}$ is a Lie algebra over $\mathbb{R}$;
2. s is an involutive automorphism of the algebra $\mathfrak{G}$;
3. the set $\mathfrak{K}$ of fixed points of the automorphism s is a compact subalgebra of the Lie algebra $\mathfrak{G}$;
4. $\mathfrak{K} \cap \mathfrak{z} = \{0\}$, where $\mathfrak{z}$ is the center of the algebra $\mathfrak{G}$.

With this definition, Proposition 4.2 implies that the manifold Λ is a Riemannian symmetrical space.

We note that the Lie algebra of the group $U(n)$ consists of skew-Hermitian matrices
$$\begin{pmatrix} a & -b \\ b & a \end{pmatrix},$$
i.e., matrices that satisfy the relation $(a + ib)^\top = -a + ib$, which means that $a^\top = -a$ and $b^\top = b$.

Let $\mathfrak{K}$ be the Lie subalgebra of matrices having the form
$$\begin{pmatrix} a & 0 \\ 0 & a \end{pmatrix},$$
where $a^\top = -a$.

Exercise. Prove that $\mathfrak{P} = \mathfrak{K}^\perp$ is given by the matrices
$$\begin{pmatrix} 0 & -b \\ b & 0 \end{pmatrix},$$
where $b^\top = b$.

In this way, we identify the tangent space to the manifold $\Lambda(\mathbb{R}^{2n})$ at the point $o = \phi(e)$ with the space of symmetrical $n \times n$ matrices b that define the matrices

$$\begin{pmatrix} 0 & -b \\ b & 0 \end{pmatrix},$$

where $b^\top = b$. The corresponding one-parameter subgroups are

$$\exp \begin{pmatrix} 0 & -bt \\ bt & 0 \end{pmatrix}.$$

2.3. The Manifold $\Lambda(\mathbb{R}^{2n})$ as a Symmetrical Space. Because we have $\Lambda(\mathbb{R}^{2n}) = \mathrm{U}(n)/\mathrm{O}(n)$, elements of $\mathbb{R}^{2n}$ are represented by unitary matrices of the form

$$\begin{pmatrix} A & -B \\ B & A \end{pmatrix},$$

where A and B are matrices such that the relations $AA^\top + BB^\top = I_n$ and $AB^\top = BA^\top$ hold by (4.7).

We consider the matrix

$$s_0 = \begin{pmatrix} I_n & 0 \\ 0 & -I_n \end{pmatrix}.$$

Obviously, $s_0^{-1} = s_0$. We consider the transformation

$$\sigma \colon \mathrm{U}(n)/\mathrm{O}(n) \to \mathrm{U}(n)/\mathrm{O}(n), \qquad \sigma(g) = s_0 g s_0.$$

This transformation transforms unitary matrices into unitary ones. It is easy to verify that

$$s_0 \begin{pmatrix} x \\ y \end{pmatrix} = \begin{pmatrix} x \\ -y \end{pmatrix}, \qquad s_0 \begin{pmatrix} A & -B \\ B & A \end{pmatrix} s_0 = \begin{pmatrix} A & B \\ -B & A \end{pmatrix},$$

i.e., the operator s_0 acts as the complex conjugation.

Exercise. Prove that the mapping σ is an involutive isometry and fixed points of σ coincide exactly with the stationary subgroup of the plane V.

§3. Riccati Equation as a Flow on the Manifold $G_n(\mathbb{R}^{2n})$

As was shown in Sec. 3 in Chap. 2, we can consider the Riccati equation as an equation that describes the evolution of n-dimensional linear subspaces under the action of symplectic transformations generated by the canonical system of ordinary differential equations

$$\begin{cases} \dot{h} = -A^{-1}C^\top h + A^{-1}p, \\ \dot{p} = (B - CA^{-1}C^\top)h + CA^{-1}p. \end{cases}$$

The Riccati equation has the form

$$\dot{W} = -B + (C + W)A^{-1}(C^{\top} + W), \tag{4.9}$$

where $A^{\top} = A$ and $B^{\top} = B$.

Because Eq. (4.9) describes the evolution of $n \times n$ matrices, it thus acts on a chart of the manifold $G_n(\mathbb{R}^{2n})$. We show that Eq. (4.9) defines a vector field on the whole Grassmann manifold. For this, we reveal what happens with Eq. (4.9) when passing to another chart.

We describe the action of transformations of the group $\mathrm{SL}(2n, \mathbb{R})$ on the manifold $G_n(\mathbb{R}^{2n})$.

Proposition 4.4. *The group* $\mathrm{SL}(2n)$ *is generated by the generators*

$$\begin{pmatrix} I & 0 \\ B & I \end{pmatrix}, \qquad \begin{pmatrix} D^{-1} & 0 \\ 0 & A \end{pmatrix}, \qquad \begin{pmatrix} 0 & I \\ -I & 0 \end{pmatrix},$$

where $\det A = \det D \neq 0$.

Proof. Indeed, the composition of the transformations

$$\begin{pmatrix} I & 0 \\ BA^{-1} & I \end{pmatrix} \begin{pmatrix} A & 0 \\ 0 & D \end{pmatrix} = \begin{pmatrix} A & 0 \\ B & D \end{pmatrix}$$

yields any block lower-triangular matrix.

The composition

$$\begin{pmatrix} 0 & I \\ -I & 0 \end{pmatrix} \begin{pmatrix} A & 0 \\ B & D \end{pmatrix} \begin{pmatrix} 0 & I \\ -I & 0 \end{pmatrix} = \begin{pmatrix} -D & B \\ 0 & -A \end{pmatrix}$$

yields any block upper-triangular matrix.

Finally, we can obtain any element

$$\begin{pmatrix} A & B \\ C & D \end{pmatrix} \in \mathrm{SL}(2n)$$

for which $\det A \neq 0$ considering the composition

$$\begin{pmatrix} I & 0 \\ CA^{-1} & DA - CA^{-1}BA \end{pmatrix} \begin{pmatrix} A & B \\ 0 & A^{-1} \end{pmatrix} = \begin{pmatrix} A & B \\ C & D \end{pmatrix}.$$

Here, $\det(DA - CA^{-1}BA) = 1$ because the product of determinants is equal to the determinant of the product.

We perform the reduction of the general case to the case $\det A \neq 0$ as follows. If $\mathrm{rk}\, A = r$, then, performing the transformation

$$\begin{pmatrix} V & 0 \\ 0 & W \end{pmatrix} \begin{pmatrix} A & B \\ C & D \end{pmatrix} \begin{pmatrix} \tilde{V} & 0 \\ 0 & \tilde{W} \end{pmatrix} = \begin{pmatrix} A' & B' \\ C' & D' \end{pmatrix},$$

we obtain

$$A' = \begin{pmatrix} I_r & 0 \\ 0 & 0 \end{pmatrix},$$

where I_r is the identity matrix of order r.

If

$$C' = \begin{pmatrix} C_1 & C_2 \\ C_3 & C_4 \end{pmatrix}$$

is the corresponding partition into blocks of the matrix C', then the rank of the matrix

$$\begin{pmatrix} C_2 \\ C_4 \end{pmatrix}$$

is equal to $n - r$; otherwise, the matrix

$$\begin{pmatrix} A' & B' \\ C' & D' \end{pmatrix}$$

is singular.

We consider the product

$$\begin{pmatrix} I & S \\ 0 & I \end{pmatrix} \begin{pmatrix} A' & B' \\ C' & D' \end{pmatrix} = \begin{pmatrix} A'' & B'' \\ C'' & D'' \end{pmatrix},$$

where

$$S = \begin{pmatrix} 0 & 0 \\ \Lambda & I \end{pmatrix}$$

and all rows of the matrix Λ are equal. Then

$$A'' = A' + SC' = \begin{pmatrix} I_r & 0 \\ * & C_4 + \Lambda C_2 \end{pmatrix}.$$

In the matrix $C_4 + \Lambda C_2$, we add linear combinations of rows of C_2 to rows of the matrix C_4. We can therefore choose Λ such that

$$\mathrm{rk}(C_4 + \Lambda C_2) = n - r.$$

This ensures the fulfillment of the condition $\det A'' \neq 0$.

It is easy to see that the above generators correspond to the following generalized linear-fractional transformations:

$$P \mapsto P + B,$$
$$P \mapsto APD, \quad \det A = \det D \neq 0,$$
$$P \mapsto -P^{-1}, \quad \det P \neq 0.$$

The transformations of the first two types do not change the quadratic character of the right-hand side of Eq. (4.9); moreover, they do not lead out of the chart considered.

We consider the transformation of the third type. Under the change of variables $W^{-1} = -Z$, Eq. (4.9) passes to the equation

$$\dot{Z} = W^{-1}\dot{W}W^{-1}$$
$$= W^{-1}[-B + (C + W)A^{-1}(C^{\top} + W)]W^{-1}$$
$$= Z(CA^{-1}C^{\top} - B)Z - ZCA^{-1} - A^{-1}C^{\top}Z + A^{-1}.$$

This is again an equation of form (4.9) written in a chart containing points that are infinitely distant for the initial chart. Because the right-hand side of Eq. (4.9) is continuous at all points of this new chart, we can consider the vector field defined by the right-hand side as a field that is defined and continuous on the whole closed manifold $G_n(\mathbb{R}^{2n})$. As a result, we obtain a possibility to consider Eq. (4.9) as an equation on $G_n(\mathbb{R}^{2n}) \times \mathbb{R}$.

We recall that in Sec. 2 in Chap. 2, it was shown that if a matrix $W(0)$ is symmetrical, then

$$W(t) = W^{\top}(t),$$

i.e., the manifold of Lagrangian planes is an integral manifold (in $G_n(\mathbb{R}^{2n})$) for the Riccati equation. Therefore, Eq. (4.9) defines a flow on the Lagrange–Grassmann manifold.

§4. Systems Associated with a Linear System of Differential Equations

In what follows, when studying the Riccati equation, we use some applications of the representation theory to ordinary differential equations. This section is devoted to presenting a method, which, in spite of its simplicity and usefulness, is not considered in the presentation of the theory of ordinary differential equations as a rule.

We consider a matrix system of ordinary differential equations

$$\dot{X} = A(t)X, \tag{4.10}$$

where X is the unknown $n \times n$ matrix and the matrix of coefficients $A(t)$ is continuous on the interval (a, b), the domain of varying the independent variable t. We let $X(t, X_0, t_0)$ denote a solution to the Cauchy problem for system (4.10) with the initial data $X(t_0) = X_0$, where $t_0 \in (a, b)$.

Proposition 4.5. *Let G be a matrix Lie group, and let $\mathfrak{G}$ be its Lie algebra. Let the coefficients $A(t)$ in the right-hand side of system of differential equations* (4.10) *belong to $\mathfrak{G}$ for all $t \in (a, b)$, and let the initial value of this system X_0 belong to G. Then the solution $X(t, X_0, t_0)$ to this system belongs to G for all $t \in (a, b)$.*

Proof. Because the group G is a matrix group, the manifold G is a smooth submanifold of the space $\mathbb{R}^{n^2}$. In the space $\mathbb{R}^{n^2}$, system (4.10) has a unique solution.

On the other hand, we can consider this system as a system of differential equations given on the group G. Indeed, for each fixed $t \in (a, b)$, the vectors $A(t)X$ in the right-hand side of Eq. (4.10) form a right-invariant vector field on the manifold G that is obtained by the right translation of the element $A(t) \in \mathfrak{G} = T_e G$ to all points $X \in G$ of the manifold G. This vector field is smooth, and system (4.10), considered as equations defined on the manifold G, has a unique solution lying in G. Obviously, these two solutions coincide.

Matrix differential equation (4.10) arises when we consider a fundamental system of solutions to the vector differential equation

$$\dot{x} = A(t)x, \tag{4.11}$$

where $x \in \mathbb{R}^n$. Slightly more general systems are Lyapunov-type systems

$$\dot{X} = L(t)X + XM(t), \tag{4.12}$$

where X, L, and M are $n{\times}n$ matrices. Proposition 4.5 is extended to this case almost literally.

Proposition 4.6. *Let the coefficients $L(t)$ and $M(t)$ in the right-hand side of system (4.12) belong to the algebra $\mathfrak{G}$ for all $t \in (a, b)$, and let $X_0 \in G$. Then $X(t, X_0, t_0) \in G$ for all $t \in (a, b)$.*

Proof. For any fixed $t \in (a, b)$, the matrices $L(t)X$ form a right-invariant vector field on the manifold G. In exactly the same way, $XM(t)$ forms a left-invariant vector field. The right-hand side of system (4.12) is the sum of tangent vectors at the point $X \in G$ and is hence also a vector tangent to the manifold G at the point X. The further argument is completely similar to the proof of Proposition 4.5.

We consider an arbitrary representation $\mathfrak{U}$ of the group G, i.e., a homomorphism of the Lie group G into the group of linear operators acting on a certain linear space V. By Proposition 4.6, the flow of solutions to Eq. (4.6) on the manifold $G \times \mathbb{R}$ induces a flow on $\mathfrak{U}(G) \times \mathbb{R}$. We must take their images under the mapping $d\mathfrak{U}$ at the point e as the coefficients L and M of the new system.

Remark. We consider the natural representation of the group $\mathrm{GL}(n)$ under which this group is always realized as a group of linear transformations of the space $\mathbb{R}^n$. The Plücker coordinates of a k-dimensional plane, being minors of order k, correspond to the kth exterior powers of the natural representation of the group $\mathrm{GL}(n)$ (the definition of an exterior power of a representation can be found in [18, 54]).

4.1. Associated Systems on Grassmann Manifolds. To describe the evolution of the Plücker coordinates of k-dimensional planes whose points are transferred according to system (4.11), we must write the derivatives of minors keeping in mind that the derivatives of the entries composing these minors satisfy system (4.11). This computation, which is needed in studying complex Riccati equations, we carry out completely in Chap. 6. Here, we restrict ourselves to considering the simplest case among nontrivial ones, the Lagrange–Grassmann manifold $\Lambda(\mathbb{R}^4)$.

Let $x = (x_1, x_2, x_3, x_4)$ and $y = (y_1, y_2, y_3, y_4)$ be two basis vectors of a Lagrangian plane W. We want to find a set of differential equations that describes the evolution of the Plücker coordinates of this Lagrangian plane in the flow described by the Hamiltonian system

$$\dot{X} = L(t)X. \tag{4.13}$$

That system (4.13) is Hamiltonian means that the 4×4 matrix $L(t)$ belongs to $\mathfrak{Sp}(2, \mathbb{R})$. Entries of the matrix L are denoted by λ_{ij}. As explained in Sec. 1, the six-dimensional vector $\xi = (\xi_{12}, \xi_{13}, \xi_{14}, \xi_{23}, \xi_{24}, \xi_{34})$, where ξ_{ij} is a minor of the matrix

$$\begin{pmatrix} x_1 & x_2 & x_3 & x_4 \\ y_1 & y_2 & y_3 & y_4 \end{pmatrix}$$

corresponding to columns of (i, j), serves as the Plücker coordinates of the plane W.

Differentiating these minors, we easily confirm that the vector ξ satisfies the linear system of ordinary differential equations with the matrix

$$\begin{pmatrix} \lambda_{11}+\lambda_{22} & \lambda_{23} & \lambda_{24} & -\lambda_{13} & -\lambda_{14} & 0 \\ \lambda_{32} & \lambda_{11}+\lambda_{33} & \lambda_{34} & \lambda_{12} & 0 & -\lambda_{14} \\ \lambda_{42} & \lambda_{43} & \lambda_{11}+\lambda_{44} & 0 & \lambda_{12} & \lambda_{13} \\ -\lambda_{31} & \lambda_{21} & 0 & \lambda_{22}+\lambda_{33} & \lambda_{34} & -\lambda_{24} \\ -\lambda_{41} & 0 & \lambda_{21} & \lambda_{43} & \lambda_{22}+\lambda_{44} & \lambda_{23} \\ 0 & -\lambda_{41} & \lambda_{31} & -\lambda_{42} & \lambda_{32} & \lambda_{33}+\lambda_{44} \end{pmatrix}. \tag{4.14}$$

We can write a similar set of differential equation for minors of the kth order of the matrix composed of k n-dimensional vectors. Then these minors (the Plücker coordinates of a k-dimensional plane in the n-dimensional space) satisfy quadratic relations (4.3), which automatically turn out to be integrals of the obtained set of differential equations. In the case of the manifold $G_2(\mathbb{R}^4)$, these quadratic relations consist of only one equation:

$$\xi_{12}\xi_{34} - \xi_{13}\xi_{24} + \xi_{14}\xi_{23} = 0. \tag{4.15}$$

Exercise. Verify by direct computation that the left-hand side of Eq. (4.15) is a first integral of the set of differential equations with matrix (4.14).

Because the plane W is Lagrangian, we have

$$\langle x, Jy \rangle = x_1 y_3 + x_2 y_4 - x_3 y_1 - x_4 y_2 = 0.$$

In terms of Plücker coordinates, this condition is rewritten in the form

$$\xi_{13} + \xi_{24} = 0.$$

The function $\xi_{13} + \xi_{24}$ is a first integral of the system with matrix (4.14) if $L(t) \in \mathfrak{Sp}(2, \mathbb{R})$.

Therefore, with each linear system of ordinary differential equations of order n, we can associate the set of systems of differential equations that describe the evolution of coordinates of the Plücker embeddings $G_k(\mathbb{R}^n) \to \mathbb{RP}^s$, where $s = C_n^k - 1$. The Plücker quadrics are integral manifolds of these systems.

These associated systems correspond to exterior powers of the natural representation of the group $GL(n)$ (see [18, 54]). We can also consider another representation. We see that the operation of separation of variables for ordinary differential equations corresponds to a decomposition of a representation into irreducible components. The inverse process is also possible: the existence of first integrals of signature zero for a linear system can justify that this system describes a translation of k-dimensional planes along a certain system of a lower order. In some cases, this allows us to reduce the dimension of a system while preserving its linearity.

Chapter 5
Matrix Double Ratio

We recall that in the study of the one-dimensional Riccati differential equation, the double ratio of four points lying on the projective line plays an important role (see the introduction and also [39, 101]). The projective nature of the manifold $G_n(\mathbb{R}^{2n})$ allows us to introduce a similar notion, which is just as useful in studying the higher-dimensional Riccati equation [60, 65, 92, 100, 110, 118, 120, 121].

Everywhere in this chapter, we assume that the charts considered on the Grassmann manifold $G_n(\mathbb{R}^{2n})$ are given by a pair of mutually orthogonal planes V_0 and V_∞.

§1. Matrix Double Ratio on the Grassmann Manifold

In the space $\mathbb{R}^{2n}$, we consider four n-dimensional planes P_1, P_2, P_3, and P_4. We chose two planes V_0 and V_∞ such that V_∞ is transversal to the planes $P_1, \ldots, P_4$. We consider the chart $\mathfrak{A}$ on the manifold $G_n(\mathbb{R}^{2n})$ that is defined by the planes V_0 and V_∞ and contains these four planes $P_1, \ldots, P_4$. The coordinates of these planes in the chart $\mathfrak{A}$ are denoted by the same letters.

Lemma 5.1. *The dimension of the intersection of two planes A and B coincides with the dimension of the kernel of the matrix $A - B$. In particular, the planes A and B are in general position iff $\det(A - B) \neq 0$.*

Proof. If the planes A and B have a nontrivial intersection, then there exists a horizontal vector $x \in V_0$, $x \neq 0$, such that $(x, Ax) = (x, Bx)$, i.e., $(A - B)x = 0$; therefore, the matrix $A - B$ is singular. The dimension of the kernel of the matrix $A - B$ is equal to the maximum number of linearly independent vectors lying in the intersection of A and B.

We assume that the planes P_2, P_3 and P_1, P_4 are pairwise in general position. By Lemma 5.1, this means that the matrices $P_2 - P_3$ and $P_1 - P_4$ are invertible. Similar to the formula for the double ratio of four points of the projective line, we compose the matrix[1]

$$\mathrm{DV} = \mathrm{DV}(P_1, P_2; P_3, P_4)$$

$$= (P_3 - P_1)(P_3 - P_2)^{-1}(P_4 - P_2)(P_4 - P_1)^{-1}. \tag{5.1}$$

We now reveal how this matrix changes under "projective transformations," i.e., under the action of $\mathrm{SL}(2n)$ on the manifold $G_n(\mathbb{R}^{2n})$.

---

[1] The abbreviation DV originates from the German term Doppelverhältniss, the *double relation*.

We proved previously (Proposition 4.4) that the group $SL(2n)$ is generated by the transformations

1. $W \mapsto W + A$;
2. $W \mapsto CWB$, $\det B = \det C$;
3. $W \mapsto -W^{-1}$.

Obviously, transformation 1 does not change DV.

We now reveal how transformations 2 act on the double ratio. Postmultiplication by B leads to the same matrix again:

$$\begin{aligned} DV(P_1B, P_2B; P_3B, P_4B) &= (P_3 - P_1)BB^{-1}(P_3 - P_2)^{-1} \\ &\quad \times (P_4 - P_2)BB^{-1}(P_4 - P_1)^{-1} \\ &= DV(P_1, P_2; P_3, P_4). \end{aligned}$$

Under premultiplication by C, we obtain

$$\begin{aligned} DV(CP_1, CP_2; CP_3, CP_4) &= C(P_3 - P_1)(P_3 - P_2)^{-1}C^{-1} \\ &\quad \times C(P_4 - P_2)(P_4 - P_1)^{-1}C^{-1} \\ &= C\,DV(P_1, P_2; P_3, P_4)C^{-1}, \end{aligned}$$

Therefore, under transformations 2, we have

$$DV \mapsto C\,DV\,C^{-1},$$

i.e., the equivalence class (with respect to conjugation) of matrices is preserved.

Applying transformation 3, we obtain

$$\begin{aligned} DV(-P_1^{-1}, -P_2^{-1}; -P_3^{-1}, -P_4^{-1}) &= (P_1^{-1} - P_3^{-1})(P_2^{-1} - P_3^{-1})^{-1} \\ &\quad \times (P_2^{-1} - P_4^{-1})(P_1^{-1} - P_4^{-1})^{-1} \\ &= P_1^{-1}(P_3 - P_1)P_3^{-1} \\ &\quad \times P_3(P_3 - P_2)^{-1}P_2P_2^{-1}(P_4 - P_2)P_4 \\ &\quad \times P_4^{-1}(P_4 - P_1)^{-1}P_1 \\ &= P_1^{-1}\,DV(P_1, P_2; P_3, P_4)P_1. \end{aligned}$$

We have thus proved that the class of matrices that are similar to the matrix $DV(P_1, P_2; P_3, P_4)$ is an invariant (which reflects the mutual disposition of the quadruple of the planes (P_1, P_2, P_3, P_4) with respect to the action of the group $SL(2n)$). In particular, by Sec. 1 of Chap. 4, this means that this class does not depend on the choice of the chart $\mathfrak{A}$, i.e., on the choice of the planes V_0 and V_∞. Namely, this class, denoted by $[DV(P_1, P_2; P_3, P_4)]$ is called the *matrix double ratio* of the quadruple of matrices P_1, P_2, P_3, and P_4.

We say that the matrix of the double ratio is *scalar* if $DV(P_1, P_2; P_3, P_4) = \sigma I_n$, $\sigma \in \mathbb{R}$. If $DV(P_1, P_2; P_3, P_4) = -I$, then this quadruple of planes is called *harmonic*. Because the class of matrices that are similar to the matrix

DV consists of one matrix σI in this case, the definitions of the scalar double ratio and the harmonic quadruple of planes do not depend on the choice of a representative from the class of similar matrices.

We consider two n-dimensional subspaces of the $2n$-dimensional Euclidean space $\mathbb{R}^{2n}$, i.e., two points A and B of the manifold $G_n(\mathbb{R}^{2n})$.

Definition 5.1. The plane A is called isoclinic to B if all vectors $p \in A$ make one and the same angle with the plane B.

Example 5.1. Isoclinic lines and planes:

1. Any two lines on the plane $\mathbb{R}^2$ are isoclinic.
2. Any two mutually orthogonal n-dimensional planes in the Euclidean space $\mathbb{R}^{2n}$ are isoclinic.

Exercise. Write equations of two isoclinic planes in the space $\mathbb{R}^4$ that make the angle $\pi/4$ with each other.

We find a condition under which two planes are isoclinic. For this, we choose a chart in the manifold $G_n(\mathbb{R}^{2n})$ that is defined by two orthogonal planes V_0 and V_∞ in $\mathbb{R}^{2n}$.

First, we consider the case where the angle made by these planes is equal to $\pi/2$.

Proposition 5.1. *The equation*

$$I_n + A^\top B = 0 \tag{5.1}$$

is a necessary and sufficient condition for the orthogonality of two planes A and B.

Proof. The planes A and B are orthogonal iff $(x, Ax) \perp (y, By)$ for any $x, y \in \mathbb{R}^n$:

$$x^\top y + x^\top A^\top By = x^\top (I_n + A^\top B)y = 0.$$

Consequently,

$$I_n + A^\top B = 0.$$

For arbitrary n-dimensional planes, we consider the matrix $I_n + A^\top B$. It is easy to verify that the following assertion holds.

Proposition 5.2. *The relation*

$$\mathrm{rk}(I_n + A^\top B) = n - k, \quad k < n, \tag{5.3}$$

holds iff A contains a k-dimensional subspace that is orthogonal to B.

We now consider the case of two isoclinic planes A and B when the angle made by these planes is equal to $\phi < \pi/2$. Then $\mathrm{rk}(I_n + A^\top B) = n$.

Proposition 5.3. *Two planes A and B are isoclinic iff*

$$(I_n + A^\top A)^{-1}(I_n + A^\top B)(I_n + B^\top B)^{-1}(I_n + B^\top A) = \sigma I_n, \qquad (5.4)$$

where $0 \leq \sigma \leq 1$. If the matrices A and B are invertible, then this expression is similar to the matrix

$$DV((-A^\top)^{-1}, (-B^\top)^{-1}; B, A).$$

Proof. We project the vector $p = (x, Ax)$ on the plane B. The projection is a vector $q = (y, By)$ such that the difference $(x - y, Ax - By)$ is orthogonal to the plane B, i.e., for any $z \in V_0$, we have

$$z^\top(x - y) + z^\top B^\top(Ax - By) = 0.$$

This means that

$$x - y + B^\top Ax - B^\top By = 0,$$

i.e.,

$$(I_n + B^\top A)x = (I_n + B^\top B)y.$$

Lemma 5.2. *For any matrix A, the matrix $I_n + A^\top A$ is symmetrical and positive definite.*

Proof. The matrix $I_n + A^\top A$ is the matrix of a positive-definite form that represents the inner square (x, Ax) of a vector belonging to the plane A.

Because $I_n + B^\top B > 0$, we have

$$y = (I_n + B^\top B)^{-1}(I_n + B^\top A)x. \qquad (5.5)$$

Let ϕ be the angle made by the vectors p and q. Then

$$\cos^2 \phi = \frac{(x^\top y + x^\top A^\top By)^2}{(x^\top x + x^\top A^\top Ax)(y^\top y + y^\top B^\top By)}.$$

Substituting (5.5) in this expression, we obtain

$$\cos^2 \phi = \frac{x^\top(I_n + A^\top B)(I_n + B^\top B)^{-1}(I_n + B^\top A)x}{x^\top(I_n + A^\top A)x}.$$

Let Γ be the arithmetic square root of the positive-definite matrix $I_n + A^\top A$, i.e., a positive-definite symmetrical matrix Γ such that $\Gamma^2 = I_n + A^\top A$. Then, letting ξ denote Γx, we obtain

$$\cos^2 \phi = \frac{\xi^\top \Gamma^{-1}(I_n + A^\top B)(I_n + B^\top B)^{-1}(I_n + B^\top A)\Gamma^{-1}\xi}{\xi^\top \xi}.$$

The angles made by the planes A and B are thus defined by the quadratic form with the matrix

$$\Gamma^{-1}(I_n + A^\top B)(I_n + B^\top B)^{-1}(I_n + B^\top A)\Gamma^{-1}.$$

The obtained matrix is similar to the matrix

$$\Gamma^{-1}\Gamma^{-1}(I_n + A^\top B)(I_n + B^\top B)^{-1}(I_n + B^\top A)\Gamma^{-1}\Gamma,$$

which is equal to

$$(I_n + A^\top A)^{-1}(I_n + A^\top B)(I_n + B^\top B)^{-1}(I_n + B^\top A). \qquad (5.6)$$

We transform the latter relation by premultiplying by $(I_n + A^\top A)$ and post-multiplying by $(I_n + A^\top A)^{-1}$. Let the matrices A and B be invertible. Then, carrying out the factors from square brackets, we obtain the matrix

$$[A^\top(B - (-A^\top)^{-1})]\,[(B - (-B^\top)^{-1})^{-1}(B^\top)^{-1}]$$
$$\times [B^\top(A - (-B^\top)^{-1})]\,[(A - (-A^\top)^{-1})^{-1}(A^\top)^{-1}],$$

which is similar to the matrix

$$\mathrm{DV}((-A^\top)^{-1}, (-B^\top)^{-1}; B, A).$$

The planes A and B are isoclinic iff $\cos^2\phi = \mathrm{const}$. Consequently, we have $\mathrm{DV}((-A^\top)^{-1}, (-B^\top)^{-1}; B, A) = \sigma I_n$, where $\sigma = \cos^2\phi$.

The quadratic form with matrix (5.6) is called the *quadratic form for determination of angles* made by planes, and, for brevity, the matrix itself is denoted by DV_{AB}.

Remark. In an arbitrary (not isoclinic) case, we consider the set $\{\xi \in V_0 \mid \|\xi\| = 1\}$. On this set, $\cos^2\phi$ is defined to be the value of the quadratic form with the matrix DV_{AB}. The method for constructing matrix (5.6) implies that the matrix DV_{AB} is symmetrical and its spectrum lies in the closed interval $[0, 1]$. This spectrum assigns critical values of the function $\cos^2\phi$ defined on the projection of the plane A, i.e., those values for which a surgery of level surfaces of the function $\cos^2\phi$ occurs [70]. In particular, the minimum and maximum eigenvalues of this form yield the minimum and maximum value of the angle made by these planes.

Isoclinic planes play an important role in studying the geometry of the Grassmann manifold. For example, paper [113] studies totally geodesic submanifolds N of the manifold $G_s(\mathbb{R}^{2n})$ that are characterized by the following property: any two distinct elements of N considered as subspaces of $\mathbb{R}^{2n}$ are in general position. It is proved that N is isometric either to the sphere or to the projective space (which is real, complex, or quaternion); moreover, any two elements of N correspond to isoclinic planes. Conversely, any set of pairwise isoclinic linear subspaces can be extended to a totally geodesic submanifold N. Also, in [113], it is proved that a subspace of the Lie algebra that corresponds to a totally geodesic submanifold has a basis consisting of matrices $S_1, \ldots, S_r$ satisfying the Hurwitz system of equations

$$S_i^\top = -S_i,$$
$$S_i S_j + S_j S_i = -2\delta_{ij} I_n. \qquad (5.7)$$

§2. Clifford Algebras

Definition 5.2. An associative algebra with the generators $e_0, e_1, \ldots, e_n$ and the relations[2]

$$e_0 e_i = e_i e_0 = e_i, \quad i = 1, \ldots, n, \tag{5.8}$$

$$e_i e_j + e_j e_i = -2\delta_{ij} e_0, \tag{5.9}$$

is called a Clifford algebra.

The monomials $e_{i_1} e_{i_2} \ldots e_{i_k}$ serve as generators of the linear space of the Clifford algebra.

Exercise. Prove that the dimension of the Clifford algebra is equal to 2^n.

The algebra generated by the relations

$$e_0 e_i = e_i e_0 = e_i, \quad i = 1, \ldots, n, \tag{5.10}$$

$$e_i e_j + e_j e_i = 2\delta_{ij} e_0, \tag{5.11}$$

is isomorphic (over the field $\mathbb{C}$) to a Clifford algebra. This isomorphism is stated via multiplication of the generators e_k, $k > 0$, by the imaginary unit i.

Clifford discovered these algebras when generalizing the construction of quaternions and was also influenced by the construction of the Grassmann algebras [20]. It is easy to see that for $n = 3$, the Clifford algebra coincides with the quaternion algebra.

From the above definition, it is obvious that Hurwitz matrix equations (5.7) define a representation of the Clifford algebra under which skew-symmetrical matrices correspond to the generators e_k, $k > 0$. Adolf Hurwitz found equations (5.10) and (5.11) when studying the decomposition problem of quadratic forms, which was later called the *Hurwitz–Radon problem*. It is formulated as follows.

Let $(x_1, \ldots, x_p)$ and $(y_1, \ldots, y_q)$ be independent variables over the field $\mathbb{R}$ or $\mathbb{C}$. It is required to determine for which values of (p, q) the biquadratic form $(x_1^2 + \ldots + x_p^2)(y_1^2 + \ldots + y_q^2)$ admits a representation in the form of a sum of squares of q bilinear forms, i.e., when the following identity holds:

$$(x_1^2 + \ldots + x_p^2)(y_1^2 + \ldots + y_q^2) \equiv (z_1^2 + \ldots + z_q^2), \tag{5.12}$$

where

$$z_i = \sum_{h=1}^{p} \sum_{j=1}^{q} a_{ihj} x_h y_j. \tag{5.13}$$

In the case of the field $\mathbb{C}$, this problem was solved by Hurwitz and was published after his death [52]. Hurwitz proved that the problem has a solution iff

---

[2] The symbol δ_{ij} is the Kronecker symbol: $\delta_{ii} = 1$, $\delta_{ij} = 0$ for $i \neq j$.

1. $p = 2r + 1$, where $r \equiv 0$ or $r \equiv 3$ (mod 4) and q is divided by 2^r;
2. $p = 2r + 1$, where $r \equiv 1$ or $r \equiv 2$ (mod 4) and q is divided by 2^{r+1};
3. $p = 2r + 2$, where $r \equiv 3$ (mod 4) and q is divided by 2^r;
4. $p = 2r + 2$, where $r \equiv 0$, $r \equiv 1$, or $r \equiv 2$ (mod 4) and q is divided by 2^{r+1}.

J. Radon obtained the same result for the case of the field $\mathbb{R}$ [85]. Later, the matrix Hurwitz equations were used by E. Stiefel in solving the problem of finding the maximum number of vector fields on spheres that are linearly independent at each of their points [102].

We show how the Hurwitz equations arise from the problem (5.12), (5.13). We write relation (5.12) in the matrix form letting A_h denote the $q \times q$ matrix with the entries a_{ihj}, where the subscript h is fixed:

$$(x_1^2 + \ldots + x_p^2) I_q = \left(\sum_{h=1}^{p} A_h x_h \right) \left(\sum_{h=1}^{p} A_h^\top x_h \right).$$

Consequently,

$$A_h A_h^\top = I_q, \qquad A_h A_k^\top + A_k A_h^\top = 0 \quad \text{for} \ \ h \neq k. \tag{5.14}$$

In particular, (5.14) implies that the matrices A_h are orthogonal and are therefore nonsingular. Now, instead of A_h, we introduce "affine coordinates" B_h by "projecting" the tuple of matrices A_h:

$$A_h = B_h A_p, \quad h = 1, \ldots, p - 1,$$

or, equivalently,

$$B_h = A_h A_p^\top. \tag{5.15}$$

The second group of Eqs. (5.14) implies in particular that the matrices B_h are skew-symmetrical. Now, substituting relations (5.15) in Eqs. (5.14) and using the relation $B_h^\top = -B_h$, we find that the Hurwitz equations are a consequence of relations (5.14).

Another important application of Clifford algebras is the construction of spinor representations of orthogonal groups, which were discovered by E. Cartan [15] in the framework of the general finite-dimensional representation theory of semisimple Lie algebras. The name *spinor representation* (originated from the word *spin*) is due to the classic works on electron theory by Dirac [26].

Although Dirac was not a mathematician and never knew the works by W. Clifford, E. Cartan, A. Hurwitz, and others, his ideas are strikingly similar to the arguments of Hurwitz presented above. Like the well-known case of discovering non-Euclidean geometry, this is one more brilliant example of a surprising phenomenon in the history of science when one and the same principally new idea is simultaneously and independently found by scientists that are completely independent of each other.

In some sense, the Dirac approach is more courageous: from the very beginning, Hurwitz had matrices in mind as a realization of its construction; Dirac did not know what mathematical objects his argument would lead to.

We outline the Dirac arguments here. The Seeman phenomenon, i.e., the splitting of spectral lines in the radiation of atoms located in a magnetic field, shows that an electron has a certain inner degree of freedom. Dirac thought about this degree of freedom as a momentum of rotation of the electron around its axis (the term *spin* originated from this).

In modern physics, no one now considers mechanical analogues as serious ones. Nevertheless, they have played an important role in the development of physics, and such linguistic relics as spin, orbit, etc., are widely used in the conventional physical terminology and even arise from it in the mathematical literature.

Dirac wanted to find an adequate mathematical technique for describing a spin. For this, he needed to extract the square root of the Laplace operator, i.e., to find a first-order differential operator whose square yields the Laplace operator. At first glance, this problem is seemed unsolvable because, as we know, a polynomial corresponding to the Laplace operator is irreducible. But when the goal is so desirable, the impossibility is overcome.

Dirac wrote the formal relation

$$\frac{\partial^2}{\partial x^2} + \frac{\partial^2}{\partial y^2} + \frac{\partial^2}{\partial z^2} = \left(a_1 \frac{\partial}{\partial x} + a_2 \frac{\partial}{\partial y} + a_3 \frac{\partial}{\partial z} \right)^2,$$

which immediately implies the following conditions on the coefficients a_1, a_2, and a_3:

$$a_i^2 = 1, \qquad a_i a_j + a_j a_i = 0, \quad i, j \in \{1, 2, 3\}.$$

The addition of the unit to these variables yields Clifford algebra (5.10), (5.11).

Using the obtained operator, Dirac wrote the following equation for an electron whose variables are spinors:

$$i \frac{\partial u}{\partial t} = a_1 \frac{\partial u}{\partial x} + a_2 \frac{\partial u}{\partial y} + a_3 \frac{\partial u}{\partial z} + mu$$

Later, Pauli found the representation of a_i in the form of matrices:

$$1 = \begin{pmatrix} 1 & 0 \\ 0 & 1 \end{pmatrix}, \quad a_1 = \begin{pmatrix} 1 & 0 \\ 0 & -1 \end{pmatrix}, \quad a_2 = \begin{pmatrix} 0 & 1 \\ 1 & 0 \end{pmatrix}, \quad a_3 = \begin{pmatrix} 0 & i \\ -i & 0 \end{pmatrix}.$$

After this historical digression, we return to the Hurwitz equation and examine how it is applied to the study of Grassmann manifolds.

§3. Totally Geodesic Submanifolds of Grassmann Manifolds

Let $\mathbb{R}^s \subset \mathbb{R}^n$, $n > s > k$. Then k-dimensional subspaces lying in $\mathbb{R}^s$ form the submanifold $G_k(\mathbb{R}^s)$ of the manifold $G_k(\mathbb{R}^n)$. This manifold is totally geodesic. We do not present a formal proof here because this fact itself is sufficiently evident: if the initial and final points are subspaces of $\mathbb{R}^m$, then a geodesic of the manifold $G_k(\mathbb{R}^n)$ does not leave the subspace $\mathbb{R}^m$, because this would increase its length. If, moreover, $2k > m$, then geodesic lines are turns of k-dimensional planes around a fixed subspace.

In this section, we discuss the most interesting case where the planes corresponding to points of a geodesic are in general position, i.e., they intersect only at the origin of the space $\mathbb{R}^n$. For simplicity, we consider only the manifold $G_n(\mathbb{R}^{2n})$.

The manifold $G_n(\mathbb{R}^{2n})$ has the structure of the Riemannian symmetrical space $G_n(\mathbb{R}^{2n}) = O(2n)/(O(n) \times O(n))$, for which $O(2n)$ is the transitive symmetry group.

Remark. As a symmetrical space, $G_n(\mathbb{R}^{2n})$ has the rank n (for the definition of the rank of a symmetrical space, see [115]). It is therefore not a two-point transitive space [115]. This means that, in general, there is no isometry ϕ that transforms a pair of points $A_1, B_1 \in G_n(\mathbb{R}^{2n})$ into any other pair $A_2, B_2 \in G_n(\mathbb{R}^{2n})$ that is equidistant to it.

We recall that two pairs of points are said to be *equidistant* if the distances between the points of each of the pairs are equal, that is, $\mathrm{dist}(A_1, B_1) = \mathrm{dist}(A_2, B_2)$.

We consider two pairs of points P, Q and S, T of the manifold $G_n(\mathbb{R}^{2n})$. We seek a criterion for the existence of an isometry of $G_n(\mathbb{R}^{2n})$ that transforms one of these pairs into the other. For this, we need the following lemmas.

Lemma 5.3. *Let the coordinates of the plane Q in a certain chart of the Grassmann manifold $G_n(\mathbb{R}^{2n})$ that correspond to the orthogonal planes V_0 and V_∞ be given by a nonsingular matrix Q. Then the orthogonal complement to the plane Q in this chart is given by the matrix $-(Q^\top)^{-1}$.*

The proof is immediately implied by the orthogonality condition of two planes (5.2).

Lemma 5.4. *The transformation that sets the orthogonal complement in correspondence to each plane Q and the action of $O(2n)$ on $G_n(\mathbb{R}^{2n})$ commute.*

Proof. Let $\mathfrak{A}$ be a chart on the manifold $G_n(\mathbb{R}^{2n})$. We consider points whose coordinates in the chart $\mathfrak{A}$ are given by nonsingular matrices (i.e., those planes Q for which Q is transversal to V_0).

We consider an element

$$g = \begin{pmatrix} A & B \\ C & D \end{pmatrix} \in O(2n).$$

We have

$$g : Q \mapsto \begin{pmatrix} A & B \\ C & D \end{pmatrix} Q = (C + DQ)(A + BQ)^{-1} = R.$$

Then $C + DQ = R(A + BQ)$. Transposing this, we obtain

$$C^\top + Q^\top D^\top = A^\top R^\top + Q^\top B^\top R^\top.$$

Consequently,

$$\begin{aligned}
(-Q^\top)^{-1} &= (B^\top R^\top - D^\top)(A^\top R^\top - C^\top)^{-1} \\
&= [B^\top + D^\top(-R^\top)^{-1}][A^\top + C^\top(-R^\top)^{-1}]^{-1} \\
&= \left(\begin{array}{cc} A^\top & C^\top \\ B^\top & D^\top \end{array} \right)(-R^\top)^{-1}.
\end{aligned}$$

Therefore,

$$(-R^\top)^{-1} = \left(\begin{array}{cc} A & B \\ C & D \end{array} \right)(-Q^\top)^{-1}.$$

If $\det Q = 0$ in the coordinate system considered, then we pass to another chart $\mathfrak{B}$ that is obtained from $\mathfrak{A}$ by applying a transformation $g \in O(2n)$.

Definition 5.3. Two pairs of points P, Q and S, T are called equivalent with respect to the action of $O(2n)$ if there exists an isometry $g \in O(2n)$ of the manifold $G_n(\mathbb{R}^{2n})$ such that $S = g(P)$ and $T = g(Q)$.

Theorem 5.1. *Two pairs of points P, Q and S, T are equivalent iff their matrix double ratios coincide:* $[\mathrm{DV}_{PQ}] = [\mathrm{DV}_{ST}]$.

Proof. The necessity is implied by Lemma 5.3 and the invariance of the matrix double ratio of a quadruple of points.

We prove the sufficiency of this condition. We carry out the proof in two stages.

1. First, let the planes P and Q be located such that none of the vectors of one of these plane is orthogonal to the other. Then the matrix DV_{PQ} (and therefore DV_{ST}) is nonsingular. Indeed, in this case, the quadratic form for angle determination that defines $\cos^2 \phi$ has no nonzero eigenvalues corresponding to directions of one of the planes that are orthogonal to the other plane.

We choose a chart $\mathfrak{A}$ in which the zero matrix serves as coordinates of the point P. The point Q necessarily lies in this chart because by Proposition 5.2, points of the manifold that do not lie in the chart $\mathfrak{A}$ consist of planes having a nontrivial intersection with the orthogonal complement of the plane P. The coordinates of the plane Q in the chart $\mathfrak{A}$ are denoted by V. Because the group $O(2n)$ acts transitively on the manifold $G_n(\mathbb{R}^{2n})$, we transform the point S into zero. In this case, the point T transforms into a point W that lies in the chart $\mathfrak{A}$ for the same reasons. By the invariance of matrix double ratios, the class $[\mathrm{DV}_{0V}]$ represented by the matrix $(\mathrm{I} + V^\top V)^{-1}$ and the class $[\mathrm{DV}_{0W}]$ represented by the matrix $(\mathrm{I} + W^\top W)^{-1}$ coincide. Therefore,

$$W^\top W = q V^\top V q^{-1}, \tag{5.16}$$

where $q \in \mathrm{SL}(n)$. Because the matrices $V^\top V$ and $W^\top W$ are symmetrical and have the same spectrum, we can assume that $q \in O(n)$.

We need to find an element belonging to the stabilizer of the zero plane (i.e., a matrix

$$\begin{pmatrix} \alpha_1 & 0 \\ 0 & \alpha_2 \end{pmatrix},$$

$\alpha_i \in O(n)$, $i = 1, 2$, that transforms V into W, i.e.,

$$\alpha_2 V \alpha_1^{-1} = W. \tag{5.17}$$

One of the solutions to Eq. (5.17) has the form $\alpha_1 = q$, $\alpha_2 = WqV^{-1}$. We verify that for $q \in O(n)$, the matrix α_2 is orthogonal. As a direct consequence of (5.16), the relation $((V^\top)^{-1}\alpha_1^\top W^\top)(W\alpha_1 V^{-1}) = I_n$ holds for $\alpha_1 = q$.

2. Now let the planes P and Q be located such that at least one of the vectors of the plane P is orthogonal to the plane Q. Then the matrix DV_{PQ} (and therefore DV_{ST}) is singular. It is easy to verify that the dimension of the linear subspace composed of vectors of the plane P that are orthogonal to the plane Q is equal to the dimension of the kernel of the matrix DV_{PQ}.

The plane Q has a nontrivial intersection with $P^\perp$. Therefore, Q cannot be in the chart in which $P = 0$. By a small deformation of the plane Q, we can make this plane be in the chart $\mathfrak{A}$. Namely, we choose a sequence of points $Q_k \to Q$ such that the matrices DV_{PQ_k} are nonsingular, i.e., all planes Q_k are transversal to the orthogonal complement of P. The fact that $Q_k \to Q$ implies that the spectrum of any of the matrices from the class $[DV_{PQ_k}]$ tends to the spectrum of the matrix DV_{PQ}. We consider the set $\mathfrak{S}_k$ consisting of matrices T_k such that $[DV_{ST_k}]$ coincides with $[DV_{PQ_k}]$. Obviously, the point T is a limit point for the tuple of the sets $\mathfrak{S}_k$. Consequently, there exists a sequence of points $T_k \in \mathfrak{S}_k$ such that $\lim_{k\to\infty} T_k = T$. By the coincidence of $[DV_{PQ_k}]$ and $[DV_{ST_k}]$ and by the already considered case, there exist transformations $g_k \in O(2n)$ such that

$$g_k(P) = S \quad \text{and} \quad g_k(Q_k) = T_k. \tag{5.18}$$

The manifold $O(2n)$ is compact; therefore, we can choose a convergent subsequence $g_k \to g \in O(2n)$ from the sequence of transformations g_k. Passing to the limit in (5.18), we confirm that the pair of planes P, Q is equivalent to the pair S, T.

We need the following simple assertion about isoclinic planes. For a suitable choice of a coordinate system, we can assume that any given k-dimensional plane has a zero matrix coordinate. Then the isoclinicity condition of two planes is reduced to the isoclinicity condition of a certain plane to the zero plane.

Proposition 5.4. *The isoclinicity condition of a plane P to the zero plane consists in the relation*

$$P^\top P = \sigma_{PP} I_n, \quad \sigma_{PP} \in \mathbb{R}. \tag{5.19}$$

To prove this assertion, it suffices to apply (5.4).

Corollary 5.1. *If a plane P is isoclinic to the zero plane and $P \neq 0$, then the matrix P is nonsingular.*

Obviously, the isoclinicity relation is symmetrical and reflexive. However, it is not transitive. It is easy to give an example of two matrices P and Q that satisfy Eq. (5.19) (which means the isoclinicity to the zero plane) and do not satisfy condition (5.4) (i.e., the planes P and Q are not isoclinic). In what follows, we are interested in those sets of planes in which any two planes are isoclinic.

Lemma 5.5. *The set Φ consisting of n-dimensional planes in the space $\mathbb{R}^{2n}$ and containing zero planes consists of mutually isoclinic planes iff relation (5.19) holds for any plane $P \in \Phi$ and we have the relations*

$$Q^\top P + P^\top Q = 2\sigma_{PQ} \, \mathrm{I}_n, \quad \sigma_{PQ} \in \mathbb{R}, \tag{5.20}$$

for any pairs of planes $P, Q \in \Phi$.

Proof. Relation (5.19) is proved above. Substituting the relations $P^\top P = \sigma_{PP} \, \mathrm{I}_n$ and $Q^\top Q = \sigma_{QQ} \, \mathrm{I}_n$ in Eq. (5.4), we easily obtain (5.20).

Definition 5.4. A set of isoclinic n-dimensional planes is called maximal if it is not a proper subset of a larger set of mutually isoclinic planes.

Lemma 5.6. *Coordinates of any maximal set of mutually isoclinic n-dimensional planes in $\mathbb{R}^{2n}$ that contains the zero plane forms a linear subspace in the space of $n \times n$ matrices.*

Proof. In the set Φ, we choose a maximal tuple of linearly independent matrices $A_0, A_1, \ldots, A_q$. We consider the linear span $\mathrm{Span}(A_0, A_1, \ldots, A_q)$ of this tuple of matrices, which coincides with the linear span of the set Φ. Let $B, C \in \mathrm{Span}\,\Phi$, i.e., $B = \sum_{i=0}^{q} b_i A_i$ and $C = \sum_{j=0}^{q} c_j A_j$. For these matrices, we compose expressions (5.19) and (5.20):

$$B^\top B = \sum_{i \leq j} b_i b_j (A_j^\top A_i + A_i^\top A_j),$$

$$C^\top B + B^\top C = \sum_{i \leq j} c_j b_i (A_j^\top A_i + A_i^\top A_j) \tag{5.21}$$

$$+ \sum_{i \leq j} b_j c_i (A_j^\top A_i + A_i^\top A_j).$$

The matrices A_i satisfy conditions (5.19) and (5.20). The identity matrices stand in the right-hand side of (5.21). Therefore, the linear span of any set of mutually isoclinic n-dimensional planes in $\mathbb{R}^{2n}$ that contains the zero plane itself consists of mutually isoclinic n-dimensional planes.

Definition 5.5. Two sets of planes Φ and Ψ in the space $\mathbb{R}^m$ are called congruent if there exists an orthogonal transformation $g \in O(m)$ that transforms the set Φ into the set Ψ.

Theorem 5.2 (Y. C. Wong). *Let Φ be the maximal set of mutually isoclinic n-dimensional planes in $\mathbb{R}^{2n}$. Then Φ is congruent to the set $\Psi = \mathrm{Span}(I_n, B_1, \ldots, B_q)$, where the real $(n \times n)$-dimensional matrices $B_1, \ldots, B_q$ form a maximal real solution of the Hurwitz equations.*

Proof. On the linear space Φ, we introduce the Euclidean structure setting $\langle A, B \rangle = \sigma_{AB}$ (see (5.20)). The inner product thus defined is obviously bilinear and symmetrical; moreover, if $A \neq 0$, then the inner square $\langle A, A \rangle = \sigma_A^2 = AA^\top$ is strictly positive by Proposition 5.4. Orthogonalizing, we pass from the basis A_i to the basis B_i, which satisfies the relations

$$B_i^\top B_i = I_n, \qquad B_j^\top B_i + B_i^\top B_j = 0 \tag{5.22}$$

for $i \neq j$, $0 \leq i \leq q$, $0 \leq j \leq q$.

Because $DV_{A_0 0} = I_n$, by Theorem 5.1, we have the existence of an orthogonal transformation g that transforms a pair of planes $A_0, 0$ into the pair $I_n, 0$. By Lemmas 5.5 and 5.6, the subspace $\Psi = g(\Phi)$ contains a basis $I_n, B_1, \ldots, B_q$ satisfying Eqs. (5.19) and (5.20), which, as we just showed, are reduced to (5.22).

Substituting $i = 0$ in (5.22), we obtain

$$B_j + B_j^\top = 0, \tag{5.23}$$

i.e., the matrices B_i are skew symmetrical. Substituting (5.23) in (5.22), we obtain the other Hurwitz equations. The obtained solution to the Hurwitz equations is maximal because the set of planes is otherwise not maximal.

Theorem 5.3 (J. Wolf). *The maximal set Φ of mutually isoclinic n-dimensional spaces in $\mathbb{R}^{2n}$ forms a totally geodesic submanifold of the Grassmann manifold $G_n(\mathbb{R}^{2n})$.*

Each connected totally geodesic submanifold with any two elements transversal to each other (considered as subspaces of $\mathbb{R}^n$) consists of mutually isoclinic planes.

Proof. In correspondence to each matrix B_i of the basis constructed in Theorem 5.2, we set the element

$$S_i = \begin{pmatrix} 0 & B_i \\ B_i & 0 \end{pmatrix}$$

of the Lie algebra of the orthogonal group $O(2n)$ (we recall that the matrix B_i is skew symmetrical). We set the subspace $\phi \subset \mathfrak{O}(2n)$ with the basis S_i in correspondence to a tuple Φ of mutually isoclinic planes.

We recall that $G_n(\mathbb{R}^{2n}) = O(2n)/(O(n) \times O(n))$. A direct calculation (which we leave to the reader as an exercise) easily shows that the subspace

ϕ is orthogonal (in the sense of the Killing form) to the Lie algebra of the stabilizer of the point e, i.e., to the subalgebra $\mathfrak{O}(n) \times \mathfrak{O}(n)$ and can therefore be considered a subspace of the space tangent to $G_n(\mathbb{R}^{2n})$.

An arbitrary element of the set ϕ has the form $S = \sum_{i=1}^{q} a_i S_i$. When computing the exponential mapping, we can suppose that $\sum_{i=1}^{q} a_i^2 = 1$. Then the Hurwitz equations imply $S^2 = -I_n$. Taking the series expansion and using the obtained relation, we easily show that $\exp(St) = (\cos t) I_n + (\sin t) S$, i.e., $\exp \phi = \Phi$.

The subspace $\phi \subset \mathfrak{O}(2n)$ is a triple Lie system. Indeed, we have

$$[[S_i, S_j], S_k] = \begin{pmatrix} 0 & [[B_i, B_j], B_k] \\ [[B_i, B_j], B_k] & 0 \end{pmatrix}.$$

Using (5.7), we easily verify that in the case where all three subscripts i, j, and k are distinct, the relation $[[B_i, B_j], B_k] = 0$ holds; in the case where some of these subscripts coincide, the commutator under consideration either vanishes or is equal to one of the matrices B_i.

Therefore, the set Φ is a totally geodesic manifold.

For the proof of the converse statement, see [113], p. 431.

§4. Curves with a Scalar Double Ratio

For three given points $P_1, P_2, P_3 \in G_n(\mathbb{R}^{2n})$, we consider the set of points $P \in G_n(\mathbb{R}^{2n})$ that are in the scalar double ratio with the points P_1, P_2, and P_3:

$$(P - P_1)(P - P_2)^{-1}(P_3 - P_2)(P_3 - P_1)^{-1} = \sigma I_n. \qquad (5.24)$$

This set is characterized by one scalar parameter σ and is called the *curve with a scalar double ratio*. We write the equation of this curve:

$$(P - P_1) = \sigma(P_3 - P_1)(P_3 - P_2)^{-1}(P - P_2).$$

Solving this equation with respect to P, we obtain

$$\begin{aligned} P &= [I_n - \sigma(P_3 - P_1)(P_3 - P_2)^{-1}]^{-1} \\ &\quad \times \{(P_1 - P_2) + [I_n - \sigma(P_3 - P_1)(P_3 - P_2)^{-1}]P_2\} \\ &= [I_n - \sigma(P_3 - P_1)(P_3 - P_2)^{-1}]^{-1}(P_1 - P_2) + P_2. \end{aligned}$$

Finally,

$$\begin{aligned} P &= J_{P_1 P_2 P_3}(\sigma) \\ &= [(P_1 - P_2)^{-1} - \sigma(P_1 - P_2)^{-1}(P_3 - P_1)(P_3 - P_2)^{-1}]^{-1} + P_2. \qquad (5.25) \end{aligned}$$

The following lemma explains the term "scalar double ratio."

Lemma 5.7. *Any four points of the curve $J_{P_1 P_2 P_3}(\sigma)$ are in the scalar double ratio.*

Proof. We set $(P_1 - P_2)^{-1} = A$, $P_2 = C$, and $(P_2 - P_1)^{-1}(P_3 - P_1)(P_3 - P_2)^{-1} = B$. Then Eq. (5.25) is written in the form

$$P = (A + \sigma B)^{-1} + C. \tag{5.26}$$

We consider four arbitrary points $Q_i = (A + \sigma_i B)^{-1} + C$ of curve (5.26) that correspond to the values σ_1, σ_2, σ_3, and σ_4 of the parameter ($\sigma_1 \neq \sigma_4$ and $\sigma_2 \neq \sigma_3$). Then

$$DV(Q_1, Q_2; Q_3, Q_4) =$$

$$= [(A + \sigma_3 B)^{-1} - (A + \sigma_1 B)^{-1}][(A + \sigma_3 B)^{-1} - (A + \sigma_2 B)^{-1}]^{-1}$$
$$\times [(A + \sigma_4 B)^{-1} - (A + \sigma_2 B)^{-1}][(A + \sigma_4 B)^{-1} - (A + \sigma_1 B)^{-1}]^{-1}$$

$$= (A + \sigma_1 B)^{-1}[(\sigma_1 - \sigma_3)B](A + \sigma_3 B)^{-1}(A + \sigma_3 B)[(\sigma_2 - \sigma_3)B]^{-1}$$
$$\times (A + \sigma_2 B)(A + \sigma_2 B)^{-1}[(\sigma_2 - \sigma_4)B](A + \sigma_4 B)^{-1}(A + \sigma_4 B)$$
$$\times [(\sigma_1 - \sigma_4)B]^{-1}(A + \sigma_1 B)$$

$$= DV(\sigma_1, \sigma_2; \sigma_3, \sigma_4) I_n.$$

As a result, $DV(Q_1, Q_2; Q_3, Q_4)$ is reduced to the following double ratio of four numbers:

$$DV(Q_1, Q_2; Q_3, Q_4) = (\sigma_3 - \sigma_1)(\sigma_3 - \sigma_2)^{-1}(\sigma_4 - \sigma_2)(\sigma_4 - \sigma_1)^{-1}$$

$$= DV(\sigma_1, \sigma_2; \sigma_3, \sigma_4).$$

Proposition 5.5. *If the curve $J_{P_1 P_2 P_3}(\sigma)$ is constructed according to three pairwise isoclinic points $P_1, P_2, P_3 \in G_n(\mathbb{R}^{2n})$, then any two points on it are isoclinic.*

Proof. By Lemma 5.7, we can consider any point of our curve as P_2. It therefore suffices to prove that any point of the curve $J_{P_1 P_2 P_3}(\sigma)$ is isoclinic to P_2. We choose a coordinate system such that $P_2 = 0$. Then (5.25) becomes

$$P = [(1 - \sigma)P_1^{-1} + \sigma P_3^{-1}]^{-1}.$$

Using (5.4), we note the isoclinicity of P_1 and 0 is equivalent to the condition $P_1^\top P_1 = \lambda I_n$, $\lambda \in \mathbb{R}^1$, and the isoclinicity of P_3 and 0 is equivalent to the condition $P_3^\top P_3 = \mu I_n$, $\mu \in \mathbb{R}^1$. Therefore, the isoclinicity of P_1 and P_3 is equivalent to the condition $P_1^\top P_3 + P_3^\top P_1 = \nu I_n$, $\nu \in \mathbb{R}^1$. We need to prove that $P^\top P = k I_n$, $k \in \mathbb{R}^1$. Passing to the inverse matrix in this relation, we have

$$[(1-\sigma)(P_1^\top)^{-1} + \sigma(P_3^\top)^{-1}][(1-\sigma)(P_1)^{-1} + \sigma(P_3)^{-1}] =$$

$$= (P_1^\top)^{-1}[(1-\sigma)P_3^\top + \sigma P_1^\top](P_3^\top)^{-1}(P_3)^{-1}[(1-\sigma)P_3 + \sigma P_1]P_1^{-1}$$

$$= (P_1^\top)^{-1}[(1-\sigma)^2 P_3^\top P_3 + \sigma(1-\sigma)(P_3^\top P_1 + P_1^\top P_3) + \sigma^2 P_1^\top P_1]P_1^{-1}$$

$$= k\, I_n;$$

this is what was required to be proved.

The property of curves with a scalar double ratio described in the next two statements is similar to the following property of circles in the Euclidean space $\mathbb{R}^n$: if three points of a circle belong to a certain plane in $\mathbb{R}^n$, then the whole circle belongs to this plane. The role of planes in the Grassmann manifold is played here by totally geodesic submanifolds of the Grassmann manifold $G_n(\mathbb{R}^{2n})$.

Proposition 5.6. *A geodesic line of $G_n(\mathbb{R}^{2n})$ consisting of pairwise iso-clinic points is a curve with a scalar double ratio.*

Proof. Any geodesic consisting of pairwise isoclinic points can be obtained by the following construction [121]. At the point $W(0) = 0$ of the manifold $G_n(\mathbb{R}^{2n})$, we consider a tangent vector that is defined by an element $A \in \mathfrak{SO}(2n)$ such that

$$A = \begin{pmatrix} 0 & -a^\top \\ a & 0 \end{pmatrix},$$

where a is an $n \times n$ matrix and $A^2 = -I_{2n}$. Then, by Theorem 3.15, a geodesic passing through the point $W(0)$ is defined by $\exp(tA)W(0)$. It is easy to see that $\exp(tA) = (\cos t) \cdot I_{2n} + (\sin t) \cdot A$; therefore,

$$(\exp(tA))\begin{pmatrix} x \\ 0 \end{pmatrix} = \begin{pmatrix} (\cos t)\,I_n & -(\sin t)a^\top \\ (\sin t)a & (\cos t)\,I_n \end{pmatrix}\begin{pmatrix} x \\ 0 \end{pmatrix} = \begin{pmatrix} (\cos t)x \\ (\sin t)ax \end{pmatrix}.$$

Therefore, we obtain

$$W(t) = (\tan t)a.$$

Because $W(t)$ is proportional to one and the same nonsingular matrix a for any t, any four points of this curve are in the scalar double ratio.

Proposition 5.7. *Let the points $P_1, P_2, P_3 \in G_n(\mathbb{R}^{2n})$ belong to a complete connected totally geodesic submanifold N consisting of those points of $G_n(\mathbb{R}^{2n})$ that (as n-dimensional planes in $\mathbb{R}^{2n}$) are pairwise in the general position. Then the curve $J_{P_1 P_2 P_3}(\sigma)$ lies entirely in N.*

Proof. We choose coordinates on $G_n(\mathbb{R}^{2n})$ such that $W(0) = P_2 = 0$. By Theorem 5.3 (see also [113], p. 431), the manifold N consists of pairwise iso-clinic planes. By the Wolf theorem, the subspace of the Lie algebra whose image under the exponential mapping yields N has a basis $S_1, \ldots, S_r \in \mathfrak{SO}(2n)$

satisfying Hurwitz conditions (5.7). Moreover, the matrices S_i can be chosen in the form

$$S_i = \begin{pmatrix} 0 & -a_i^\top \\ a_i & 0 \end{pmatrix}, \qquad i = 1, \ldots, r.$$

In terms of the matrices a_i, the Hurwitz relations become

$$a_i^\top a_i = I_n, \qquad a_i^\top a_j + a_j^\top a_i = 0, \quad i \neq j.$$

We consider the corresponding geodesics

$$\exp(tS_i) = (\cos t) I_{2n} + (\sin t) S_i. \tag{5.27}$$

The mapping $\exp(tS_i)$ transforms the matrix $W(0)$ into $(\tan t)a_i$.
    Let

$$P_1 = (\tan t)a_1, \qquad P_3 = (\tan \tau)a_2. \tag{5.28}$$

By (5.27), the manifold N consists of matrices having the form

$$N = \left\{ \alpha_0 I_n + \sum_{i=1}^{r} \alpha_i a_i \,\middle|\, \alpha_i \in \mathbb{R}^1 \right\}.$$

We need to verify that there exist numbers α_i such that

$$J_{P_1 P_2 P_3}(\sigma) = [(1-\sigma)(a_1 \tan t)^{-1} + \sigma(a_2 \tan \tau)^{-1}]^{-1} = \alpha_0 I_n + \sum_{i=1}^{r} \alpha_i a_i.$$

We set $\alpha_i = 0$ for $i \geq 3$. It suffices to show that the expression $(\lambda_1 a_1^{-1} + \lambda_2 a_2^{-1})^{-1}$ for any $\lambda_1, \lambda_2 \in \mathbb{R}^1$ is a linear combination of a_1 and a_2. To show this, we choose λ_1 and λ_2 such that $\lambda_1^2 + \lambda_2^2 = 1$. Then, by the Hurwitz equations, we have

$$(\lambda_1 a_1^\top + \lambda_2 a_2^\top)(\lambda_1 a_1 + \lambda_2 a_2) = \lambda_1^2 a_1^\top a_1 + \lambda_1 \lambda_2 (a_1^\top a_2 + a_2^\top a_1) + \lambda_2^2 a_2^\top a_2 = I_n.$$

Therefore,

$$(\lambda_1 a_1^{-1} + \lambda_2 a_2^{-1})^{-1} = (\lambda_1 a_1^\top + \lambda_2 a_2^\top)^{-1} = \lambda_1 a_1 + \lambda_2 a_2;$$

this is what was required to be proved.

    The formulas

$$DV(P_1, 0; P, P_3) = \sigma I_n,$$
$$P = [(1-\sigma)P_1^{-1} + \sigma P_3^{-1}]^{-1}, \tag{5.29}$$

can also be considered for complex $n \times n$ matrices P_i and complex numbers σ. Then they set a one-to-one and bicontinuous correspondence between the extended complex plane $\sigma \in \widehat{\mathbb{C}}^1$ and its image in $G_n(\mathbb{C}^{2n})$ under mapping (5.29). Therefore, complex curves with a scalar double ratio on the manifold $G_n(\mathbb{C}^{2n})$ are smooth embeddings of Riemannian spheres in $G_n(\mathbb{C}^{2n})$; this stresses once more an analogy between real curves with a scalar double ratio and circles.

§5. Fourth Harmonic as a Geodesic Symmetry

Definition 5.6. A point $P \in G_n(\mathbb{R}^{2n})$ is called the fourth harmonic of the points $P_1, P_2, P_3 \in G_n(\mathbb{R}^{2n})$ if the quadruple of points P_1, P_2, P_3, and P is harmonic.

The manifold $G_n(\mathbb{R}^{2n})$ is a Riemannian symmetrical space. This means that for any point $Q \in G_n(\mathbb{R}^{2n})$, there exists an involutive isometry for which the point Q is an isolated fixed point. We present an explicit formula for this isometry.

Theorem 5.4. *For any point $Q \in G_n(\mathbb{R}^{2n})$, the mapping $P \mapsto \phi(P)$ that sets the fourth harmonic of the points Q, $(-Q^\top)^{-1}$, and P in correspondence to each point P is an involutive isometry of the manifold $G_n(\mathbb{R}^{2n})$ with respect to the point Q.*

Proof. Let a point P be sufficiently close to the point Q. By Theorem 5.1, the pairs of points $Q, (-Q^\top)^{-1}$ and $I_n, (-I_n)$ are equivalent because the corresponding matrices of double ratios vanish (or, equivalently, both pairs of planes corresponding to points of the Grassmann manifold are mutually orthogonal). Therefore, it suffices to consider the case where $Q = I_n$ and $(-Q^\top)^{-1} = -I_n$. Then

$$\mathrm{DV}(I, -I; P, \phi(P)) = (P - I)(P + I)^{-1}(\phi + I)(\phi - I)^{-1} = -I,$$
$$\phi + I = -(P - I)^{-1}(P + I)\phi + (P - I)^{-1}(P + I),$$
$$\phi = (I + (P - I)^{-1}(P + I))^{-1}((P - I)^{-1}(P + I) - I)$$
$$= ((P - I) + (P + I))^{-1}(P - I)(P - I)^{-1}((P + I) - (P - I)) = P^{-1}.$$

The transformation ϕ is given by the orthogonal matrix

$$\begin{pmatrix} 0 & I_n \\ I_n & 0 \end{pmatrix}$$

and thus defines an isometry of $G_n(\mathbb{R}^{2n})$. The point I_n is obviously an isolated fixed point of this isometry.

Theorem 5.4 gives us a possibility to study the global properties of involutive isometries. Namely, fixed points of the involution $\phi(P)$ are solutions to the equation $P^2 = I_n$. This implies that the matrix P is diagonalizable and has eigenvalues equal to ± 1. The orbit of the group $O(n)$ that passes through the point Π_k corresponding to the diagonal matrix with eigenvalues ± 1 and signature k is denoted by O_k. The orbits O_k remain fixed under the involution ϕ.

Let $N \subset G_n(\mathbb{R}^{2n})$. We consider the set of geodesics of the Grassmann manifold that are orthogonal to the submanifold N and a point $P \in N$. Let $\gamma = \mathrm{Exp}(tX)$, where $t \in \mathbb{R}$ and $X \in TN^\perp$, be the equation of a certain geodesic from the set considered that passes through the point P for $t = 0$. Then the *geodesic symmetry with respect to the manifold N* is the mapping that sets the point $\mathrm{Exp}(-sX)$ in correspondence to each point $\mathrm{Exp}(sX)$.

Proposition 5.8. *In a neighborhood of any point of the orbit O_k, the mapping ϕ is a geodesic symmetry with respect to this orbit.*

Proof. We note that a matrix double ratio is continuous. Let $B \in O_k$. Then for $A_k \to B$, the sequence $\phi(A_k)$ converges to B. Therefore, a small neighborhood of the orbit O_k contains a point A such that $\phi(A)$ also belongs to this neighborhood. Through the points A and $\phi(A)$, we draw a shortest geodesic γ. Because ϕ is an isometry, the image γ under the mapping ϕ is a geodesic again; because each pair of points $A, \phi(A)$ passes to itself, we have $\phi(\gamma) = (\gamma)$. Because ϕ is an isometry, γ and O_k are orthogonal.

Conversely, under the isometry ϕ, any geodesic that is orthogonal to O_k passes into itself. Points that lie on this geodesic and are equidistant from O_k pass to each other.

A quadruple of points P_1, P_2, P_3, and P_4 is called *biharmonic* if $P_2 = (-P_1^{\top})^{-1}$, $P_4 = (-P_3^{\top})^{-1}$, and $DV(P_1, P_2; P_3, P_4) = -I_n$.

As was shown, the study of biharmonic quadruples can be reduced to the case where $P_3 = I_n$ and $P_4 = -I_n$. Then the biharmonicity condition means that $DV((-P^{\top})^{-1}, P; I_n, -I_n) = -I_n$, and hence the matrix P is skew symmetrical, i.e., $P^{\top} = -P$.

5.1. Manifold of Isotropic Planes. In a fixed chart of the manifold $G_n(\mathbb{R}^{2n})$, we consider the set of points whose coordinates are skew-symmetrical matrices. We let $\mathbb{S}$ denote the manifold[3] that is the closure of this set. We study certain geometric properties of the manifold $\mathbb{S}$.

Proposition 5.9. *The manifold $\mathbb{S}$ is the locus of points of $G_n(\mathbb{R}^{2n})$ that are equidistant from I_n and $(-I_n)$ and are isoclinic to these planes.*

Proof. We compute the quadratic form for angle determination for the matrix I_n and an arbitrary skew-symmetrical matrix P:

$$DV_{PI} = (I_n - P)(I_n + I_n)^{-1}(I_n + P)(I_n - P^2)^{-1}$$
$$= (1/2)(I_n - P)(I_n + P)(I_n - P^2)^{-1} = I_n/2.$$

The plane P is therefore isoclinic to the plane I_n and the angle made by them is equal to $\pi/4$. Therefore, P is isoclinic and orthogonal to the plane $(-I_n)$ and is also inclined to it by the angle $\pi/4$.

Conversely, P is isoclinic to I_n and $(-I_n)$ and equidistant from them iff the quadratic form for angle determination between P and I_n has the form $I_n/2$:

$$DV_{PI} = (I_n + P^{\top})(2 I_n)^{-1}(I_n + P)(I_n + P^{\top}P)^{-1} = I_n/2.$$

But then $(I_n + P^{\top})(I_n + P) = I_n + P^{\top}P$, and therefore $P^{\top} = -P$.

---

[3] Below, in Chap. 6, the properties of the manifold $\mathbb{S}$ are used to isolate those classes of Riccati equations whose complexification defines flows on Cartan–Siegel homogeneity domains.

Under an orthogonal transformation of $\mathbb{R}^{2n}$, i.e., under an isometry of $G_n(\mathbb{R}^{2n})$, the cone of planes that are isoclinic to I_n and $(-I_n)$ and make the angle $\pi/4$ with them passes to the same cone with respect to two other mutually orthogonal planes. The obtained manifold is isometric to $\mathbb{S}$. However, it is no longer represented by skew-symmetrical matrices.

We prove the following invariant property of the manifold $\mathbb{S}$.

Theorem 5.5. *The set $\mathbb{S}$ described by the equation $P^{\mathsf{T}} = -P$ is a totally geodesic submanifold of $G_n(\mathbb{R}^{2n})$.*

Proof. We consider the decomposition of the Lie algebra $\mathfrak{SO}(2n)$ into the subalgebra corresponding to the stationary subgroup of the point $o = \phi(e) \in G_n(\mathbb{R}^{2n}) = O(2n)/(O(n) \times O(n))$ and its orthogonal complement. We then have

$$\mathfrak{SO}(2n) = \mathfrak{P} + \mathfrak{K},$$

where

$$\mathfrak{K} = \left\{ \begin{pmatrix} a & 0 \\ 0 & b \end{pmatrix} \,\middle|\, a^{\mathsf{T}} = -a, b^{\mathsf{T}} = -b \right\}$$

is the subalgebra corresponding to the stabilizer K of the plane $o = \phi(e)$ and

$$\mathfrak{P} = \left\{ \begin{pmatrix} 0 & b \\ -b^{\mathsf{T}} & 0 \end{pmatrix} \right\}$$

is the complement to $\mathfrak{K}$ in $\mathfrak{SO}(2n)$, which is invariant under the action of $\operatorname{Ad} K$. We compute the exponential mapping for $\mathfrak{P}$:

$$\exp \begin{pmatrix} 0 & b \\ -b^{\mathsf{T}} & 0 \end{pmatrix} t = \begin{pmatrix} \cos bt & \sin bt \\ -\sin bt & \cos bt \end{pmatrix}.$$

Trigonometric functions of a matrix argument are defined using the Taylor series.

The geodesic of the manifold $G_n(\mathbb{R}^{2n})$ is given by

$$W(t) = \phi \exp \begin{pmatrix} 0 & b \\ -b^{\mathsf{T}} & 0 \end{pmatrix} = -\tan bt.$$

Taking the Taylor series of the obtained relation, we easily verify that the matrix $W(t) = -\tan bt$ is skew symmetrical iff the matrix b is skew symmetrical. Therefore, $T_{W(0)}\mathbb{S}$ (the tangent plane to the manifold $\mathbb{S}$ at the point $W(0) = o = \phi(e)$) consists of the matrices

$$\begin{pmatrix} 0 & b \\ b & 0 \end{pmatrix}, \quad \text{where } b^{\mathsf{T}} = -b.$$

We verify that this subspace is a triple Lie system. We have

$$\left[\begin{pmatrix} 0 & b \\ b & 0 \end{pmatrix}, \begin{pmatrix} 0 & c \\ c & 0 \end{pmatrix}\right] = \begin{pmatrix} [b,c] & 0 \\ 0 & [b,c] \end{pmatrix},$$

$$\left[\begin{pmatrix} [b,c] & 0 \\ 0 & [b,c] \end{pmatrix}, \begin{pmatrix} 0 & a \\ a & 0 \end{pmatrix}\right] = \begin{pmatrix} 0 & [[b,c],a] \\ [[b,c],a] & 0 \end{pmatrix}.$$

Because skew-symmetrical matrices form a Lie algebra, we see that $a^\top = -a$, $b^\top = -b$, and $c^\top = -c$ imply $[[b,c],a]^\top = -[[b,c],a]$.

Therefore, $\mathbb{S}$ is a totally geodesic submanifold of $G_n(\mathbb{R}^{2n})$.

The manifold $\mathbb{S}$ also admits the following invariant description. We consider the group $O(n,n)$ consisting of matrices that leave the nondegenerate quadratic form in $2n$ variables of signature 0 invariant:

$$\zeta((x,y),(x,y)) = \sum_{i=1}^{n} x_i y_i = \langle x,y \rangle.$$

(Here and in what follows, the angle brackets denote the standard Euclidean inner product of n-dimensional vectors.) The plane A is called *isotropic* if $\zeta|_A = 0$. It is easy to show that the maximum dimension of an isotropic plane is equal to n.

Proposition 5.10. *A point of the manifold $G_n(\mathbb{R}^{2n})$ whose first coordinate is equal to a matrix A is isotropic iff $A^\top = -A$.*

Proof. Indeed, the isotropy condition for the plane $y = Ax$ consists in the relation $\zeta((x,Ax),(x,Ax)) = 0$ for any $x \in \mathbb{R}^n$, i.e., $\langle x, Ax \rangle = 0$. This implies $A^\top = -A$.

In $G_n(\mathbb{R}^{2n})$, we consider a basis that is defined by the following two planes: the plane P_0, the maximal subspace such that the restriction of the form ζ to it is positive definite, and the plane P_∞, the maximal subspace such that the restriction of the form ζ to it is negative definite. Because the signature of ζ is zero, both planes are n-dimensional and the form ζ in these coordinates becomes $\zeta((x,y),(x,y)) = \langle x,x \rangle - \langle y,y \rangle$. The plane A is isotropic if for any vector $x \in P_0$, the relation $\langle x,x \rangle - \langle Ax, Ax \rangle = 0$ holds. Therefore, in the basis P_0, P_∞, the manifold $\mathbb{S}$ consists of orthogonal matrices.

§6. Clifford Parallels

We consider the group Spin(3) of unit quaternions, which is the three-dimensional sphere S^3 as a manifold; Spin(3) is a semisimple compact Lie group, and a two-side invariant Riemannian metric (corresponding to angles made by rays going to points of S^3 from the center of the sphere) exists on it.

The group of motions in Spin(3) is the group of left and right multiplications by quaternions whose module is equal to one.

To imagine S^3 visually, we use the projections from the center of the standard three-dimensional sphere $\sum_{k=0}^3 x_k^2 = 1$ on the three-dimensional space tangent to the sphere S^3 at its north pole, i.e., at the point $x_0 = 1$. We assume that the north pole is the identity of the group. Then the plane $x_0 = 1$ becomes the space $\mathbb{RP}^3$ whose infinitely distant plane $\mathbb{RP}^2$ corresponds to the section of the sphere S^2 by the plane $x_0 = 0$. In the space $\mathbb{RP}^3$, we introduce homogeneous coordinates $(x_0 : x_1 : x_2 : x_3)$ defined with an accuracy up to multiplication by a nonzero constant. The invariant metric on S^3 induces a metric on the plane $L = \{x_0 = 1\}$ and transforms it into a Riemannian manifold (more precisely, to one of its charts whose closure $\mathbb{RP}^3$ is SO(3), which is double covered by the group Spin(3)).

Such a plane is called the *elliptic space* Ell(3). The group of motions in Ell(3) is the projection of the group of rotations of the sphere S^3. Therefore, Ell(3) is a realization of Spin(3) with the induced invariant metric. We describe this realization and, in particular, left-invariant and right-invariant vector fields on Ell(3).

We know that geodesics in S^3 are great circles. Under the projection from the center, they pass to lines of the plane L, which thus serve as geodesics in Ell(3). We first consider a certain specific left-invariant vector field on S^3. We know that such a field can be obtained from a fixed unit vector in T_e Spin(3) $= \mathfrak{SO}(3)$ if we apply all left translations on Spin(3) to it. To be more specific, we choose a vector that is tangent to the great circle $\{a+bi \mid a^2+b^2 = 1\}$. The exponential mapping of the line spanned by this vector forms a one-parameter subgroup $P = \{\cos\phi + i\sin\phi \mid \phi \in \mathbb{R}\}$. A vector tangent to it at the point $\phi = 0$ is $(-\sin\phi + i\cos\phi)|_{\phi=0} = i$.

We consider left translations of this tangent vector under the action of the group Spin(3). To compute the translation to the point $g \in$ Spin(3), we need to postmultiply g by i. Under multiplication by i, we have $1 \mapsto i$, $i \mapsto -1$, and $j \mapsto -k$, $k \mapsto j$. Therefore, the matrix of the multiplication operator by i is equal to

$$A = \begin{pmatrix} 0 & -1 & 0 & 0 \\ 1 & 0 & 0 & 0 \\ 0 & 0 & 0 & 1 \\ 0 & 0 & -1 & 0 \end{pmatrix}.$$

The matrix A is simultaneously skew symmetrical and orthogonal. This corresponds to the fact that the point i can be also considered as a point of Spin(3) as well as a vector belonging to the Lie algebra $\mathfrak{SO}(3)$.

The image of this vector at the point $g = (a_0 + a_1 i + a_2 j + a_3 k)$ is

$$(g)(i) = Ag = \begin{pmatrix} a_1 \\ -a_0 \\ -a_3 \\ a_2 \end{pmatrix}.$$

Because A is orthogonal, $\|Ag\| = 1$ for any g; because A is skew symmetrical, $Ag \perp g$, i.e., Ag is a tangent vector at the point g.

We consider the translation of the manifold S^3 along the trajectories of the following differential equation corresponding to the constructed vector field:

$$\begin{aligned} \dot{x}_0 &= x_1 & \dot{x}_2 &= -x_3 \\ \dot{x}_1 &= -x_0 & \dot{x}_3 &= x_2. \end{aligned} \qquad (5.30)$$

The trajectories of this field are obtained via a translation by elements of the one-parameter subgroup P and are therefore geodesics in S^3, i.e., for any initial state $X(0) = (x_0(0), x_1(0), x_2(0), x_3(0))$, the corresponding points circumscribe the great circle with unit speed.

Exercise. Integrate system (5.30) and find the trajectory passing through the point $X(0) = (x_0(0), x_1(0), x_2(0), x_3(0))$.

The flow of Eq. (5.30) forms a one-parameter group of isometries of the unit sphere. In particular, a pair of points $X^1(0), X^2(0)$ passes to a pair $X^1(\tau), X^2(\tau)$. Moreover, $\mathrm{dist}(X^1(0), X^2(0)) = \mathrm{dist}(X^1(\tau), X^2(\tau))$ and $\mathrm{dist}(X^1(0), X^1(\tau)) = \mathrm{dist}(X^2(0), X^2(\tau))$.

Definition 5.7. A transformation F of a Riemannian manifold M under which $\mathrm{dist}(x, F(x)) = \mathrm{const}$ for any x is called a Clifford translation.

In the Euclidean space, a Clifford translation is a usual translation. We have no such translations on the sphere S^2.

Above, we constructed the Clifford translation on S^3. We now find its action on Ell(3). Choosing an appropriate coordinate system in Ell(3), we can assume that the line $x_1 = x_2 = 0$ corresponds to the subgroup P. It corresponds to the trajectory of system (5.30) with the initial conditions $x_1(0) = x_2(0) = 0$. System (5.30) admits two first integrals $x_0^2 + x_1^2 = C_1$ and $x_2^2 + x_3^2 = C_2$. This means that the trajectories lie on the two-dimensional surface

$$\begin{aligned} x_0^2 + x_1^2 &= \rho, \\ x_2^2 + x_3^2 &= 1 - \rho, \end{aligned} \qquad (5.31)$$

which is a torus on S^3. The image of this surface under the projection on Ell(3) is the surface

$$\begin{aligned} x_0^2 + x_1^2 + x_2^2 + x_3^2 &= 1, \\ (1 - \rho)(x_0^2 + x_1^2) - \rho(x_2^2 + x_3^2) &= 0. \end{aligned} \qquad (5.32)$$

The first equation in (5.32) is a norming condition of the homogeneous coordinates. The second equation assigns the one-parameter family of surfaces

$$\Gamma_\rho = \{(1 - \rho)(x_0^2 + x_1^2) - \rho(x_2^2 + x_3^2) = 0\}.$$

For $\rho = 1$, we obtain $\Gamma_1 = L$. For $\rho = 0$, we obtain $\Gamma_0 = \{x_0 = x_1 = 0\}$, which is a line in the infinitely distant part of Ell(3) (the corresponding two circles on S^3 are linked). For $0 < \rho < 1$, this surface in our chart is a hyperboloid of one sheet,

$$\Gamma_\rho = \{(1 - \rho) = \rho x_2^2 + \rho x_3^2 - (1 - \rho)x_1^2\}.$$

Therefore, the family of tori (5.31) passes to the family of hyperboloids of one sheet under the projection. Because the trajectories of system (5.30) are geodesics, they should coincide with rectilinear rulings of the hyperboloids. By the theorem on the continuous dependence on initial data for system (5.30), the left-invariant fields corresponds to one of the two families of rectilinear rulings. Obviously, the right-invariant vector field corresponds to the other family.

One asks how to realize the left Clifford translation in Ell(3)? It is required to draw a hyperboloid Γ_ρ through each point of X and a ruling of one of the families on Γ_ρ. Under the translation by τ, each point passes one and the same distance along the corresponding ruling (in the sense of the metric of Ell(3)). Because this transformation is an isometry, each geodesic passes to a geodesic.

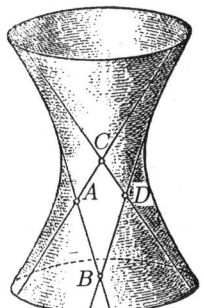

Fig. 5.1

This implies that lines of the second family pass to each other. We choose any two points A and B lying on one and the same line and translate them by τ. We obtain two points C and D lying on another line (see Fig. 5.1). We have $AB = CD$ and $AC = BD$. Further, because the considered mapping is an isometry, the sum of the angles CAB and DCA equals π. Further, the line AB turns out to be Clifford parallel to CD (in the sense of some other translation in S^3). That is, the spatial quadrangle $ABDC$ has all the properties of a parallelogram, i.e., its opposite sides are equal and Clifford parallel, and the sum of the corresponding angles is equal to π (however, its diagonals do not intersect each other) [55].

It is clear that we can take the great circle of S^3 as the initial curve L. We can choose the lines AB and AC to be orthogonal to each other, and then we obtain a Clifford rectangle. This implies that Clifford parallels are lines that are distant from each other by a constant value. It turns out that Clifford parallels are directly related to the geometry of Grassmann manifolds and the Riccati equation.

§7. Connection Between Clifford Parallels and Isoclinic Planes

As in the case of the space Ell(3), we use the central projection from the origin O of the space $\mathbb{R}^{2n}$ on the hyperplane $x_{2n} = 1$ to describe Ell($2n - 1$). Under this projection, we set a point of its intersection on the hyperplane $x_{2n} = 1$ in correspondence to each line passing through O. In correspondence to lines that are parallel to this hyperplane, we set improper elements of the hyperplane that correspond to the direction of these lines. Such a projection transforms the hyperplane $x_{2n} = 1$ into the real projective space $\mathbb{RP}^{2n-1}$. The semisimple compact Lie group $O(2n)$ leaving the angular metric on the sphere invariant acts on the standard $(2n-1)$-dimensional sphere $S^{2n-1} \subset \mathbb{R}^{2n}$. This metric induces a metric g on the plane $x_{2n} = 1$. A projective (linear-fractional) transformation of the plane $x_{2n} = 1$ preserving the induced metric corresponds to each orthogonal transformation of $\mathbb{R}^{2n}$. The plane $x_{2n} = 1$ equipped with the metric g is called an *elliptic space* and is denoted by Ell($2n - 1$). Because great circles are geodesics on the sphere, ordinary lines serve as geodesic lines in the space Ell($2n - 1$). For the same reason, ordinary planes are totally geodesic submanifolds.

The elliptic space Ell($2n - 1$) is compact, and the maximum distance between any of its points is equal to π.

A $(n-1)$-dimensional (proper or improper) plane $\widetilde{A} \subset$ Ell($2n - 1$), which is the intersection $\widetilde{A} = A \cap \{x_{2n} = 1\}$ of A with $x_{2n} = 1$, corresponds to any n-dimensional subspace $A \subset \mathbb{R}^{2n}$.

A two-dimensional plane K is called *semiorthogonal* to a plane A if $K \cap A$ is one-dimensional and there exists a vector $p \in K$ such that $p \perp A$ (see Fig. 5.2).

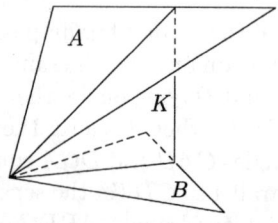

Fig. 5.2

Proposition 5.11. *Two planes A and B are isoclinic iff it is possible to draw a hyperplane through each one-dimensional direction in A (or B) that is orthogonal to A as well as to B.*

Proof. We consider the projection of the vector (x, Ax) of the plane A on the plane B. We have shown in Sec. 1 that the vector (y, By), where

$$y = (I_n + B^\top B)^{-1}(I_n + B^\top A)x,$$

is such a projection. If, in turn, we consider the projection of the obtained vector (y, By) on the plane A, then we obtain the vector (z, Az), where

$$z = (I_n + A^\top A)^{-1}(I_n + A^\top B)y. \qquad (5.33)$$

Joining (5.5) and (5.33), we obtain the following statement.

Lemma 5.8. *A vector z is collinear to a vector x iff it is an eigenvector of the matrix*

$$(I_n + A^\top A)^{-1}(I_n + A^\top B)(I_n + B^\top B)^{-1}(I_n + B^\top A). \qquad (5.34)$$

The collinearity of the vectors x and z means that the two-dimensional plane $\mathrm{Span}\{(x, Ax), (y, By)\}$ is orthogonal to the plane A as well as to the plane B. We now note that matrix (5.34) occurring in Lemma 5.8 is a matrix for angle determination. Therefore, each vector is an eigenvector of this matrix iff this matrix is scalar, i.e., when A and B are isoclinic.

Let A and B be two n-dimensional subspaces in $\mathbb{R}^{2n}$, and let $\widetilde{A}$ and $\widetilde{B}$ be their projections on $\mathrm{Ell}(2n - 1)$.

Two $(n-1)$-dimensional planes of the space $\mathrm{Ell}(2n - 1)$ are called *Clifford parallel* if the distance from any point on one to the other plane is constant.

Theorem 5.6 (J. C. Wong). *Two spaces A and B are isoclinic iff $\widetilde{A}$ and $\widetilde{B}$ are Clifford parallel.*

Proof. Let A and B be isoclinic, and let O be a point of its intersection. We choose an arbitrary point $x \in \widetilde{A}$ and draw a plane K through x that is semiorthogonal to the plane A as well as to the plane B. Let $y = K \cap B$. The angle made by the vector Ox and the plane B is equal to the angle xOy. This angle is the minimum angle among all angles of the form xOz, where $z \in B$; therefore the distance from the point x to the plane $\widetilde{B}$ (in the metric of the space $\mathrm{Ell}(2n - 1)$) is equal to the angle xOy. This angle does not depend on the choice of the point $x \in \widetilde{A}$ because the planes A and B are isoclinic. Therefore, the planes $\widetilde{A}$ and $\widetilde{B}$ are Clifford parallel.

The above argument is obviously reversible. If the distance from the point $x \in \widetilde{A}$ to the plane $\widetilde{B}$ in the space $\mathrm{Ell}(2n - 1)$ does not depend on the choice of the point x, then this means that the angle made by the vector $Ox \in A$ and the plane B does not depend on the choice of the point $x \in A$.

§8. Matrix Double Ratio on the Lagrange–Grassmann Manifold

We begin by recalling some facts (see Chap. 4). Let the linear space $\mathbb{R}^{2n}$ of variables (x, p) be equipped with the standard symplectic structure $\omega = \sum_{i=1}^{n} dp_i \wedge dx_i$. The group of transformations leaving the form ω invariant is denoted by $\mathrm{Sp}(n, \mathbb{R})$. Its Lie algebra is denoted by $\mathfrak{Sp}(\mathfrak{n}, \mathbb{R})$. A plane l is called *Lagrangian* if it is n-dimensional and $\omega|_l = 0$. The set of all Lagrangian subspaces $\mathbb{R}^{2n}$ is called the *Lagrange–Grassmann manifold* and is denoted by $\Lambda(\mathbb{R}^{2n})$. Under the identification $\Lambda(\mathbb{R}^{2n}) = U(n)/O(n)$, it is equipped with the structure of a Riemannian symmetrical space. In a local coordinate system defined by a pair of orthogonal Lagrangian planes P_0 and P_∞, points of $\Lambda(\mathbb{R}^{2n})$ are parametrized by the set of symmetrical matrices. The concept of the matrix double ratio is also extended to the Lagrange–Grassmann manifold if we consider this manifold as a submanifold of the manifold $G_n(\mathbb{R}^{2n})$.

Theorem 5.7 ([121]). *The submanifold $\Lambda(\mathbb{R}^{2n})$ is a totally geodesic submanifold of $G_n(\mathbb{R}^{2n})$.*

Proof. The decomposition of the algebra $\mathfrak{SO}(2n)$ at the point $V = (x, 0)$ has the form $\mathfrak{SO}(2n) = \mathfrak{P} + \mathfrak{K}$, where

$$\mathfrak{K} = \left\{ \begin{pmatrix} a & 0 \\ 0 & b \end{pmatrix} \middle| a, b \in \mathfrak{SO}(n) \right\}$$

is the Lie algebra of the stabilizer $K = O(n) \times O(n)$ of the plane $V = (x, 0)$ and

$$\mathfrak{P} = \left\{ \begin{pmatrix} 0 & b \\ -b^\top & 0 \end{pmatrix} \right\}$$

is the direct complement of $\mathfrak{K}$ (in the sense of the Killing form), which is invariant with respect to the action of $\text{Ad}\, K$. The differential of the natural projection $\phi\colon O(2n) \to G_n(\mathbb{R}^{2n})$ at a point V transforms a matrix g into $d\phi(g)$. We thus obtain an identification of $\mathfrak{P}$ with the tangent space of $G_n(\mathbb{R}^{2n})$ at the point V that is invariant with respect to the action of K. To realize $\Lambda(\mathbb{R}^{2n})$ as a submanifold of $G_n(\mathbb{R}^{2n})$, we consider an embedding $U(n) \to O(2n)$ under which the matrix $A + iB \in U(n)$ passes to

$$\begin{pmatrix} A & -B \\ B & A \end{pmatrix} \in SO(2n).$$

Under this embedding, the stabilizer V is equal to

$$K_1 = \left\{ \begin{pmatrix} A & 0 \\ 0 & A \end{pmatrix} \right\},$$

and the decomposition of the Lie algebra $\mathfrak{U}$ becomes $\mathfrak{U}(n) = \mathfrak{P}_1 + \mathfrak{K}_1$, where $\mathfrak{P}_1 \subset \mathfrak{P}$:

$$\mathfrak{P}_1 = \left\{ \begin{pmatrix} 0 & b \\ -b & 0 \end{pmatrix} \middle| b^\top = b \right\}.$$

Therefore, $\{W \mid W \in \Lambda(\mathbb{R}^{2n})\}$ is a result of the action of matrices of the form $\exp A_b$, where $A_b \in \mathfrak{P}_1$, on the $n \times n$ matrix $o = \phi(e)$, i.e., the plane $\{x \mid x \in V_0\}$. We find an explicit expression for the matrices W:

$$\exp A_b = \exp \begin{pmatrix} 0 & b \\ -b & 0 \end{pmatrix} = I_n + \begin{pmatrix} 0 & b \\ -b & 0 \end{pmatrix} + \cdots = \begin{pmatrix} \cos b & \sin b \\ -\sin b & \cos b \end{pmatrix};$$

further,

$$\exp A_b(V_0) = \left(\begin{array}{cc} \cos b & \sin b \\ -\sin b & \cos b \end{array} \right) \left(\begin{array}{c} x \\ 0 \end{array} \right) = \left(\begin{array}{c} (\cos b)x \\ -(\sin b)x \end{array} \right).$$

Therefore, $W_b = \phi(\exp A_b) = -\tan b$. Considering matrix Taylor series and their analytic continuations [29], we easily show that if a matrix b is symmetrical, the matrix W_b is also symmetrical. Vice versa, any symmetrical matrix W can be represented in the form $W_b = -\tan b$, where b is a certain symmetrical matrix.

To complete the proof, it remains to verify that $\mathfrak{P}_1$ is a triple Lie system, i.e., $[\mathfrak{P}_1, [\mathfrak{P}_1, \mathfrak{P}_1]] \subset \mathfrak{P}_1$. Indeed,

$$\left[\left(\begin{array}{cc} 0 & b \\ -b & 0 \end{array} \right), \left(\begin{array}{cc} 0 & c \\ -c & 0 \end{array} \right) \right] = -\left(\begin{array}{cc} bc - cb & 0 \\ 0 & bc - cb \end{array} \right) = \left(\begin{array}{cc} \delta & 0 \\ 0 & \delta \end{array} \right),$$

where $\delta^\mathsf{T} = -\delta$. Further,

$$\left[\left(\begin{array}{cc} \delta & 0 \\ 0 & \delta \end{array} \right), \left(\begin{array}{cc} 0 & a \\ -a & 0 \end{array} \right) \right] = \left(\begin{array}{cc} 0 & \delta a - a\delta \\ -(\delta a - a\delta) & 0 \end{array} \right) \in \mathfrak{P}_1.$$

This completes the proof.

Lemma 5.9. *If three points P_1, P_2, and P_3 lie in $\Lambda(\mathbb{R}^{2n})$, then a curve with the double scalar ratio $J_{P_1,P_2,P_3}(\sigma)$ belongs entirely to $\Lambda(\mathbb{R}^{2n})$.*

To prove this lemma, it suffices to transport the origin to the point P_2 and use (5.26), which implies that P is symmetrical provided P_1 and P_3 are symmetrical matrices.

Theorem 5.7 and Lemma 5.9 allow us to extend the main results of Secs. 4 and 5 to the case of a Lagrange–Grassmann manifold. In particular, the following statements are direct consequences of Propositions 5.5 and 5.6.

Proposition 5.12. *A geodesic line of the manifold $\Lambda(\mathbb{R}^{2n})$ with pairwise isoclinic elements is a curve with a scalar double ratio.*

Proposition 5.13. *Let three points $P_1, P_2, P_3 \in \Lambda(\mathbb{R}^{2n})$ lie on a complete connected totally geodesic submanifold N whose points, considered as subspaces of $\mathbb{R}^{2n}$, are pairwise in general position. Then all points of the curve $J_{P_1,P_2,P_3}(\sigma)$ belong to N.*

However, the criterion for equivalence of pairs of points with respect to isometries of $\Lambda(\mathbb{R}^{2n})$ does not coincide with the criterion for equivalence presented in Theorem 5.1 for the manifold $G_n(\mathbb{R}^{2n})$.

We showed in Theorem 5.1 that the double ratio DV_{AB} yields a complete set of invariants with respect to the action of the group $O(2n)$. In the case of the Lagrange–Grassmann manifold, the coincidence of matrix double ratios certainly remains a necessary condition for the equivalence of two pairs of matrices, but it is no longer sufficient.

Indeed, as in the proof of Theorem 5.1, the question of the equivalence of two pairs of matrices with the same double ratio is reduced to verification of the equivalence of pairs of matrices $(0, V)$ and $(0, W)$ under the condition that the matrices V^2 and W^2 are similar. In this case, the stabilizer of the zero plane in the group of transformations that preserve Lagrangian planes consists of matrices

$$\begin{pmatrix} \alpha & 0 \\ 0 & -\alpha \end{pmatrix}.$$

Because $\alpha \in O(n)$, the equivalence of the pairs $(0, V)$ and $(0, W)$ means that $W = \alpha V \alpha^{-1}$. But the similarity of the squares of the matrices does not imply the similarity of the matrices themselves. Therefore, the matrix double ratio does not yield a complete set of invariants with respect to the group of isometries of the Lagrange–Grassman manifold.

In the next section, we introduce one more invariant of the action of the isometry group on the Lagrange–Grassmann manifold. For this, we use the Morse–Maslov–Arnol'd index in the form of Leray–Kashiwara [68].

§9. Morse–Maslov–Arnol'd Index in the Leray–Kashivara Form

We recall that the *Morse index* of a quadratic functional is the maximum dimension of a subspace on which this functional is negative definite. Morse proved that in the case of the integral functional

$$\int_{t_0}^{t_1} [\dot{h}^\top(t) A(t)\dot{h}(t) + 2h^\top(t)C(t)\dot{h}(t) + h^\top(t)B(t)h(t)]\, dt$$

satisfying the strengthened Legendre condition (i,e,, when the matrix $A(t)$ is positive definite for all $t \in (t_0, t_1)$), this dimension is equal to the number of points that are conjugate to the point t_0 on the interval (t_0, t_1) when counted with their multiplicity (for the proof see [70], Sec. 15).

Later, far-reaching generalizations of the Morse theorem were found, which, in essence, led to the creation of symplectic geometry as an independent mathematical branch. Various symplectic-invariant definitions of the Morse index were proposed. Here, we follow the Kashiwara definition ([68], Sec. 1.5).

Let A_1, A_2, and A_3 be three Lagrange subspaces of a symplectic space, i.e., the space $\mathbb{R}^{2n}$ equipped with a nondegenerate skew-symmetrical form ω with the matrix J. We consider the $3n$-dimensional space $Z = A_1 \oplus A_2 \oplus A_3$ consisting of triples of vectors

$$(x_1, x_2, x_3), \qquad x_1 \in A_1, \quad x_2 \in A_2, \quad x_3 \in A_3,$$

and the quadratic form Q on Z defined by

$$Q = \langle x_1, Jx_2 \rangle + \langle x_2, Jx_3 \rangle + \langle x_3, Jx_1 \rangle. \tag{5.35}$$

Definition 5.8. The Morse index $\tau(A_1, A_2, A_3)$ of a triple of Lagrangian planes (A_1, A_2, A_3) is the signature of the quadratic form (5.35).

We recall that the *signature* $\operatorname{sign} Q$ of a *quadratic form* Q is the difference between the number of positive squares and the number of negative squares in the basis in which the form Q is diagonal. Because the signature is invariant under linear transformations, $\tau(gA_1, gA_2, gA_3) = \tau(A_1, A_2, A_3)$ for any $g \in \operatorname{Sp}(n, \mathbb{R})$. Moreover, because the matrix J is skew-symmetrical, we have

$$\tau(A_1, A_2, A_3) = -\tau(A_2, A_1, A_3) = -\tau(A_1, A_3, A_2).$$

First, we assume that the planes A_1 and A_3 are transversal; the operator of projection of the space $\mathbb{R}^{2n}$ on A_1 parallel to A_3 is denoted by π_{13}, and the operator of projection on A_3 parallel to A_1 is denoted by π_{31}.

Lemma 5.10. *If A_1 and A_3 are transversal, the index $\tau(A_1, A_2, A_3)$ is equal to the signature of the quadratic form $\widetilde{Q}$ given by*

$$\widetilde{Q}(x) = \langle \pi_{13}x, J\pi_{31}x \rangle \tag{5.36}$$

on the space A_2.

Proof. The condition of the lemma implies that there is a unique decomposition of the space $\mathbb{R}^{2n}$ into the direct sum $\mathbb{R}^{2n} = A_1 \oplus A_3$. Let $x_2 \in A_2$. Then

$$x_2 = \pi_{13}x_2 + \pi_{31}x_2.$$

That the planes A_1 and A_3 are Lagrangian means that for any pair of vectors $x, \tilde{x} \in A_i$, the skew-inner product $\langle x, J\tilde{x} \rangle$ vanishes. This implies

$$
\begin{aligned}
Q &= \langle x_1, Jx_2 \rangle + \langle x_2, Jx_3 \rangle + \langle x_3, Jx_1 \rangle \\
&= \langle x_1, J(\pi_{31}x_2) \rangle + \langle \pi_{13}x_2, Jx_3 \rangle + \langle x_3, Jx_1 \rangle \\
&= \langle \pi_{13}x_2, \pi_{31}x_2 \rangle - \langle (x_1 - \pi_{13}x_2), J(x_3 - \pi_{31}x_2) \rangle.
\end{aligned}
$$

We set

$$y_1 = x_1 - \pi_{13}x_2, \qquad y_2 = x_2, \qquad y_3 = x_3 - \pi_{31}x_2.$$

In these coordinates, $Q = \langle \pi_{13}y_2, J\pi_{31}y_2 \rangle - \langle y_1, Jy_3 \rangle$, where the first summand in the right-hand side is a quadratic form on A_2 and the second is a quadratic form on $A_1 \oplus A_3$. Therefore,

$$\tau(A_1, A_2, A_3) = \operatorname{sign}\langle \pi_{13}y_2, J\pi_{31}y_2 \rangle + \operatorname{sign}\langle y_1, Jy_3 \rangle.$$

It is easy to show that $\operatorname{sign}\langle y_1, Jy_3 \rangle = 0$ because this quadratic form does not contain the squares of coordinates and subscripts of coordinates in different terms of this form are not repeated.

We note that the bilinear form $\langle \pi_{13}x, J\pi_{31}y \rangle$ on A_2 is symmetrical because for $x, y \in A_2$, by the Lagrangian property of A_1 and A_3, we have

$$\langle x, Jy \rangle = 0 = \langle \pi_{13}x + \pi_{31}x, J(\pi_{13}y + \pi_{31}y) \rangle$$
$$= \langle \pi_{13}x, J\pi_{31}y \rangle + \langle \pi_{31}x, J\pi_{13}y \rangle.$$

Further, we define the Maslov index $\tau(A_1, A_2, A_3, A_4)$ for a quadruple of n-dimensional Lagrangian planes by

$$\tau(A_1, A_2, A_3, A_4) = \tau(A_1, A_2, A_3) + \tau(A_1, A_3, A_4). \tag{5.37}$$

We now consider a pair of Lagrangian planes A and B and define

$$\theta_{AB} = \tau(-A^{-1}, -B^{-1}, A, B)$$

for it (we recall that the matrices A and B are symmetrical). The value θ_{AB} is not an invariant of a pair of planes with respect to the action of the group $\mathrm{Sp}(n, \mathbb{R})$, because the transformation $Q \mapsto -Q^{-1}$ does not commute with the action of $\mathrm{Sp}(n, \mathbb{R})$, because the orthogonal complement is not preserved under an arbitrary symplectic transformation.

Theorem 5.8. *The value θ_{AB} is a unitary invariant of a pair of planes. Therefore, the relation $\theta_{AB} = \theta_{CD}$ is a necessary condition for the equivalence of two pairs of planes A, B and C, D.*

Proof. By Lemma 5.4, the transformation $Q \mapsto -Q^{-1}$ commutes with the action of the group $\mathrm{U}(n)$, which also preserves the Maslov index of a quadruple of planes.

We find a simple explicit formula for the constructed invariant θ_{AB} in a certain specific coordinate system. Using (5.37), we obtain

$$\theta_{AB} = \tau(-A^{-1}, -B^{-1}, A, B)$$
$$= \tau(-A^{-1}, -B^{-1}, A) + \tau(-A^{-1}, A, B). \tag{5.38}$$

First, we consider the first summand in the right-hand side. By (5.36), it is equal to

$$\mathrm{sign}\langle \pi_{13}y, J\pi_{31}y \rangle, \tag{5.39}$$

where $y \in -B^{-1}$. Because the planes A and $-A^{-1}$ are orthogonal, π_{13} is the orthogonal projection on $-A^{-1}$, and π_{31} is the orthogonal projection on A.

We consider a chart in which the plane $-B^{-1}$ has zero coordinates. Let the plane A have coordinates V in this chart. To compute the invariant θ_{AB}, we must first project the vector $y = (x, 0) \in (-B^{-1})$ on the planes A and $-A^{-1}$, i.e., decompose it into the sum of two vectors of the form (x_1, Vx_1) and $(x_2, -V^{-1}x_2)$. To do this, we must solve the system

$$x = x_1 + x_2,$$
$$0 = Vx_1 - V^{-1}x_2. \tag{5.40}$$

Solving it in x_1 and x_2, we obtain

$$x_1 = (I + V^2)^{-1}x,$$
$$x_2 = V^2(I + V^2)^{-1}x.$$

We now find the signature of quadratic form (5.39). Because the matrix V is symmetrical, by Lemma 5.2, the matrix $1 + V^2$ is nonsingular. Therefore,

$$\pi_{13}y = ((I + V^2)^{-1}x, V(I + V^2)^{-1}x),$$
$$\pi_{31}y = (V^2(I + V^2)^{-1}x, -V(I + V^2)^{-1}x),$$
$$J\pi_{31}y = (-V(I + V^2)^{-1}x, -V^2(I + V^2)^{-1}x).$$

Quadratic form (5.39) becomes

$$\langle -V(I + V^2)^{-2}x, x \rangle - \langle V^3(I + V^2)^{-2}x, x \rangle = \langle -V(I + V^2)^{-1}x, x \rangle. \qquad (5.41)$$

Because the matrix $(I + V^2)^{-1}$ is positive definite and commutes with the matrix V, the index of form (5.39) is equal to the index of the form with the matrix $-V$.

The second summand in the right-hand side of (5.38) is computed similarly. Because $y = (0, x)$, instead of system (5.40), we obtain the system

$$0 = x_1 + x_2,$$
$$x = Vx_1 - V^{-1}x_2.$$

The solution to this system has the form

$$x_1 = V(1 + V^2)^{-1}x,$$
$$x_2 = -V(I + V^2)^{-1}x.$$

Therefore,

$$\pi_{13}y = (V(I + V^2)^{-1}x, V^2(I + V^2)^{-1}x),$$
$$\pi_{31}y = (-V(I + V^2)^{-1}x, (I + V^2)^{-1}x),$$
$$J\pi_{31}y = ((I + V^2)^{-1}x, V(I + V^2)^{-1}x).$$

Therefore, the quadratic form corresponding to the second summand in (5.38) becomes

$$\langle V(I + V^2)^{-2}x, x \rangle + \langle V^3(I + V^2)^{-2}x, x \rangle = \langle V(I + V^2)^{-1}x, x \rangle.$$

Therefore,

$$\tau(-A^{-1}, -B^{-1}, A) = -\tau(-A^{-1}, B, A).$$

Therefore, θ_{AB} in the considered coordinate system is equal to the doubled index of the matrix V.

§10. Fourth Harmonic as an Isometry of the Lagrange–Grassmann Manifold

To prove the theorem on the fourth harmonic for the Lagrange–Grassmann manifold, we need a particular case of the equivalence of pairs of planes.

Proposition 5.14. *For any matrix $Q \in \Lambda(\mathbb{R}^{2n})$, the pair of planes Q and $-Q^{-1}$ is equivalent to the pair of planes I_n and $-I_n$.*

Indeed, both pairs consist of mutually orthogonal planes; therefore, if we transform Q into I_n by a unitary transformation, then $-Q^{-1}$ passes to the orthogonal complement of I_n, i.e., the plane $-I_n$.

Theorem 5.9. *For any point $Q \in \Lambda(\mathbb{R}^{2n})$, the mapping $\sigma : P \mapsto \sigma(P)$ that sets the fourth harmonic of the triple $Q, -Q^{-1}, P$ in correspondence to each point $P \in \Lambda(\mathbb{R}^{2n})$ is an involutive isometry of $\Lambda(\mathbb{R}^{2n})$ with respect to the point Q.*

Proof. Proposition 5.14 allows us to reduce the theorem to consideration of the case where $Q = I_n$ and $-Q^{-1} = -I_n$. As in the proof of Theorem 5.4, we obtain $\sigma(P) = P^{-1}$. The mapping σ is involutive; the point Q is its isolated fixed point. By Theorem 1.4, the unitary group acts transitively on the set of Lagrangian planes. The stationary subgroup of a certain real plane V consists of those orthogonal transformations $\mathfrak{A}$ that leave V fixed. The canonical identification $\Lambda(\mathbb{R}^{2n}) = U(n)/O(n)$ shows that the isometries of $\Lambda(\mathbb{R}^{2n})$ are elements of the group $U(n)$. However, the group $U(n)$ does not exhaust all isometries of the manifold $\Lambda(\mathbb{R}^{2n})$. The mapping σ given by the matrix

$$\begin{pmatrix} 0 & I_n \\ I_n & 0 \end{pmatrix}$$

is also an isometry, although it does not belong to the group $U(n)$. (Indeed, the mapping σ is multiplication by $-i$ with the subsequent complex conjugation, which is not a unitary transformation.)

We are now ready to show the application of the matrix double ratio to the study of the Riccati equation.

§11. Application of the Matrix Double Ratio to the Study of the Riccati Equation

We consider three arbitrary solutions $P_1(t)$, $P_2(t)$, and $P_3(t)$ to the Riccati equation

$$\dot{W} = B - (W + C)A^{-1}(W + C^{\mathsf{T}}). \tag{5.42}$$

For any $n \times n$ matrix P, $a_i(t, P)$ denotes the coefficients of the characteristic polynomial of the matrix $\mathrm{DV}(P, P_1, P_2, P_3)$. These coefficients are obviously independent of the choice of a representative of the class $[\mathrm{DV}]$.

Theorem 5.10. *The functions $a_i(t, P)$ are first integrals of the Riccati equation.*

Proof. In Sec. 3 of Chap. 2, we showed that Riccati equation (5.42) describes the transport of n-dimensional planes under linear mappings of the space $\mathbb{R}^{2n}$ along the trajectories of a linear canonical Hamiltonian system. This implies that for each t, the mapping $W(t_0) \mapsto W(t)$ is represented by a generalized linear-fractional transformation of matrices. Under this transformation, the matrix double ratio is not changed, and this means that the coefficients of the characteristic polynomial of the matrix DV remain constant.

In other words, any solution to the Riccati equation does not leave orbits of the action of the adjoint representation of the group $SL(n)$ on the set of n-dimensional matrices $\mathfrak{M}_n$.[4]

Let $P_1(t)$, $P_2(t)$, and $P_3(t)$ be three solutions to the Riccati equation. For each $\sigma \in \mathbb{R}^1$, we construct a matrix-valued function $P_4(t)$ such that for each t, the double ratio of the obtained quadruple of matrices is equal to σI_n. The condition

$$DV(P_1(t), P_2(t), P_3(t), P_4(t)) = \sigma I_n \qquad (5.43)$$

uniquely defines the function $P_4(t)$. Indeed, we solve Eq. (5.43) for P_4. We obtain

$$P_4 = \{I_n - \sigma(P_3 - P_2)(P_3 - P_1)^{-1}\}^{-1}[P_2 - \sigma(P_3 - P_2)(P_3 - P_1)^{-1}P_2].$$

Proposition 5.15. *The function $P_4(t)$ is a solution to the Riccati equation.*

Proof. The function $P_4(t)$ is defined for those values of t for which $1/\sigma$ is not an eigenvalue of the matrix $(P_3 - P_2)(P_3 - P_1)^{-1}$. Otherwise, $P_4(t)$ is not defined, i.e., at these points, the trajectory leaves the chart considered, and in order to determine the coordinates of this plane, we must pass to another chart.

We consider a solution $P(t)$ to the Riccati equation that coincides with $P_4(t)$ for a certain $t = t_0$. By Theorem 5.1, the matrix double ratio of the quadruple of points $P_1(t)$, $P_2(t)$, $P_3(t)$, and $P(t)$ is constant and identical to σI_n. Therefore, by the theorem on the uniqueness of a solution to the Cauchy problem for ordinary differential equations, we have $P_4(t) \equiv P(t)$.

Riccati equation (5.42) assigns a transport of planes under the action of the flow defined by the canonical system of ordinary differential equations (see Sec. 3 of Chap. 2) for the functional

$$\int \langle A\dot{h}, \dot{h} \rangle + 2\langle C\dot{h}, h \rangle + \langle Bh, h \rangle. \qquad (5.44)$$

---

[4] The determinant of the matrix via which the conjugation of similar matrices is realized can be normalized to one.

We fix two arbitrary planes and find conditions under which Riccati equation (5.42) preserves the quadratic form for determination of angles made by planes.

Theorem 5.11. *Let coefficients of functional (5.44) be connected by the relations*

$$CA^{-1}C^{\mathsf{T}} - B = A^{-1},$$
$$A^{-1}C^{\mathsf{T}} = -CA^{-1}. \tag{5.45}$$

Then the quadratic form for determination of angles made by any two solutions to the Riccati equation remains constant.

Proof. Let $X_1(t)$ and $X_2(t)$ be two solutions to Eq. (5.42). Letting V denote $(-X^{\mathsf{T}})^{-1}$, we find that the Riccati equation for $-(X_1^{\mathsf{T}})^{-1}$ and $-(X_2^{\mathsf{T}})^{-1}$ has the form

$$-\dot{V} = A^{-1} - VCA^{-1} - A^{-1}C^{\mathsf{T}}V + V(CA^{-1}C^{\mathsf{T}} - B)V. \tag{5.46}$$

Condition (5.45) implies that Eq. (5.42) coincides with (5.46). Therefore, all four matrices X_1, X_2, $-(X_1^{\mathsf{T}})^{-1}$, and $-(X_2^{\mathsf{T}})^{-1}$ are solutions to Eq. (5.42). This implies that the double ratio of these matrices is preserved and the quadratic form for determination of the angle made by the planes X_1 and X_2 is hence preserved.

We note that conditions (5.45) are the skew-symmetry conditions for the matrix of coefficients of the corresponding linear Hamiltonian system. Therefore, the mapping along trajectories of the Hamiltonian system is not only symplectic but also orthogonal (see Sec. 4 of Chap. 4); therefore, it preserves the inner product and hence the quadratic form for determination of angles made by any two planes.

Example 5.2. We consider the functional

$$\int (\langle \dot{h}, \dot{h} \rangle + 2\langle C\dot{h}, h \rangle - \langle (I_n + C^2)h, h \rangle)\, dt,$$

where $C^{\mathsf{T}} = -C$.

The Riccati equation for this functional has the form

$$\dot{W} = W^2 + CW - WC + I_n.$$

We write the Euler equation

$$\dot{X} = CX + Y,$$
$$\dot{Y} = -X + CY. \tag{5.47}$$

The preservation of the Euclidean and symplectic structure of the space $\mathbb{R}^{2n}$ under the mapping along trajectories of Euler equation (5.47) implies that the Hermitian structure of this space that arises when the space is identified with

$\mathbb{C}^n$ is preserved. Therefore, we can write Eq. (5.47) as a linear equation in the complex $n \times n$ matrix $Z = X + iY$ whose coefficients are determined by a certain complex $n \times n$ matrix $M = K + iL$:

$$\dot{Z} = \dot{X} + i\dot{Y} = (K + iL)(X + iY). \qquad (5.48)$$

We find K and L. From system (5.48), we obtain

$$\dot{X} = KX - LY, \qquad \dot{Y} = LX + KY$$

$$K = C, \qquad L = -I_n.$$

Therefore,

$$\dot{Z} = (C - iI_n)Z. \qquad (5.49)$$

The complex form of the Riccati equation gives us a possibility to answer the question about conjugate points of functional (5.44).

We first suppose that $C = $ const. Then a solution to Eq. (5.49) has the form $Z = \exp((C + iI_n)t)$. The matrices C and I_n commute, and therefore

$$Z = \exp(Ct)\exp(-it) = \exp(Ct)\cos t - i\exp(Ct)\sin t.$$

Because $\det \exp(Ct) \neq 0$, the conjugate points are zeros of the function $\cos t$ (in this case, the plane corresponding to the solution of the Riccati equation is vertical, and the matrix $\Re Z$ is singular). An analogue of this formula also holds for a variable matrix $C(t)$.

Let a matrix Γ be the fundamental solution to the equation

$$\dot{X} = C(t)X,$$

i.e., $\dot{\Gamma} = C\Gamma$, where $\det \Gamma \neq 0$. Then $Z = \Gamma \cos t - i\Gamma \sin t$ is a solution to Eq. (5.49). Indeed,

$$\dot{Z} = C\Gamma \cos t - iC\Gamma \sin t - \Gamma \sin t - i\Gamma \cos t,$$

$$(C - i)(\Gamma \cos t - i\Gamma \sin t) = C\Gamma \cos t - iC\Gamma \sin t - i\Gamma \cos t - \Gamma \sin t.$$

The conjugate point corresponds to the vertical position of the plane that is defined by the Riccati equation. Introducing the complex variable $Z = X + iY$, we see that canonical variables q correspond to the matrix X and canonical variables p correspond to the matrix Y. Therefore, the condition that the plane contains a vertical direction becomes $\det X = 0$. But $X = \Gamma \cos t$, where $\det \Gamma \neq 0$. Therefore, the distance between conjugate points is equal to π.

Therefore, the existence of an additional symmetry of the Riccati equation allows us to reduce this equation to a complex ordinary differential equation with a quadratic right-hand side, which is again called the *complex Riccati equation*. An advantage of this approach to the study of the Riccati equation consists in the fact that its (complex) dimension is half the dimension of the initial equation. This simple remark shows the importance of studying complex Riccati equations.

Chapter 6
Complex Riccati Equations

In this chapter, we show that depending on the coefficients, the complex Riccati equation has one or another homogeneity domain of the space of several complex variables as its integral manifold. A domain $D \subset \mathbb{C}^n$ is called a *homogeneity domain* if there exists an infinite group of analytic automorphisms of this domain onto itself. Homogeneity domains (Cartan–Siegel domains) often occur in many important branches of calculus (for example, in analytic number theory, automorphic function theory, etc. [16, 50, 82, 99]). We present the main facts related to the homogeneity domains.

§1. Cartan–Siegel Domains

Homogeneity domains first appeared in the studies by E. Cartan [16] as domains D of the space $\mathbb{C}^n$ that possess an infinite group G of analytic automorphisms.

For $n = 1$, the unit disk $|z| < 1$ is a unique (up to analytic isomorphisms) bounded homogeneity domain. The group G of its analytic automorphisms consists of linear-fractional transformations

$$z \mapsto \frac{(az + b)}{(\bar{b}z + \bar{a})},$$

where a and b are complex numbers such that $a\bar{a} - b\bar{b} = 1$.

For $n \geq 2$, the description of all homogeneity domains is a difficult question. The answer becomes comprehensible under the additional assumption that the group G acts *transitively* on the domain D (i.e., G contains transformations that transform any point of D into another point of D). The group G of all analytic automorphisms of a bounded domain $D \subset \mathbb{C}^n$ turns out to be a real Lie group (see [16], p. 134), and the domain D can therefore be described as a quotient space of the Lie group G by the stationary subgroup. Therefore, the domain D becomes an (analytic) homogeneous space. Because of this, such a domain is called a *homogeneity domain* of the space of several complex variables.

We note that to describe all homogeneity domains, it suffices to find only one domain from each equivalence class of domains (with respect to analytic mappings); moreover, it suffices to consider only irreducible classes of homogeneity domains, i.e., the classes that are not direct products of homogeneity domains of lower dimensions.

E. Cartan showed (see [16]) that for $n = 2$, any bounded irreducible homogeneity domain is equivalent to the domain $|z_1|^2 + |z_2|^2 < 1$. Moreover, he conjectured that for $n = 3$, any bounded irreducible homogeneity domain is equivalent to one of the following two domains:

$$|z_1|^2 + |z_2|^2 + |z_3|^2 < 1$$

or

$$Z\overline{Z} < \mathrm{I}_2, \quad Z = \begin{pmatrix} z_1 & z_2 \\ z_2 & z_3 \end{pmatrix}, \quad \mathrm{I}_2 = \begin{pmatrix} 1 & 0 \\ 0 & 1 \end{pmatrix}.$$

For $n \geq 3$, E. Cartan introduced an additional requirement. Namely, he assumed that the domain D is symmetrical in the sense that for each point $z \in D$, the group G contains an involution with a unique fixed point z. (For $n \leq 3$, this assumption holds automatically; see [16].) This assumption allows a classification of all bounded symmetrical domains D.

Remark. In Chap. 4, we considered Riemannian symmetrical spaces corresponding to compact Lie groups. Symmetrical homogeneity domains, *Hermitian symmetrical spaces*, correspond to noncompact Lie groups. (For example, the unit disk $|z| < 1$ corresponds to the noncompact Lie group of its analytic automorphisms. Indeed, this group coincides with $\mathrm{SL}(2, \mathbb{R})$ because the unit disk is analytically isomorphic to the upper half-plane [96]).

Bounded nonsymmetrical homogeneity domains were first described by I. I. Pyatetskii–Shapiro in [83].

Among symmetrical homogeneous domains, we find four types of irreducible domains. Moreover, there is an exceptional domain in each of the spaces $\mathbb{C}^{16}$ and $\mathbb{C}^{27}$, which we do not describe here. Further, when describing these four types of irreducible bounded symmetrical domains, we follow the Siegel presentation [99] (with certain modifications).

Type I. Let p and q be two positive integers, $p + q = m$. We consider the group of complex linear transformations $G_{\mathrm{I}} = \mathrm{U}(p, q)$ of the space $\mathbb{C}^m$ that leave the invariant Hermitian form with p negative and q positive squares invariant:

$$\Omega(t) = -|t_1|^2 - \cdots - |t_p|^2 + |t_{p+1}|^2 + \cdots + |t_m|^2 = t^\top H \bar{t},$$

where

$$t = \begin{pmatrix} t_1 \\ \vdots \\ t_m \end{pmatrix}, \quad H = \begin{pmatrix} -\mathrm{I}_p & 0 \\ 0 & \mathrm{I}_q \end{pmatrix}.$$

In other words, we consider the linear transformations with the matrix $g \in \mathfrak{M}_m(\mathbb{C})$ satisfying the relations $g^T H \bar{g} = H$. We agree to represent the number of rows and columns of the matrix with double superscripts (the first is the number of rows, and the second is the number of columns). We set

$$g = g^{(m,m)} = \begin{pmatrix} A^{(p,p)} & B^{(p,q)} \\ C^{(q,p)} & D^{(q,q)} \end{pmatrix} = \begin{pmatrix} A & B \\ C & D \end{pmatrix}.$$

It is easy to see that $g^{\mathsf{T}} H \bar{g} = H$ iff

$$A^{\mathsf{T}} \bar{A} - C^{\mathsf{T}} \bar{C} = I_p, \qquad D^{\mathsf{T}} \bar{D} - B^{\mathsf{T}} \bar{B} = I_q, \qquad A^{\mathsf{T}} \bar{B} = C^{\mathsf{T}} \bar{D}.$$

In the space $\mathbb{C}^m$, we consider the set $\mathfrak{M}$ of all q-dimensional linear subspaces consisting of vectors for which the Hermitian form Ω assumes positive values. The set $\mathfrak{M}$ can be considered as a subset of the complex Grassmann manifold $G_q(\mathbb{C}^m)$. Moreover, $\mathfrak{M}$ is invariant with respect to the action of the group $U(p, q)$ because this group leaves the form Ω invariant. We consider a chart $\mathfrak{A}$ in the Grassmann manifold $G_q(\mathbb{C}^m)$ in which the q-dimensional plane corresponding to the coordinates $(t_{p+1}, \ldots, t_m)$ serves as the horizontal plane V_0, and the p-dimensional coordinate plane corresponding to the coordinates $(t_1, \ldots, t_p)$ serves as the vertical plane V_∞. The set $\mathfrak{M}$ lies entirely in the chart $\mathfrak{A}$ because any plane $A \in \mathfrak{M}$ does not contain vectors that belong to V_∞ ($\Omega(x)$ is negative for such vectors). We consider an $m \times q$ matrix

$$U = \begin{pmatrix} V^{(p,q)} \\ W^{(q,q)} \end{pmatrix} = \begin{pmatrix} V \\ W \end{pmatrix},$$

composed of vectors of a basis of the plane A (columns of the matrix U are composed of these vectors). Then the Hermitian form with the matrix $U^{\mathsf{T}} H \bar{U}$ is positive definite:[1]

$$U^{\mathsf{T}} H \bar{U} > 0. \tag{6.1}$$

It is easy to see that in terms of the matrices W and V condition (6.1) can be rewritten as

$$W^{\mathsf{T}} \bar{W} > V^{\mathsf{T}} \bar{V}. \tag{6.2}$$

Lemma 6.1. *For any plane $A \in \mathfrak{M}$, the matrix W is nonsingular.*

Proof. If there exists a vector $t \in V_0$ such that $Wt = 0$, then

$$0 = t^{\mathsf{T}} W^{\mathsf{T}} \bar{W} \bar{t} > t^{\mathsf{T}} V^{\mathsf{T}} \bar{V} \bar{t} = (Vt)^{\mathsf{T}} \overline{(Vt)} \geq 0,$$

which is impossible.

On the manifold $G_q(\mathbb{C}^m)$, we consider standard matrix coordinates of q-dimensional planes in the chart $\mathfrak{A}$:

$$Z^{(p,q)} = VW^{-1}.$$

Then condition (6.2) can be written in the form

$$Z^{\mathsf{T}} \bar{Z} < I_q. \tag{6.3}$$

---

[1] We recall that $A > 0$ means that the Hermitian form with the matrix A is positive definite and $A > B$ means that the matrix $A - B$ is positive definite.

Indeed, applying the similarity transformation by the matrix $(W^{-1})^\top$ to the matrix $V^\top \overline{V}$, we obtain

$$(W^{-1})^\top (V^\top \overline{V})(\overline{W})^{-1} = (VW^{-1})^\top (\overline{V}\,\overline{W}^{-1}) = Z^\top \overline{Z}.$$

Lemma 6.2. *The set of points Z satisfying inequality* (6.3) *is bounded in the space* $\mathbb{C}^{q^2}$.

Proof. Inequality (6.3) implies that the Hermitian inner squares of the columns of the matrix Z do not exceed one; hence, the sum of squares of the modules of all entries of the matrix Z does not exceed q.

As for real Grassmann manifolds, a linear automorphism of the space $\mathbb{C}^{p+q}$,

$$\begin{pmatrix} V \\ W \end{pmatrix} \mapsto \begin{pmatrix} A & B \\ C & D \end{pmatrix} \begin{pmatrix} V \\ W \end{pmatrix},$$

induces a transformation of the Grassmann manifold $G_q(\mathbb{C}^m)$,

$$Z \mapsto (AZ + B)(CZ + D)^{-1}, \qquad Z = VW^{-1}. \tag{6.4}$$

Therefore, the bounded domain (6.3) is invariant with respect to the group of generalized linear-fractional transformations (6.4), where

$$A^\top \overline{A} - C^\top \overline{C} = I_p, \qquad D^\top \overline{D} - B^\top \overline{B} = I_q, \qquad A^\top \overline{B} = C^\top \overline{D}. \tag{6.5}$$

It can be shown (see [100]) that this group acts transitively on $\mathfrak{M}$. Therefore, the set $\mathfrak{M}$ is a homogeneity domain. Because the group $U(p,q)$ contains the involution $Z \mapsto -Z$, the domain $\mathfrak{M}$ is a symmetrical domain.

For $q = 1$, the domain $\mathfrak{M}$ is the unit ball of the space $\mathbb{C}^m$.

Type II. We assume that $p = q \geq 2$. We define the subgroup G_{II} of the group $U(p,p)$ consisting of linear transformations of the space $\mathbb{C}^{2p}$ as follows. Let elements of the subgroup G_{II} leave not only the Hermitian form

$$\eta(t) = -|t_1|^2 - \cdots - |t_p|^2 + |t_{p+1}|^2 + \cdots + |t_{2p}|^2 = t^\top H\bar{t}$$

but also the quadratic form

$$\kappa(t) = 2t_1 t_{p+1} + 2t_2 t_{p+2} + \cdots + 2t_p t_{2p} = t^\top K t \tag{6.6}$$

invariant, where

$$K = \begin{pmatrix} 0 & I_p \\ I_p & 0 \end{pmatrix}. \tag{6.7}$$

We thus consider complex linear transformations whose matrices g satisfy the conditions

$$g^\top H \bar{g} = H, \qquad g^\top K g = K. \tag{6.8}$$

If we consider the Hermitian form with the matrix

$$J = K^{-1}H = \begin{pmatrix} 0 & I_p \\ I_p & 0 \end{pmatrix}\begin{pmatrix} -I_p & 0 \\ 0 & I_p \end{pmatrix} = \begin{pmatrix} 0 & I_p \\ -I_p & 0 \end{pmatrix},$$

then conditions (6.8) imposed on the matrix g can be replaced by

$$g^{\mathsf{T}}H\overline{g} = H, \qquad g^{-1}J\overline{g} = J. \tag{6.9}$$

Indeed, (6.8) implies

$$g^{-1}J\overline{g} = g^{-1}K^{-1}H\overline{g} = (g^{\mathsf{T}}Kg)^{-1}(g^{\mathsf{T}}H\overline{g}) = K^{-1}H = J,$$

and (6.9) implies

$$g^{\mathsf{T}}Kg = g^{\mathsf{T}}HJ^{-1}g = (g^{\mathsf{T}}H\overline{g})(g^{-1}J\overline{g})^{-1} = HJ^{-1} = K.$$

We represent the matrix g in the block form

$$g = \begin{pmatrix} A^{(p,p)} & B^{(p,p)} \\ C^{(p,p)} & D^{(p,p)} \end{pmatrix}.$$

It is easy to see that relations (6.9) hold iff

$$B = -\overline{C}, \qquad D = \overline{A}, \qquad A^{\mathsf{T}}\overline{A} - C^{\mathsf{T}}\overline{C} = I_p, \qquad A^{\mathsf{T}}C = -C^{\mathsf{T}}A. \tag{6.10}$$

As in the description of domains of Type I, in the space $\mathbb{C}^{2p}$, we consider the set $\mathfrak{N}$ of all p-dimensional linear subspaces consisting of vectors on which the Hermitian form η assumes positive values. Then $\mathfrak{N}$ can be considered as a subset of the complex Grassmann manifold $G_p(\mathbb{C}^{2p})$. The set $\mathfrak{N}$ is invariant with respect to the action of the group G_{II} because this group leaves the form η invariant. Further, we note that the quadratic cone $t^{\mathsf{T}}Kt = 0$ in the space $\mathbb{C}^{2p}$ contains the family of p-dimensional linear subspaces. For example, this family contains a subspace defined by the set of equations $t_1 = t_2 = \cdots = t_p = 0$. We let $\mathfrak{P}$ denote the submanifold of the manifold $G_p(\mathbb{C}^{2p})$ consisting of elements of the family considered. As a homogeneity domain of type II, we take the intersection

$$\mathfrak{Q} = \mathfrak{P} \cap \mathfrak{N}.$$

We consider a chart $\mathfrak{A}$ on the manifold $G_p(\mathbb{C}^{2p})$ in which the plane V_0 corresponding to the coordinates $(t_{p+1}, \ldots, t_{2p})$ serves as the horizontal plane and the plane V_∞ corresponding to the coordinates $(t_1, \ldots, t_p)$ serves as the vertical plane. The set $\mathfrak{N}$ lies entirely in the chart $\mathfrak{A}$ because, by definition, any plane $A \in \mathfrak{N}$ does not contain vectors belonging to V_∞.

We consider the $p \times p/2$ matrix

$$U = \begin{pmatrix} V^{(p,p)} \\ W^{(p,p)} \end{pmatrix} = \begin{pmatrix} V \\ W \end{pmatrix}$$

composed of vectors of a basis of the plane A (these vectors form the columns of the matrix U). Then the Hermitian form with the matrix $U^{\mathsf{T}}H\overline{U}$ is positive definite,

$$U^\top H \overline{U} > 0,$$

and the matrix $U^\top K U$ equals zero. In terms of the matrices W and V, these conditions mean that

$$W^\top \overline{W} > V^\top \overline{V}, \qquad V^\top W = -W^\top V. \tag{6.11}$$

Obviously, the mappings $U \mapsto gU$ preserve the matrix $U^\top H \overline{U}$ as well as the matrix $U^\top K U$, i.e., transform the considered set of matrices U into itself. By Lemma 6.1, the matrix W is nonsingular, and we can introduce the matrix coordinates $Z = VW^{-1}$ in the domain $\mathfrak{Q}$ of the manifold $G_p(\mathbb{C}^{2p})$. Relations (6.11) yield the following conditions on the matrix $Z \in \mathfrak{Q}$:

$$Z^\top \overline{Z} < I_p, \qquad Z^\top = -Z. \tag{6.12}$$

In accordance with Lemma 6.2, relations (6.12) define a bounded domain $\mathfrak{Q}$ in the $(p(p-1)/2)$-dimensional space of entries of complex skew-symmetrical matrices Z. The domain $\mathfrak{Q}$ is invariant with respect to the group G of analytic transformations

$$Z \mapsto (AZ + B)(CZ + D)^{-1},$$

where A, B, C, and D satisfy relations (6.10). It can be proved that this group acts transitively on $\mathfrak{Q}$ [100]. The domain $\mathfrak{Q}$ is symmetrical because the group G contains the involution $Z \mapsto -\overline{Z}$.

Type III. This type of domain is of the greatest importance because exactly this type corresponds to the complex Riccati equations arising from problems in the calculus of variations.

Let $p = q$. We consider the following subgroup G_{III} of the group $U_{p,p}$. Elements of the subgroup G_{III} leave both the Hermitian form

$$\eta(t) = -|t_1|^2 - \cdots - |t_p|^2 + |t_{p+1}|^2 + \cdots + |t_{2p}|^2 = t^\top H \bar{t}$$

and the skew-symmetrical form

$$(t_1 u_{p+1} - u_1 t_{p+1}) + \cdots + (t_p u_{2p} - u_p t_{2p}) = t^\top J u \tag{6.13}$$

defining the symplectic structure in the space $\mathbb{C}^{2p}$ invariant.

We therefore consider linear transformations with the matrix g satisfying the relations

$$g^\top H \bar{g} = H, \qquad g^\top J g = J. \tag{6.14}$$

For the matrix g, which is written in the block form

$$g = \begin{pmatrix} A^{(p,p)} & B^{(p,p)} \\ C^{(p,p)} & D^{(p,p)} \end{pmatrix},$$

relations (6.14) are equivalent to

$$B = \overline{C}, \qquad D = \overline{A}, \qquad A^\top \overline{A} - C^\top \overline{C} = I_p, \qquad A^\top C = C^\top A. \tag{6.15}$$

As above, let

$$U = \begin{pmatrix} V^{(p,p)} \\ W^{(p,p)} \end{pmatrix}$$

range over the set of all matrices having $2p$ rows and p columns such that $U^\top H \overline{U} > 0$ and $U^\top J U = 0$. In other words, the matrix U is formed by a basis of the Lagrangian plane for whose vectors the Hermitian form η assumes positive values.

Rewriting these conditions in terms of the matrices V and W, we obtain $W^\top \overline{W} > V^\top \overline{V}$ and $V^\top W = W^\top V$. The mappings $U \mapsto gU$ transform the set of such matrices U into itself. By Lemma 6.1, the matrix W is nonsingular, and the matrix coordinates $Z = VW^{-1}$ on the corresponding Lagrangian manifold are varied in a bounded (by Lemma 6.2) domain in the $(p(p+1)/2)$-dimensional complex space whose coordinates are entries of the complex symmetrical matrix Z. This domain is denoted by $\mathfrak{R}$. Therefore,

$$\mathfrak{R} = \{Z^\top \overline{Z} < I_p, \ Z^\top = Z\}. \tag{6.16}$$

The group G of analytic automorphisms

$$Z \mapsto (AZ + B)(CZ + D)^{-1}$$

of the domain $\mathfrak{R}$ is described by conditions (6.15). It can be proved that this group acts transitively on $\mathfrak{R}$ [100]. The domain $\mathfrak{R}$ is symmetrical because the group G contains the involution $Z \mapsto -Z$.

The domain $\mathfrak{R}$ is called the *generalized unit disk*. For $p = 1$, the matrix Z is one-dimensional. Therefore, the condition $Z^\top = Z$ holds automatically, and the domain $\mathfrak{R}$ is the ordinary unit disk in the complex plane $\mathbb{C}^1$. We show below that the domain $\mathfrak{R}$ is analytically equivalent to the generalized Siegel half-plane.

Type IV. Let p be a positive integer. We consider the group G of linear transformations with real coefficients that leave the quadratic form

$$\zeta(t) = t_1^2 + t_2^2 - t_3^2 - \cdots - t_{p+2}^2 = t^\top H t$$

invariant; here

$$t = \begin{pmatrix} t_1 \\ t_2 \\ \vdots \\ t_{p+2} \end{pmatrix}, \qquad H = \begin{pmatrix} I_2 & 0 \\ 0 & -I_p \end{pmatrix}.$$

In other words, we consider the group of linear transformations whose matrices g satisfy the equations $g = \overline{g}$ and $g^\top H g = H$. This group is usually denoted by $O(p, 2)$. The action of the group $O(p, 2)$ on the $(p+2)$-dimensional complex space $\mathbb{C}^{p+2}$ leaves the cone

$$t^\top H t = t_1^2 + t_2^2 - t_3^2 - \cdots - t_{p+2}^2 = 0 \tag{6.17}$$

and the domain

$$t^\top H\bar{t} > 0 \tag{6.18}$$

invariant.

The points lying in the set (6.17), (6.18) are such that the complex variables t_1 and t_2, considered as points of a two-dimensional plane, are linearly independent over $\mathbb{R}$. Indeed, otherwise, $|t_1|^2 + |t_2|^2 = |t_1^2 + t_2^2|$ and

$$|t_3|^2 + \cdots + |t_{p+2}|^2 < |t_3^2 + \cdots + t_{p+2}^2|,$$

which is impossible.

Therefore, the imaginary part of the ratio t_1/t_2 is different from zero, and the transformations of the group $O(p, 2)$ either preserve the sign of this imaginary part for all points of the set (6.17), (6.18) or change it to the opposite sign for all the points. The linear transformations preserving the sign of the imaginary part of t_1/t_2 therefore form a subgroup G of index 2 in $O(p, 2)$. This subgroup acts on the set defined by

$$t^\top Ht = 0, \qquad t^\top H\bar{t} > 0, \qquad \Im(t_1/t_2) > 0 \tag{6.19}$$

and is denoted by G_{IV} in what follows.

We introduce new coordinates on the set defined by relations (6.19). To do this, we divide the first relation in (6.19) by $(t_1 + it_2)^2$. We then obtain

$$\left(\frac{t_3}{t_1 + it_2}\right)^2 + \cdots + \left(\frac{t_{p+2}}{t_1 + it_2}\right)^2 = \frac{t_1 - it_2}{t_1 + it_2}. \tag{6.20}$$

Dividing the second relation in (6.19) by $|t_1 + it_2|^2$ and noting that

$$2(|t_1|^2 + |t_2|^2) = |t_1 - it_2|^2 + |t_1 + it_2|^2$$

(the sum of the squares of the diagonals of a parallelogram is equal to the sum of the squares of its sides), we obtain

$$\left|\frac{t_3}{t_1 + it_2}\right|^2 + \cdots + \left|\frac{t_{p+2}}{t_1 + it_2}\right|^2 < \frac{1}{2}\left(1 + \left|\frac{t_1 - it_2}{t_1 + it_2}\right|^2\right).$$

The third relation in (6.19) means that

$$\left|\frac{t_1 - it_2}{t_1 + it_2}\right| < 1.$$

We make the change of variables

$$z_k = \frac{t_{k+2}}{t_1 + it_2}, \qquad z_{p+1} = \frac{t_1 - it_2}{t_1 + it_2}, \qquad k = 1, \ldots, p. \tag{6.21}$$

The sense of this change consists in the following. In the space $\mathbb{C}^{p+2}$, relations (6.19) isolate a part of the manifold $t^\top Ht = 0$ that lies in the domain

$$t^\top H \bar{t} > 0, \qquad \Im(t_1/t_2) > 0.$$

We note that relations (6.19) are homogeneous in the variables t; therefore, introducing the variables $t_1 - it_2$ and $t_1 + it_2$ instead of t_1 and t_2 and dividing all coordinates by $t_1 + it_2$, we in fact introduce affine coordinates in the complex projective space $\mathbb{CP}^{p+1}$, and condition (6.20) isolates a certain hypersurface in this affine space whose equation in the new coordinates becomes

$$z_{p+1} = z_1^2 + \cdots + z_p^2. \tag{6.22}$$

Under change of variables (6.21), the set defined by relations (6.19) transforms into the bounded domain $\mathfrak{T} \in \mathbb{CP}^p$,

$$|z_1|^2 + \cdots + |z_p|^2 < \frac{1}{2}\{1 + |z_1^2 + \cdots + z_p^2|^2\} < 1; \tag{6.23}$$

moreover, the group G_{IV} leaves manifold (6.22) invariant. Equation (6.22) defines z_{p+1} as a single-valued function of the other coordinates. We can therefore assume that G_{IV} acts on the domain $\mathfrak{T}$ given by inequalities (6.23). Indeed, let $g \in G_{\mathrm{IV}}$, and let $\pi \colon \mathfrak{T} \to \mathbb{CP}^p$ be the projection of manifold (6.22) that ignores the last coordinate, that is,

$$\pi \colon (z_1, \ldots, z_p, z_{p+1}) \mapsto (z_1, \ldots, z_p).$$

Then the group consisting of the mappings $\pi \circ g \circ \pi^{-1}$ is well defined and acts on $\mathfrak{T}$. The obtained group is transitive, and the mapping $z \mapsto -z$ transforms $\mathfrak{T}$ into a symmetrical homogeneity domain.

§2. Klein–Poincaré Upper Half-Plane and Generalized Siegel Upper Half-Plane

We return to the homogeneity domain of type III, which is the most interesting to us. We begin by recalling the generalized Klein–Poincaré half-plane.

As is known, the simplest homogeneity domain in the space $\mathbb{C}^1$, the unit disk, can be conformally mapped onto the upper half-plane $\Im z > 0$. In this case, the group of analytic automorphisms of the unit disk passes to the group $\mathrm{SL}(2, \mathbb{R})$, which acts by linear-fractional transformations that transform the upper half-plane into itself.

In Chap. 1, we considered the upper half-plane with the metric

$$ds^2 = \frac{dx^2 + dy^2}{y^2}. \tag{6.24}$$

This metric is invariant under the action of the group $\mathrm{SL}(2, \mathbb{R})$. It is easy to verify that the Gaussian curvature of metric (6.24) is constant and negative; therefore, we can consider the upper half-plane a model of the Lobachevsky plane. This model is usually called the *Klein–Poincaré half-plane*. Metric (6.24) is the real part of the invariant Hermitian metric $dz\,d\bar{z}/y^2$, whose imaginary part vanishes.

Therefore, the Klein–Poincaré upper half-plane is the set

$$P = \{z \in \mathbb{C} \mid \Im z > 0\}$$

with the metric

$$ds^2 = \frac{dx^2 + dy^2}{y^2} = \frac{dz\,d\bar{z}}{y^2}.$$

The group of linear-fractional transformations $\mathrm{SL}(2, \mathbb{R})$ (this is also $\mathrm{Sp}(1, \mathbb{R})$) acts on the Klein–Poincaré half-plane,

$$\begin{pmatrix} \alpha & \beta \\ \gamma & \delta \end{pmatrix}(z) = \frac{\gamma + \delta z}{\alpha + \beta z},$$

which transforms the upper half-plane into itself and preserves the metric ds^2.

The line $y = 0$ (which does not belong to the Lobachevsky plane) is called its *absolute*.

As is known [96], the group of linear-fractional transformations is the group of analytic automorphisms of the Riemann sphere. The upper half-plane can be considered the image of the hemisphere η under the stereographic projection. Under this mapping, the group of linear-fractional transformations, which transform the hemisphere η into itself, passes to the group $\mathrm{SL}(2, \mathbb{R})$. The restriction of the group acting on the sphere to the boundary of the hemisphere induces the action of the group $\mathrm{SL}(2, \mathbb{R})$ on the absolute.

In this case, each transformation $g \in \mathrm{SL}(2, \mathbb{R})$ is everywhere defined, except for one point at which the numerator of the linear-fractional transformation vanishes (under the action on the sphere, the image of the corresponding point is the north pole).

Therefore, when the action of $\mathrm{SL}(2, \mathbb{R})$ on the absolute is considered, it is natural to add an infinitely distant point to it, thus transforming the absolute into the projective line. Then any linear-fractional transformation is a one-to-one mapping of this projective line onto itself.

The set $\mathrm{SL}(2, \mathbb{Z})$ of second-order matrices whose entries are integers and whose determinant is equal to one is a subgroup of the group $\mathrm{SL}(2, \mathbb{R})$ because the inverse matrices are also integer valued.

The significance of the group $\mathrm{SL}(2, \mathbb{Z})$ was determined in the famous K. Gauss paper "Disquisitiones arithmeticae." When solving quadratic Diophantine equations, more precisely, when studying the problems of the representability of integers by values of a quadratic form in integer arguments, we can proceed as follows. We associate a lattice $\mathfrak{D}$ on the plane with a given quadratic form (which is in general scalene); the problem then consists in the following: does a circle with an integer radius centered at one of the nodes of the lattice touch the nodes of this lattice? Gauss considered such a lattice as a lattice in $\mathbb{C}$. This geometric interpretation of complex numbers as points of the complex plane, which is customary now, was discovered by Gauss specifically for these purposes. He set and solved the problem of the classification of lattices from the arithmetical viewpoint. We take one of the nodes of the lattice as the origin. Then we choose one of the nodes ω_1 that is nearest to the origin and direct the real axis to it. As the identity, we take the distance from the origin to this node. Also, we choose one more node that does not belong to the real axis and represent it by ω_2. Therefore, a lattice is defined by a complex number $\omega = \omega_1/\omega_2$, but the choice of ω is not unique for a given lattice. First, we can suppose that $\Im \omega > 0$. Second, we can consider other base nodes of the lattice $\mathfrak{O}$ that have the form

$$\widetilde{\omega}_1 = \alpha \omega_1 + \beta \omega_2, \qquad \widetilde{\omega}_2 = \gamma \omega_1 + \delta \omega_2,$$

where

$$\det \begin{pmatrix} \alpha & \beta \\ \gamma & \delta \end{pmatrix} = 1, \quad \alpha, \beta, \gamma, \delta \in \mathbb{Z}.$$

These nodes define the same lattice. In other words, the group $\mathrm{SL}(2, \mathbb{Z})$ leaves the lattice $\mathfrak{O}$ invariant. Therefore, the set of nonequivalent lattices or, equivalently, the set of distinct elliptic curves can be identified with the quotient group $\mathrm{SL}(2, \mathbb{R})/\mathrm{SL}(2, \mathbb{Z})$. This quotient group is itself a smooth curve, which is called a *modular curve* because it is a moduli manifold of elliptic curves.

Points of the modular curve are parameterized by the parameter $\omega = \omega_1/\omega_2$ such that $\Im \omega > 0$; moreover, $\widetilde{\omega}$ yields the same lattice as ω whenever

$$\widetilde{\omega} = \frac{\alpha \omega + \beta}{\gamma \omega + \delta}, \quad \det \begin{pmatrix} \alpha & \beta \\ \gamma & \delta \end{pmatrix} = 1, \quad \alpha, \beta, \gamma, \delta \in \mathbb{Z}.$$

We can associate with each point of the modular curve, or what is the same, with each lattice the following *modular functions* of the complex parameter ω:

$$g_2(\omega) = 60 \sum_{\substack{\zeta_i \in \mathfrak{O}, \\ \zeta_i \neq 0}} \frac{1}{\zeta_i^4}, \qquad g_3(\omega) = 140 \sum_{\substack{\zeta_i \in \mathfrak{O}, \\ \zeta_i \neq 0}} \frac{1}{\zeta_i^6}.$$

Obviously, these functions are invariant under the action of the group $\mathrm{SL}(2, \mathbb{Z})$. We introduce the Weierstrass function

$$\wp(z; \omega) = \frac{1}{z^2} + \sum_{\substack{\zeta_i \in \mathfrak{O}, \\ \zeta_i \neq 0}} \left[\frac{1}{(z + \zeta_i)^2} - \frac{1}{\zeta_i^2} \right].$$

The function $\wp$ is double periodic and meromorphic (at the nodes of the lattice $\mathfrak{O}$, the Weierstrass function has poles of the second order).

Proposition 6.1. *The function $\wp(z, \omega)$ satisfies the differential equation*

$$\wp'^2 - 4\wp^3 + g_2\wp + g_3 = 0. \tag{6.25}$$

Proof. Differentiating the function $\wp$ in z, we obtain

$$\frac{d\wp}{dz}(z,\omega) = -\frac{2}{z^3} - \sum_{\substack{\zeta_i \in \mathfrak{D}, \\ \zeta_i \neq 0}} \frac{2}{(z + \zeta_i)^3}.$$

We expand the functions $\wp$ and $\wp'$ into the Laurent series at zero. Substituting these series in the left-hand side of Eq. (6.25), we easily show that the coefficients of the leading terms of the obtained expansion vanish. Therefore, the left-hand side of Eq. (6.25) is a double-periodic analytic function that has no singularities at the period parallelogram. Therefore, this function is bounded and is constant by the Liouville theorem. Because it vanishes at zero, Eq. (6.25) holds identically.

Therefore, the mapping

$$z \mapsto \big(\wp(z), \wp'(z)\big)$$

transforms the period parallelogram into a complex curve given by the equation

$$y^2 = 4x^3 - g_2 x - g_3. \tag{6.26}$$

In this case, the nodes of the period parallelogram pass to an infinitely distant point of this curve, Therefore, the compactification of the curve (6.26) is a torus.

2.1. Generalized Siegel Upper Half-Plane. We return to the main line of our discussion. To reveal the nature of the complex Riccati equation, we need another realization of the Cartan–Siegel homogeneity domain of type III, an unbounded domain in the space of several complex variables, which is a natural generalization of the Klein–Poincaré upper half-plane.

The *generalized Siegel upper-half plane* is the set of complex symmetrical $n \times n$ matrices with a positive imaginary part,

$$\mathfrak{S}_n = \{X + iY \mid X^\top = X, \, Y^\top = Y, \, Y > 0\}.$$

This set forms an open domain $\mathfrak{S}_n$ in the space of complex symmetrical matrices $\mathbb{C}^{n(n+1)/2}$ and is therefore an analytic manifold. The boundary of this manifold consists of those $z \in \mathbb{C}^{n(n+1)/2}$ for which the condition $Y > 0$ is first violated when we approach these points along any curve lying in the domain $\mathfrak{S}_n$. At these points, the matrix Y becomes *semidefinite*, i.e., positive semidefinite and, moreover, degenerate. The order of degeneration can be arbitrary.

In accordance with this, the boundary $s = \partial \mathfrak{S}_n$ of the manifold $\mathfrak{S}_n$ turns out to be a *stratified manifold*. It consists of a finite number of pieces of different dimensions, which are called *strata*. Strata adjoin each other such that the boundary of each stratum consists of strata of lower dimensions. For

$\mathfrak{S}_n$, each stratum s_k, $k = 0, 1, \ldots, n - 1$, consists of complex symmetrical matrices of rank k with nonnegative imaginary parts:

$$s_k = \{X + iY \mid X^\top = X,\ Y^\top = Y,\ Y \geq 0,\ \operatorname{rk} Y = k\}.$$

2.2. Siegel Half-Plane as a Symmetrical Space.

Theorem 6.1. *The generalized Siegel half-plane $\mathfrak{S}_n$ is a homogeneous symmetrical Hermitian manifold.*

To prove Theorem 6.1, we need several auxiliary assertions.

First, we find a group of analytic automorphisms that act transitively on $\mathfrak{S}_n$. We consider the action of generalized linear-fractional transformations on $\mathfrak{S}_n$. Similar to the one-dimensional case, where the generalized upper half-plane coincides with the usual upper half-plane $\{z \mid \Im z > 0\}$, we define the action of the block matrix

$$\begin{pmatrix} A & B \\ C & D \end{pmatrix}$$

on elements ζ of the space $\mathfrak{S}_n$ as

$$\begin{pmatrix} A & B \\ C & D \end{pmatrix} (\zeta) = (C + D\zeta)(A + B\zeta)^{-1}.$$

In order that real matrices ζ pass to real matrices, we consider real $n \times n$ matrices as $A, B, C,$ and D.

We consider three forms of such transformations:

Form 1.
$$\begin{pmatrix} A & B \\ C & D \end{pmatrix} = \begin{pmatrix} 1 & 0 \\ P & 1 \end{pmatrix}$$

Under this transformation, $\zeta \mapsto \zeta + P$. The domain $\mathfrak{S}_n$ passes to itself iff the matrix P is symmetrical: $P^\top = P$.

Form 2.
$$\begin{pmatrix} A & B \\ C & D \end{pmatrix} = \begin{pmatrix} (V^\top)^{-1} & 0 \\ 0 & V \end{pmatrix}, \quad V \in \operatorname{GL}(n, \mathbb{R})$$

In this case, $\zeta \mapsto V\zeta V^\top$. Obviously, under this action, symmetrical matrices pass to symmetrical ones, and the imaginary part of the matrix ζ remains positive definite, i.e., $\mathfrak{S}_n$ passes to $\mathfrak{S}_n$.

Form 3.
$$\begin{pmatrix} A & B \\ C & D \end{pmatrix} = J = \begin{pmatrix} 0 & I_n \\ -I_n & 0 \end{pmatrix}$$

In this case, $\zeta \mapsto -\zeta^{-1}$.

Lemma 6.3. *A transformation of form 3 is defined on the entire $\mathfrak{S}_n$. The image of this mapping lies in $\mathfrak{S}_n$.*

Proof. Indeed, let $X + iY \in \mathfrak{S}_n$. Because $Y > 0$, there exists a matrix $V \in \mathrm{GL}(n)$ such that $VYV^\top = \mathrm{I}$. Then $V\zeta V^\top = T + i\,\mathrm{I}$.

We begin with the identity

$$T^2 + \mathrm{I} = (T + i\,\mathrm{I})(T - i\,\mathrm{I}). \tag{6.27}$$

By Lemma 5.2, the matrix $T^2 + \mathrm{I}$ is positive definite and is hence invertible. Therefore, both factors in the right-hand side of Eq. (6.27) are nonsingular; hence, the matrix ζ is nonsingular. Further,

$$
\begin{aligned}
-\zeta^{-1} &= (-V^{-1}(T + i\,\mathrm{I})(V^\top)^{-1})^{-1} \\
&= -V^\top (T + i\,\mathrm{I})^{-1} V = V^\top (i\,\mathrm{I} - T)(T^2 + \mathrm{I})^{-1} V.
\end{aligned}
$$

Therefore, the matrix $(-\zeta^{-1})$ is symmetrical, and its imaginary part is positive definite.

Proposition 6.2. *The group generated by transformations of forms 1–3 coincides with the real symplectic group* $\mathrm{Sp}(n, \mathbb{R})$.

Before proving this assertion, we recall that $\mathrm{Sp}(n, \mathbb{R})$ is the group that leaves the skew-symmetrical quadratic form with the matrix

$$J = \begin{pmatrix} 0 & \mathrm{I}_n \\ -\mathrm{I}_n & 0 \end{pmatrix}$$

invariant. This means that

$$g = \begin{pmatrix} A & B \\ C & D \end{pmatrix} \in \mathrm{Sp}(n, \mathbb{R})$$

iff $gJg^\top = J$. We rewrite the latter relation in terms of $n \times n$ blocks of the matrix g:

$$\begin{pmatrix} A & B \\ C & D \end{pmatrix} J \begin{pmatrix} A^\top & C^\top \\ B^\top & D^\top \end{pmatrix} = \begin{pmatrix} -BA^\top + AB^\top & -BC^\top + AD^\top \\ -DA^\top + CB^\top & -DC^\top + CD^\top \end{pmatrix}.$$

From this, we obtain the following necessary and sufficient conditions for a matrix g to belong to the group $\mathrm{Sp}(n, \mathbb{R})$:

$$AB^\top = BA^\top, \tag{6.28}$$

$$CD^\top = DC^\top, \tag{6.29}$$

$$AD^\top - BC^\top = \mathrm{I}_n. \tag{6.30}$$

Proof of Proposition 6.2. Let G denote the group generated by transformations of forms 1–3. It is easy to verify that the matrix of each of these transformations is symplectic.

We consider an arbitrary matrix

$$g = \begin{pmatrix} A & B \\ C & D \end{pmatrix} \in \mathrm{Sp}(n, \mathbb{R}).$$

To prove the assertion, it suffices to show that premultiplying or postmultiplying g by matrices of forms 1–3, we obtain matrices of forms 1–3. We divide our actions into three stages.

Stage I. We apply a transformation of form 2 to the matrix g from the left and right:

$$\begin{pmatrix} (V^\top)^{-1} & 0 \\ 0 & V \end{pmatrix} \begin{pmatrix} A & B \\ C & D \end{pmatrix} \begin{pmatrix} (V_1^\top)^{-1} & 0 \\ 0 & V_1 \end{pmatrix}.$$

Under this transformation, the left upper corner A of the matrix g is premultiplied and postmultipled by nonsingular matrices. We can choose the matrices V and V_1 such that the matrix g becomes the matrix

$$g_1 = \begin{pmatrix} A_1 & B_1 \\ C_1 & D_1 \end{pmatrix},$$

where

$$A_1 = \begin{pmatrix} I_r & 0 \\ 0 & 0 \end{pmatrix}$$

and r is the rank of the matrix A.

Stage II. Let

$$B_1 = \begin{pmatrix} \beta_1 & \beta_2 \\ \beta_3 & \beta_4 \end{pmatrix}$$

be the partition of the matrix B_1 into blocks of the same dimension as in the matrix A_1. We have

$$A_1 B_1^\top = \begin{pmatrix} \beta_1^\top & \beta_3^\top \\ 0 & 0 \end{pmatrix}, \qquad B_1 A_1^\top = \begin{pmatrix} \beta_1 & 0 \\ \beta_3 & 0 \end{pmatrix}.$$

Because $g_1 \in \mathrm{Sp}(n, \mathbb{R})$,

$$\beta_3 = 0, \qquad \beta_1 = \beta_1^\top \tag{6.31}$$

follows from (6.28). It follows from (6.31) that

$$\det \beta_4 \neq 0 \tag{6.32}$$

because otherwise the first n rows of the matrix g_1 would be linearly dependent.

We use a transformation of form 1 and postmultiply g_1 by the matrix

$$\begin{pmatrix} I_n & 0 \\ \lambda I_n & I_n \end{pmatrix},$$

where λ is a real number. We obtain the matrix g_2 whose left upper minor has the form

$$A_2 = \begin{pmatrix} I_r + \lambda\beta_1 & \lambda\beta_2 \\ 0 & \lambda\beta_4 \end{pmatrix}.$$

By (6.32), for a sufficiently small λ, we have $\det A_2 \neq 0$. Repeating the arguments of stage I, we reduce the matrix g_2 to the form

$$g_3 = \begin{pmatrix} A_3 & B_3 \\ C_3 & D_3 \end{pmatrix},$$

where $A_3 = I_n$. Applying relation (6.28) to the matrix g_3, we find that the matrix B_3 is symmetrical.

Stage III. Because B_3 is symmetrical, we have

$$J^{-1} \begin{pmatrix} I_n & 0 \\ B_3 & I_n \end{pmatrix} J = \begin{pmatrix} I_n & -B_3 \\ 0 & I_n \end{pmatrix} \in G. \tag{6.33}$$

We postmultiply

$$g_3 = \begin{pmatrix} I_n & B_3 \\ C_3 & D_3 \end{pmatrix}$$

by (6.33). We obtain the matrix

$$g_4 = \begin{pmatrix} I_n & 0 \\ C_4 & D_4 \end{pmatrix}. \tag{6.34}$$

Sequentially applying relations (6.30) and (6.29) to the matrix g_4, we obtain $D_4 = I_n$ and $C_4^\top = C_4$.

Therefore, as a result of applying transformations of forms 1–3 to the matrix g, we obtain a matrix of form 1; this completes the proof of the assertion.

We have thus defined the action of the group $\mathrm{Sp}(n, \mathbb{R})$ on the generalized Siegel upper half-plane $\mathfrak{S}_n$.

Lemma 6.4. *The group* $\mathrm{Sp}(n, \mathbb{R})$ *acts transitively on* $\mathfrak{S}_n$.

Proof. It suffices to show that the point $i\,I_n \in \mathfrak{S}_n$ can be transformed into any point $X + iY \in \mathfrak{S}_n$.

First, we note that the condition $Y > 0$ implies that the quadratic form with the matrix Y can be reduced to the sum of squares, i.e., there exists a nonsingular matrix A such that $Y = AA^\top$. Then the application of a transformation of form 2, where $V = A$, transforms the matrix $i\,I_n$ into iY.

Using a transformation of form 1, we transform the real part of the matrix iY into the matrix X.

Therefore, $\mathrm{Sp}(n, \mathbb{R})$ is a transitive group of analytic transformations of the space $\mathfrak{S}_n$. To represent $\mathfrak{S}_n$ in the form of a homogeneous space, we must find the stabilizer of a certain point of $\mathfrak{S}_n$. It is easiest to find the stabilizer of the point $i\,I$.

Lemma 6.5. *The stationary subgroup of the group* $\mathrm{Sp}(n, \mathbb{R})$ *at the point* $i\,\mathrm{I}$ *coincides with*

$$\mathrm{Sp}(n) = \mathrm{SO}(2n) \cap \mathrm{Sp}(n, \mathbb{R}). \tag{6.35}$$

Group (6.35) is called the *unitary restriction* of the symplectic group or the *compact symplectic group*.

Proof. Let the matrix

$$\begin{pmatrix} A & B \\ C & D \end{pmatrix}$$

leave the point $i\,\mathrm{I} \in \mathfrak{S}_n$ fixed:

$$\begin{pmatrix} A & B \\ C & D \end{pmatrix} \circ (i\,\mathrm{I}) = (C + iD)(A + iB)^{-1} = i\,\mathrm{I}.$$

Then $C + iD = Ai - B$; therefore, $C = -B$ and $D = A$. Therefore, the stabilizer of the point $i\,\mathrm{I}$ consists of matrices of the form

$$\begin{pmatrix} A & B \\ -B & A \end{pmatrix}.$$

Because these matrices are symplectic, by (6.28)–(6.30), we have

$$AA^\top + BB^\top = \mathrm{I}, \qquad AB^\top = BA^\top.$$

These conditions can be rewritten in the matrix form

$$\begin{pmatrix} A & -B \\ B & A \end{pmatrix} \begin{pmatrix} A^\top & B^\top \\ -B^\top & A^\top \end{pmatrix} = \begin{pmatrix} \mathrm{I} & 0 \\ 0 & \mathrm{I} \end{pmatrix}.$$

We have thus proved that matrices that leave the point $i\,\mathrm{I}$ fixed are orthogonal. The determinant of any symplectic matrix is equal to one.

Proof of Theorem 6.1. Proposition 6.2 and Lemmas 6.3–6.5 imply that the generalized Siegel upper half-plane is the homogeneous space

$$\mathfrak{S}_n = \mathrm{Sp}(n, \mathbb{R})/(\mathrm{SO}(2n) \cap \mathrm{Sp}(n, \mathbb{R})).$$

To prove that $\mathfrak{S}_n$ is a homogeneous Hermitian space, we describe the Hermitian metric on the space $\mathfrak{S}_n$ that is invariant with respect to the action $\mathrm{SU}(n)$. This metric was first described by C. Siegel in [100].

We consider two points S_1 and S_2 of the space $\mathfrak{S}_n$. We define the expression

$$\sigma(S_1, S_2) = (\overline{S}_1 - S_2)(\overline{S}_1 - \overline{S}_2)^{-1}(S_1 - \overline{S}_2)(S_1 - S_2)^{-1}.$$

We note that this expression is similar to matrix double ratio (5.1), which is applied to the matrices S_1, S_2, $\overline{S}_1$, and $\overline{S}_2$. However, the function $\sigma(S_1, S_2)$ is not the double ratio of the quadruple of points S_1, $\overline{S}_1$, S_2, and $\overline{S}_2$ of the Siegel half-plane because if $S_1, S_2 \in \mathfrak{S}_n$, then $\overline{S}_1, \overline{S}_2 \notin \mathfrak{S}_n$.

Siegel proved in [100] that the expression $\sigma(S_1, S_2)$ is invariant with respect to the action of the group $\mathrm{Sp}(n, \mathbb{R})$ on $\mathfrak{S}_n$. We note that when S_2 tends to S_1, the factors $\overline{S}_1 - \overline{S}_2$ and $S_1 - S_2$ tend to zero, and the factors $\overline{S}_1 - S_2$ and $S_1 - \overline{S}_2$ tend to $-2\Im S_1$ and $2\Im S_1$ respectively.

We let $Z = X + iY$ denote the matrix coordinates of the current point in $\mathfrak{S}_n$. We introduce the Hermitian form[2]

$$ds^2 = \mathrm{Tr}\{Y^{-1}(d\overline{Z})Y^{-1}(dZ)\}. \tag{6.36}$$

For $n = 1$, formula (6.36) assigns the metric

$$ds^2 = \frac{dx^2 + dy^2}{y^2} = \frac{dz \wedge d\overline{z}}{y^2}$$

on the Lobachevsky plane in the Klein–Poincaré model.

Proposition 6.3. (See [100].) *The action of* $\mathrm{Sp}(n, \mathbb{R})$ *on* $\mathfrak{S}_n$ *leaves the Hermitian form* ds^2 *invariant.*

Proof. We recall that by Proposition 6.2, symplectic transformations are generated by transformations of three different forms. We show that metric (6.36) is not changed under these transformations.

1. The mapping $X + iY \to (X + P) + iY$, $P^{\mathsf{T}} = P$, obviously does not change ds^2.

2. Under the mapping $Z \mapsto A(X + iY)A^{\mathsf{T}}$, we have

$$dZ \mapsto A\,dZ\,A^{\mathsf{T}}, \qquad d\overline{Z} \mapsto A\,d\overline{Z}\,A^{\mathsf{T}},$$
$$Y \mapsto AYA^{\mathsf{T}}, \qquad Y^{-1} \mapsto (A^{\mathsf{T}})^{-1}Y^{-1}A^{-1}.$$

Therefore,

$$ds^2 \mapsto \mathrm{Tr}\{(A^{\mathsf{T}})^{-1}Y^{-1}A^{-1}A\,d\overline{Z}\,A^{\mathsf{T}}(A^{\mathsf{T}})^{-1}Y^{-1}A^{-1}A\,dZ\,A^{\mathsf{T}}\}$$
$$= \mathrm{Tr}\{(A^{\mathsf{T}})^{-1}Y^{-1}\,d\overline{Z}\,Y^{-1}dZ\,A^{\mathsf{T}}\},$$

and ds^2 is hence preserved.

3. Under the mapping $X + iY \mapsto W = -(X + iY)^{-1}$, we have $Z = -W^{-1}$. Further,

$$dZ = W^{-1}(dW)W^{-1}, \qquad d\overline{Z} = \overline{W}^{-1}\,d\overline{W}\,\overline{W}^{-1}. \tag{6.37}$$

Therefore,

$$Y = \frac{1}{2i}(Z - \overline{Z}) = \frac{1}{2i}[\overline{W}^{-1} - W^{-1}]. \tag{6.38}$$

Let $W = U + iV$. We use two different methods to transform expression (6.38). First,

$$\frac{1}{2i}[\overline{W}^{-1}(W - \overline{W})W^{-1}] = \frac{1}{2i}\overline{W}^{-1}2iVW^{-1} = \overline{W}^{-1}VW^{-1}.$$

---

[2] Here, dZ is the matrix $dX + i\,dY$ whose entries are differential 1-forms.

In this case,
$$Y^{-1} = WV^{-1}\overline{W}. \qquad (6.39)$$

Second,
$$\frac{1}{2i}[W^{-1}(W - \overline{W})\overline{W}^{-1}] = \frac{1}{2i}(W^{-1}2iV\overline{W}^{-1}).$$

In this case,
$$Y^{-1} = \overline{W}V^{-1}W. \qquad (6.40)$$

Substituting (6.37), (6.39), and (6.40) in (6.36), we obtain
$$\begin{aligned}
ds^2 &= \text{Tr}\{WV^{-1}\overline{W}\,\overline{W}^{-1}\,d\overline{W}\,\overline{W}^{-1}\overline{W}V^{-1}WW^{-1}(dW)W^{-1}\} \\
&= \text{Tr}\{V^{-1}\,d\overline{W}\,V^{-1}dW\}.
\end{aligned}$$

Therefore, ds^2 is a Hermitian metric that is invariant with respect to the action of $\text{Sp}(n, \mathbb{R})$. Proposition 6.3 is proved.

Proposition 6.4. *The manifold $\mathfrak{S}_n$ is a symmetrical space.*

Proof. We show that a transformation $W \mapsto -W^{-1}$ of form 3 is an involution of the manifold $\mathfrak{S}_n$. Indeed, this transformation is involutive because $-(-W^{-1})^{-1} = W$. Fixed points are defined by the relation $-W^{-1} = W$ or $W^2 = -\text{I}$.

The point $W = i\,\text{I}$ is a unique fixed point in $\mathfrak{S}_n$ because in order to obtain a positive-definite coefficient for the imaginary part, it is necessary to extract the arithmetical square root of I, which is uniquely defined. Proposition 6.4 is proved.

Therefore, Theorem 6.1 is completely proved.

Remark. The Hermitian metric defines the Riemannian structure $\mathfrak{R}(ds^2)$, as well as the symplectic structure $\omega = \mathfrak{I}(ds^2)$, on the manifold $\mathfrak{S}_n$, i.e., this manifold is simultaneously Riemannian and symplectic. If, moreover, the differential form ω is closed, i.e., $d\omega = 0$, then the manifold is called *Kählerian*.

Exercise. Show that the form $\mathfrak{I}(ds^2)$ is closed.

2.3. Action of $\text{Sp}(n, \mathbb{R})$ on the Boundary of the Siegel Half-Plane. As in the case of the Klein–Poincaré upper half-plane, the action of the group $\text{Sp}(n, \mathbb{R})$ defines a compactification of each of the strata of the boundary of the manifold $\mathfrak{S}_n$. To verify this, we prove the following statement.

Proposition 6.5. *The action of the group $\text{Sp}(n, \mathbb{R})$ on the generalized Siegel upper half-plane and on its boundary (if it is defined, i.e., if the corresponding inverse matrix exists) does not change the rank of the imaginary part of the matrix Z.*

Proof. We use Theorem 6.1. A transformation of form 1 does not change the imaginary part of the matrix Z. A transformation of form 2 transforms the matrix Y into its adjoint. In this case, the rank of Y is not changed. It remains to verify that under the passage to the inverse matrix, the rank of the imaginary part is preserved. Let $\operatorname{rk} Y = k$. We apply a transformation of form 3. Because $Y^\top = Y$, there exists $\alpha \in \mathrm{GL}(n)$ such that

$$B = \alpha Y \alpha^\top = \begin{pmatrix} \mathrm{I}_k & 0 \\ 0 & 0 \end{pmatrix},$$

where I_k is the identity matrix of order k. We set $A = \alpha X \alpha^\top$. If the matrix $A^2 + B$ is not singular, then there exists the inverse matrix $(A + iB)^{-1}$, which is equal to

$$(A + iB)^{-1} = (A - iB)(A^2 + B)^{-1} \tag{6.41}$$

(we recall that $B^2 = B$). The obtained relation implies that the rank of the imaginary part of the matrix $(A + iB)^{-1}$ is equal to k because the coefficient of the imaginary part, the matrix B, is multiplied by a nonsingular matrix in (6.41). If $A^2 + B$ is singular, then the matrix $A + iB$ is also singular; therefore, the transformation of form 3 is not defined for such a matrix.

In the case where a linear-fractional transformation is not defined, similar to the projective space, we can redefined it by passing to another chart. The passage to another chart means the addition of those elements which are not in the initial chart but belong to its closure. The added points define a compactification of a stratum s_k, which is denoted by $\hat{s}_k$.

The manifold $\hat{s}_0$, being a compactification of the set of real symmetrical matrices with respect to the action of the group of generalized linear-fractional transformations, coincides with the Lagrange–Grassmann manifold $\Lambda(\mathbb{R}^{2n})$.

2.4. Cayley Transform. The transformation

$$z_1 = i\,\frac{1 + z}{1 - z}$$

transforms the unit disk $|z| < 1$ into the upper half-plane of the complex plane $\mathbb{C}$. The inverse transformation is given by the formula

$$z = \frac{z_1 - i}{z_1 + i}.$$

The generalization of these formulas to the Siegel upper half-plane is called the *Cayley transform*.

Let Z and ζ be two $n \times n$ matrices, and let $\det(1 - Z) \neq 0$. The *Cayley transform* is the passage from the matrix Z to the matrix ζ (or vice versa) given by the formulas

$$\zeta = i(\mathrm{I} + Z)(\mathrm{I} - Z)^{-1}, \qquad Z = (\zeta - i\,\mathrm{I})(\zeta + i\,\mathrm{I})^{-1}.$$

Lemma 6.6. *Let a matrix Z belong to the Siegel domain of type* III, *i.e.,* $Z \in \Re_n$. *Then* $\det(I - Z) \neq 0$.

Proof. We find the kernel of the matrix $I - Z$. We have $(I - Z)x = 0$; this implies $Zx = x$. Then

$$x^\top Z^\top = x^\top, \qquad \overline{Z}\,\overline{x} = \overline{x}.$$

We compose the quadratic expression

$$x^\top (I - Z\overline{Z})\,\overline{x} = x^\top \overline{x} - x^\top Z\overline{Z}\,\overline{x} = 0.$$

By the definition of the generalized unit disk (6.16), the inequality $I - Z\overline{Z} > 0$ holds. Therefore, $x = 0$.

Proposition 6.6. *The generalized Siegel upper half-plane $\mathfrak{S}_n$ is analytically equivalent to the homogeneity domain of type* III, *the generalized unit disk*

$$\Re_n = \{ Z \in M(n, \mathbb{C}) \mid Z^\top = Z,\ I - Z^\top \overline{Z} > 0 \}.$$

The correspondence between $\Re_n$ and $\mathfrak{S}_n$ is given by the Cayley transform.

Proof. Let $Z \in \Re_n$. By Lemma 6.6, the expression $\zeta = i(I + Z)(I - Z)^{-1}$ is defined. Obviously, $\zeta^\top = \zeta$ (because the matrices $(I - Z)$ and $(I + Z)$ commute). We prove that $\Im \zeta > 0$. Indeed,

$$\Im \zeta = \frac{1}{2i}(\zeta - \overline{\zeta}) = \frac{1}{2i}\{ i(I + Z)(I - Z)^{-1} + i(I + \overline{Z})(I - \overline{Z})^{-1} \}$$

$$= \frac{1}{2}\{ (I - Z)^{-1}[(I + Z)(I - \overline{Z}) + (I - Z)(I + \overline{Z})](I - \overline{Z})^{-1} \}$$

$$= (I - Z)^{-1}[I - Z\overline{Z}]\overline{(I - Z)^{-1}} > 0.$$

The latter inequality holds because the obtained expression coincides with the result of passing to new variables for the positive-definite Hermitian form $I - Z\overline{Z} > 0$. This passage consists in premultiplying $I - Z\overline{Z}$ by the matrix $(I - Z)^{-1}$ and postmultiplying it by the complex-conjugate transposed matrix. (We recall that the matrix Z is symmetrical.) Such a transformation does not change the positive definiteness of the form. Therefore, $\zeta \in \mathfrak{S}_n$.

Conversely, let $\zeta \in \mathfrak{S}_n$. Then the matrix Z is symmetrical. Indeed, we have

$$Z^\top = (\zeta^\top + i\,I)^{-1}(\zeta^\top - i\,I). \tag{6.42}$$

Because the matrix ζ is symmetrical and the summands in (6.42) commute, we have $Z^\top = Z$.

It remains to show that $I - Z\overline{Z} > 0$. We have

$$I - Z\overline{Z} = I - (\zeta + i\,I)^{-1}(\zeta - i\,I)(\overline{\zeta} + i\,I)(\overline{\zeta} - i\,I)^{-1}$$

$$= (\zeta + i\,I)^{-1}[(\zeta + i\,I)(\overline{\zeta} - i\,I) - (\zeta - i\,I)(\overline{\zeta} + i\,I)](\overline{\zeta} - i\,I)^{-1}$$

$$= (\zeta + i\,\mathrm{I})^{-1}\frac{2}{i}(\zeta - \overline{\zeta})(\overline{\zeta} - i\,\mathrm{I})^{-1}$$

$$= 4(\zeta + i\,\mathrm{I})^{-1}\frac{1}{i}(\Im\zeta)(\overline{\zeta} - i\,\mathrm{I})^{-1} > 0.$$

The latter inequality holds because the passage to new variables in the positive-definite Hermitian form does not violate its positive definiteness.

Therefore, the matrix Z belongs to the generalized unit disk.

§3. Complexified Riccati Equation as a Flow on the Generalized Siegel Upper Half-Plane

Let $W = X + iY$ be a complex $n \times n$ matrix with the real and imaginary parts X and Y. If W satisfies the Riccati equation

$$\dot{W} = (C + W)A^{-1}(C^{\mathsf{T}} + W) - B$$

with the real matrices A, B, and C, where A and B are symmetrical matrices, then we obtain the set of equations

$$\dot{X} = -B + (X + C)A^{-1}(X + C^{\mathsf{T}}) - YA^{-1}Y,$$
$$\dot{Y} = YA^{-1}(C^{\mathsf{T}} + X) + (X + C)A^{-1}Y \tag{6.43}$$

for the matrices X and Y. We show that system (6.43) defines a flow on the generalized Siegel upper half-plane.

Theorem 6.2. *Let $W(t) = X(t) + iY(t)$ be a solution to system (6.43) with the initial condition $W(t_0) \in \mathfrak{S}_n$. Then $W(t) \in \mathfrak{S}_n$ for all t.*

Proof. By Lemma 2.1, the matrix $W(t)$ remains symmetrical for all values of t.

Let $L(t)$ be the matrix $L(t) = A^{-1}(t)[X(t) + C^{\mathsf{T}}(t)]$ with entries denoted by λ_{ij}. If $X(t), Y(t)$ is a solution to system (6.43), then the matrix $Y(t)$ satisfies the Lyapunov-type linear equation

$$\dot{Y} = YL + L^{\mathsf{T}}Y. \tag{6.44}$$

We prove a formula that describes the evolution of minors of the matrix $Y(t)$, which is a solution to this equation. In other words (see Sec. 4 in Chap. 4), we explicitly write a system of ordinary differential equations that is associated with the initial system for exterior powers of the trivial representation (under which the $n \times n$ matrix group is realized as a group of linear transformations of the space $\mathbb{R}^n$).

Let I or J be a tuple of m distinct elements of the set $1, 2, \ldots, n$, and let $\mathcal{L}_{j\alpha}$ be the operator on the set of these tuples that replaces the element in the jth position with the element α, i.e., $\mathcal{L}_{j\alpha}(i_1, \ldots, i_j, \ldots, i_m) = (i_1, \ldots, \alpha, \ldots, i_m)$. Let $\Delta(I, J)$ be the minor of the matrix $Y(t)$ composed of rows with numbers from the tuple I and columns with numbers from the tuple J.

Lemma 6.7. *The following formula holds:*

$$\frac{d}{dt}\Delta(I, J) = \sum_{j \in J} \sum_{\alpha=1}^{n} \lambda_{\alpha j} \Delta(I, \mathcal{L}_{j\alpha}(J)) + \sum_{i \in I} \sum_{\beta=1}^{n} \lambda_{\beta i} \Delta(\mathcal{L}_{i\beta}(I), J). \tag{6.45}$$

Proof. We can differentiate the determinant Δ by two different methods. In the first case, we consider the sum of determinants in which the columns of Δ are successively differentiated, and in the second case, its rows.

The right-hand side of system (6.44) consists of two sums:

$$\dot{y}_{ij} = \sum_{\alpha=1}^{n} \lambda_{\alpha j} y_{i\alpha} + \sum_{\beta=1}^{n} \lambda_{\beta i} y_{\beta j}.$$

When computing Δ', we sequentially substitute the sum $\sum_{\alpha=1}^{n} \lambda_{\alpha j} y_{i\alpha}$ for each of the columns of Δ and the sum $\sum_{\beta=1}^{n} \lambda_{\beta i} y_{\beta j}$ for each of its rows. We then obtain

$$\frac{d}{dt}\Delta(I, J) = \sum_{j \in J} \begin{vmatrix} y_{i_1 j_1} & \cdots & \sum_{\alpha=1}^{n} \lambda_{\alpha j} y_{i_1 \alpha} & \cdots & y_{i_1 j_s} \\ \vdots & \ddots & \vdots & \ddots & \vdots \\ y_{i_s j_1} & \cdots & \sum_{\alpha=1}^{n} \lambda_{\alpha j} y_{i_s \alpha} & \cdots & y_{i_s j_s} \end{vmatrix}$$

$$+ \sum_{i \in I} \begin{vmatrix} y_{i_1 j_1} & \cdots & y_{i_1 j_s} \\ \vdots & \ddots & \vdots \\ \sum_{\beta=1}^{n} \lambda_{\beta i} y_{\beta j_1} & \cdots & \sum_{\beta=1}^{n} \lambda_{\beta i} y_{\beta j_s} \\ \vdots & \ddots & \vdots \\ y_{i_s j_1} & \cdots & y_{i_s j_s} \end{vmatrix}$$

$$= \sum_{j \in J} \sum_{\alpha=1}^{n} \lambda_{\alpha j} \Delta(I, \mathcal{L}_{j\alpha}(J)) + \sum_{i \in I} \sum_{\beta=1}^{n} \lambda_{\beta i} \Delta(\mathcal{L}_{i\beta}(I), J). \tag{6.46}$$

The obtained relation proves the lemma.

Corollary 6.1. *When $m = n$, we obtain*

$$(\det Y)' = 2(\operatorname{Tr} L) \det Y \tag{6.47}$$

for the equation $\dot{Y} = YL + L^{\top}Y$.

Therefore, if $\det Y(t_0) \neq 0$, then $\det Y(t)$ is different from zero for any t, i.e., when moving along the Siegel upper half-plane, the trajectory of system (6.43) cannot attain the boundary s of the Siegel domain in a finite time.

We prove that the trajectory $W(t)$ of system (6.46) cannot go to infinity in a finite time. All infinitely distant points of the manifold $\mathfrak{S}_n$ belong to its boundary $\hat{s}$. We can steer any infinitely distant point $\hat{s}$ to a finite point belonging to s using a generalized linear-fractional transformation. As was shown in Chap. 2, under such a transformation, system (6.43) again passes to a system of form (6.43). This implies that a solution to system (6.43) cannot go to infinity in a finite time; this completes the proof of the theorem.

Therefore, the compactified Riccati equation defines a (nonstationary) flow on the generalized Siegel upper half-plane. In other words, the Siegel upper half-plane is an integral manifold of system (6.43). We show that each stratum $\hat{s}_k$ of the boundary of $\mathfrak{S}_n$ is an integral manifold of this system.

Theorem 6.3. *Equations* (6.43) *define a smooth flow on each of the manifolds* $\hat{s}_k$, $k = 0, 1, \ldots, n - 1$.

Proof. We consider Eq. (6.43). We note that if $\operatorname{rk} Y(t_0) = k$, then there exists a minor of the kth order that is different from zero, and all minors of larger order vanish. By Lemma 6.7, the vector composed of all minors of a fixed order m satisfies the homogeneous linear system of equations (6.46). By the uniqueness theorem, this vector either is identically equal to zero or never vanishes. Therefore, the rank of the matrix $Y(t)$ is preserved, i.e., $\operatorname{rk} Y(t) = k$. Therefore, the solutions to system (6.43) do not leave the stratum s_k. However, in contrast to the solutions lying inside the generalized Siegel upper half-plane $\mathfrak{S}_n$, the solutions lying on its boundary s can go to infinity in a finite time because outside the chart considered (at infinity), we have no points of $\mathfrak{S}_n$, but there are points of $\hat{s}$. We show that even when passing through infinity, the trajectory cannot leave the stratum $\hat{s}_k$.

Indeed, by Proposition 6.2, the transformations $\Gamma \in \operatorname{Sp}(n, \mathbb{R})$ transform $\hat{s}_k$ into itself. Because under the compactification, the stratum s_k contains the image of all its points under the action of $\operatorname{Sp}(n, \mathbb{R})$, we can steer the infinitely distant point $a \in \hat{s}_k$ to its finite point using a certain element from $\operatorname{Sp}(n, \mathbb{R})$. As was shown in Chap. 2, this transformation does not change the form of system (6.43), and the trajectory passing through a does not leave $\hat{s}_k$.

§4. Flow on Cartan–Siegel Homogeneity Domains

In the preceding section, we considered the Riccati equation

$$\dot{W} - W A^{-1} W - C A^{-1} W - W A^{-1} C^\mathsf{T} - C A^{-1} C^\mathsf{T} + B = 0, \qquad (6.48)$$

in which the matrices A and B are real and symmetrical. This case corresponds to the Riccati equation considered in the classical calculus of variations, namely, to the Riccati equation for the minimization problem of the functional

$$I = \frac{1}{2} \int_{t_0}^{t_1} [\langle A(t)\dot{h},\, \dot{h} \rangle + 2\langle C(t)\dot{h},\, h \rangle + \langle B(t)h,\, h \rangle]\, dt. \qquad (6.49)$$

Riccati equation (6.48) is obtained from the canonical Hamiltonian system of ordinary differential equations

$$\dot{h} = -A^{-1}C^{\mathsf{T}}h + A^{-1}p, \qquad \dot{p} = (-CA^{-1}C^{\mathsf{T}} + B)h + CA^{-1}p$$

whose matrix

$$\begin{pmatrix} -A^{-1}C^{\mathsf{T}} & A^{-1} \\ -CA^{-1}C^{\mathsf{T}} + B & CA^{-1} \end{pmatrix}$$

belongs to the Lie algebra $\mathfrak{Sp}(n, \mathbb{R})$ of the Lie group $\mathrm{Sp}(n, \mathbb{R})$ for all t.

In Sec. 3, we showed that the complex Riccati equation of this form defines a flow on the generalized Siegel upper half-plane (i.e., on the Siegel homogeneity domain of type III) and on each of the strata of its boundary. We show here that for other conditions on the coefficients, the complex Riccati equation defines a flow on other Cartan–Siegel homogeneity domains.

4.1. Riccati-Type Equation for a Linear System Whose Matrix Belongs to a Given Lie Algebra.
Let $\mathfrak{G}$ be the Lie algebra of a Lie group G.[3] We consider the matrix linear system of ordinary differential equations

$$\dot{\Gamma} = B(t)\Gamma. \tag{6.50}$$

We assume that for all values of the independent variable t, the matrix $B(t)$ assumes its values in the Lie algebra $\mathfrak{G}$. Then, according to Proposition 4.5, $\Gamma(t) \in G$ whenever $\Gamma(0) \in G$.

Further let $G_k(\mathbb{R}^n)$ (or $G_k(\mathbb{C}^n)$ when the main field of constants is $\mathbb{C}$) be the Grassmann manifold of k-dimensional planes with the canonical set of charts that were introduced in Sec. 1 in Chap. 4. For the initial condition $\Gamma(0) \in G$, system (6.50) defines a one-parameter system of linear transformations of the space $\mathbb{R}^n$ (or $\mathbb{C}^n$) under which each transformation belongs to the group G. These transformations, being nonsingular, transform any k-dimensional subspace of the space $\mathbb{R}^n$ (or $\mathbb{C}^n$) into a k-dimensional subspace again and therefore generate a flow of transformations of the corresponding Grassmann manifold. As usual, we consider $\Gamma(0) = \mathrm{I}$ as the initial condition. In this case, the evolution begins with the identity transformation.

We find a differential equation that describes the evolution of these transformations on the Grassmann manifold. For this, we introduce a partition of the matrices B and Γ into blocks. As before, a pair of superscripts denotes the number of rows and columns in the corresponding submatrix. Therefore, let

$$B = \begin{pmatrix} B_{11}^{kk} & B_{12}^{k(n-k)} \\ B_{21}^{(n-k)k} & B_{22}^{(n-k)(n-k)} \end{pmatrix}, \qquad \Gamma = \begin{pmatrix} \Gamma_{11}^{kk} & \Gamma_{12}^{k(n-k)} \\ \Gamma_{21}^{(n-k)k} & \Gamma_{22}^{(n-k)(n-k)} \end{pmatrix}.$$

Let W be the $(n-k) \times k$ matrix that defines the coordinates of a point $W \in G_k(\mathbb{R}^n)$ (or $G_k(\mathbb{C}^n)$). We prove the following generalization of Theorem 2.3.

---

[3] The results of this subsection can be applied to the cases where the coefficients of the Riccati equation and its solutions belong to the field $\mathbb{R}$ or the field $\mathbb{C}$.

Theorem 6.4. *Under the action of the flow of solutions to Eq. (6.50), the evolution of the matrix $W(t)$ is described by the equation*

$$\dot{W} = B_{21}^{(n-k)k} + B_{22}^{(n-k)(n-k)} W - W B_{11}^{kk} - W B_{12}^{k(n-k)} W. \tag{6.51}$$

Proof. As was shown in Sec. 1 in Chap. 4, under the transformation $\Gamma(t)$, the matrix W is transformed by the linear-fractional transformation

$$W(t) = (\Gamma_{21}^{(n-k)k} + \Gamma_{22}^{(n-k)(n-k)} W_0)(\Gamma_{11}^{kk} + \Gamma_{12}^{k(n-k)} W_0)^{-1}. \tag{6.52}$$

We omit superscripts in the subsequent computations.

Differentiating (6.52) and using differential equation (6.50), we obtain

$$\begin{aligned}
\dot{W} = &(B_{21}\Gamma_{11} + B_{22}\Gamma_{21} + B_{21}\Gamma_{12}W_0 + B_{22}\Gamma_{22}W_0)(\Gamma_{11} + \Gamma_{12}W_0)^{-1} \\
&- (\Gamma_{21} + \Gamma_{22}W_0)(\Gamma_{11} + \Gamma_{12}W_0)^{-1} \\
&\times (B_{11}\Gamma_{11} + B_{12}\Gamma_{21} + B_{11}\Gamma_{12}W_0 + B_{12}\Gamma_{22}W_0)(\Gamma_{11} + \Gamma_{12}W_0)^{-1}.
\end{aligned}$$

We transform the obtained equation to isolate the coefficients of B_{ij}:

$$\begin{aligned}
\dot{W} = &B_{21}(\Gamma_{11} + \Gamma_{12}W_0)(\Gamma_{11} + \Gamma_{12}W_0)^{-1} \\
&+ B_{22}(\Gamma_{21} + \Gamma_{22}W_0)(\Gamma_{11} + \Gamma_{12}W_0)^{-1} \\
&- (\Gamma_{21} + \Gamma_{22}W_0)(\Gamma_{11} + \Gamma_{12}W_0)^{-1} \\
&\times [B_{11}(\Gamma_{11} + \Gamma_{12}W_0)(\Gamma_{11} + \Gamma_{12}W_0)^{-1} \\
&+ B_{12}(\Gamma_{21} + \Gamma_{22}W_0)(\Gamma_{11} + \Gamma_{12}W_0)^{-1}].
\end{aligned}$$

Substituting the expression for $W(t)$ from (6.52), we finally obtain

$$\dot{W} = B_{21} + B_{22}W - WB_{11} - WB_{12}W. \tag{6.53}$$

The theorem is proved.

Remark. Any solution to Riccati equation (6.53) is obtained from the solution to linear system (6.50) by the "Grassmannization" method, i.e., using the ratio of two matrices that defines the coordinates on the Grassmann manifold. The general solution to the linear system is a linear combination of a finite number of its solutions. Therefore, we can write explicit formulas that express the dependence of the general solution to the Riccati equation on a finite number of its solutions. In the literature on Riccati equation theory, this dependence is called the *chain rule* (see [89, 112]).

The remainder of Sec. 4 is devoted to specifications of Theorem 6.4.

4.2. Flow on the Siegel Homogeneity Domain of Type I. We recall that the Siegel homogeneity domain of type I is the set of rectangular $q \times p$ matrices W with complex entries satisfying the condition

$$W^{\mathsf{T}} \overline{W} < I \tag{6.54}$$

(as usual, the inequality sign between matrices means that the difference between the right- and left-hand sides is a positive-definite Hermitian matrix, and the bar denotes complex conjugation). The boundary of domain (6.54) is a stratified manifold whose strata ζ_k are given by the conditions

$$\zeta_k = \{W \mid W^{\mathsf{T}} \overline{W} \leq I, \ \mathrm{rk}(I - W^{\mathsf{T}} \overline{W}) = k\};$$

moreover, $\zeta_0 = \{W \mid W^{\mathsf{T}} \overline{W} = I\}$ is the Shilov boundary (see [96], Vol. 2) of domain (6.54).

We consider the equation describing the transport of p-dimensional planes using mappings defined by a linear $(p+q)$-dimensional system of differential equations whose matrix of coefficients belongs to the Lie algebra of the group acting on the Siegel homogeneity domain of type I for all values of the argument.

The general group of analytic automorphisms of domain (6.54) is denoted by $U(p,q)$ and is defined as the group of $(p+q) \times (p+q)$ matrices with complex entries that preserve the nondegenerate Hermitian form in $p+q$ variables of signature $q-p$. In other words, the group $U(p,q)$ consists of those matrices G for which

$$G^{\mathsf{T}} H \overline{G} = H, \tag{6.55}$$

where

$$H = \begin{pmatrix} -I_p & 0 \\ 0 & I_q \end{pmatrix}$$

(as usual, I_p denotes the square p-dimensional identity matrix). If

$$G = \begin{pmatrix} A & B \\ C & D \end{pmatrix}$$

is the corresponding block partition of the matrix G, then condition (6.55) becomes

$$A^{\mathsf{T}} \overline{A} - C^{\mathsf{T}} \overline{C} = I_p, \qquad D^{\mathsf{T}} \overline{D} - B^{\mathsf{T}} \overline{B} = I_q, \qquad A^{\mathsf{T}} \overline{B} = C^{\mathsf{T}} \overline{D}.$$

Differentiating these relations, we verify that the Lie algebra $\mathfrak{u}(p,q)$ of the group $U(p,q)$ consists of the matrices

$$\mathfrak{u}(p,q) = \left\{ \begin{pmatrix} \mathfrak{p} & \mathfrak{q} \\ \mathfrak{q}^{\mathsf{T}} & \mathfrak{r} \end{pmatrix} \middle| \ \mathfrak{p}^{\mathsf{T}} = -\overline{\mathfrak{p}}, \ \mathfrak{r}^{\mathsf{T}} = -\overline{\mathfrak{r}} \right\}.$$

We consider the following set of ordinary differential equations (written in the block form) with coefficients in the Lie algebra $\mathfrak{u}(p,q)$:

$$\dot{x} = \mathfrak{p}(t)x + \mathfrak{q}(t)y, \qquad \dot{y} = \overline{\mathfrak{q}}^{\top}(t)x + \mathfrak{r}(t)y, \tag{6.56}$$

where $x \in \mathbb{R}^p$, $y \in \mathbb{R}^q$, $\mathfrak{p}^{\top}(t) = -\overline{\mathfrak{p}}(t)$, and $\mathfrak{r}^{\top}(t) = -\overline{\mathfrak{r}}(t)$. We set

$$B(t) = \begin{pmatrix} \mathfrak{p}(t) & \mathfrak{q}(t) \\ \overline{\mathfrak{q}}^{\top}(t) & \mathfrak{r}(t) \end{pmatrix}$$

and consider the transformation $\Gamma(t): \mathbb{C}^{p+q} \to \mathbb{C}^{p+q}$ satisfying Eq. (6.50).

By Theorem 6.4, the equation satisfied by the coordinates W of p-dimensional planes under translations by time t along trajectories of system (6.56) has the form

$$\dot{W} = \overline{\mathfrak{q}}^{\top} + \mathfrak{r}W - W\mathfrak{p} - W\mathfrak{q}W. \tag{6.57}$$

Theorem 6.5. *Solutions to Eq. (6.57) define a flow on the Siegel homogeneity domain of type I. Each stratum of the boundary of this domain is an integral manifold of Eq. (6.57).*

Proof. We have

$$\begin{aligned}
\frac{d}{dt}(W^{\top}\overline{W} - \mathrm{I}) &= W^{\top}(\mathfrak{q}^{\top} + \overline{\mathfrak{r}}\,\overline{W} - \overline{W}\,\overline{\mathfrak{p}} - \overline{W}\,\overline{\mathfrak{q}}\,\overline{W}) \\
&\quad + (\overline{\mathfrak{q}} - W^{\top}\overline{\mathfrak{r}} + \overline{\mathfrak{p}}\,W^{\top} - W^{\top}\mathfrak{q}^{\top}W^{\top})\overline{W} \\
&= W^{\top}\mathfrak{q}^{\top} + W^{\top}\overline{\mathfrak{r}}\,\overline{W} - (W^{\top}\overline{W} - \mathrm{I})\overline{\mathfrak{p}} - \overline{\mathfrak{p}} \\
&\quad - (W^{\top}\overline{W} - \mathrm{I})\overline{\mathfrak{q}}\,\overline{W} - \overline{\mathfrak{q}}\,\overline{W} + \overline{\mathfrak{q}}\,\overline{W} - W^{\top}\overline{\mathfrak{r}}\,\overline{W} \\
&\quad + \overline{\mathfrak{p}}\,(W^{\top}\overline{W} - \mathrm{I}) + \overline{\mathfrak{p}} - W^{\top}\mathfrak{q}^{\top}(W^{\top}\overline{W} - \mathrm{I}) \\
&\quad - W^{\top}\mathfrak{q}^{\top}. \tag{6.58}
\end{aligned}$$

Let $Y = W^{\top}\overline{W} - \mathrm{I}$ and $E = \overline{\mathfrak{p}} - W^{\top}\mathfrak{q}^{\top}$. Then (6.58) becomes

$$\dot{Y} = EY + Y\overline{E}^{\top}. \tag{6.59}$$

We note that if $Y(t)$ is a solution to system (6.59) and $\operatorname{rk}Y(t_0) = k$, then there exists a minor of order k that is different from zero, and all other minors of larger order vanish.

By Lemma 6.7, the vector composed of all minors of a fixed order m satisfies the homogeneous linear system of equations (6.46), where $\Lambda = \overline{E}^{\top}$ and $M = E$. Using the uniqueness theorem, we conclude that this vector either is identically equal to zero or never vanishes. Therefore, the rank of the matrix $Y(t)$ is preserved: $\operatorname{rk}Y(t) = k$. In particular, for $k = p$, we obtain solutions to system (6.57) that emanate inside domain (6.54) and remain inside this domain for all t, and solutions emanating from the stratum s_k do not leave it at least while they remain in the considered chart of the Grassmann manifold. Of course, being solutions to an equation with a quadratic right-hand side, they can go to infinity in a finite time, i.e., they leave the given chart.

We show that when passing through any infinitely distant point a, a solution to Eq. (6.57) also does not leave the stratum s_k.

The complex automorphisms of the Siegel homogeneity domain preserve the stratification of the boundary [82]. Therefore, we can choose an automorphism $g \in U(p,q)$ that transforms the infinitely distant point a into the finite point ga of the same stratum s_k. To describe the action of g on Eq. (6.57), we proceed as follows. Instead of performing a linear-fractional change in Eq. (6.57) itself (which leads to cumbersome computations), we find the action of the element g on the corresponding linear system (6.56) written in the matrix form $\dot{X} = BX$, where

$$B(t) = \left(\begin{array}{cc} \mathfrak{p}(t) & \mathfrak{q}(t) \\ \overline{\mathfrak{q}}^{\mathsf{T}}(t) & \mathfrak{r}(t) \end{array} \right).$$

Setting $X = gY$, we obtain

$$\dot{Y} = g^{-1}BgY \tag{6.60}$$

The matrix of coefficients of this system $g^{-1}Bg$ also belongs to $\mathfrak{U}(p,q)$. Indeed, because $g \in U(p,q)$ and $B \in \mathfrak{U}(p,q)$, we have $g^{-1}Bg = \mathrm{Ad}_g B \in \mathfrak{U}(p,q)$.

The matrix equation constructed by system (6.60) obviously coincides with the equation that is obtained as a result of the linear-fractional action of the element g on Eq. (6.57). Therefore, the Riccati equation obtained as a result of the change of variables has the same structure as Eq. (6.57). Therefore, when passing through the point ga, its solution does not leave the stratum s_k.

4.3. Flow on the Siegel Homogeneity Domain of Type II.

We recall that the Siegel homogeneity domain of type II is the set of complex $n \times n$ matrices W for which

$$W^{\mathsf{T}}\overline{W} < I, \qquad W^{\mathsf{T}} = -W. \tag{6.61}$$

We consider the $2n$-dimensional system of differential equations

$$\dot{X} = \mathfrak{A}(t)X \tag{6.62}$$

whose matrix of coefficients $\mathfrak{A}(t)$ belongs to the Lie algebra of the group acting on the Siegel homogeneity domain of type II for all t. We find the equation satisfied by the coordinates of n-dimensional planes in the flow of Eq. (6.62).

The general group G of analytic automorphisms of domain (6.61) is defined by the relations

$$g^{\mathsf{T}}H\overline{g} = H, \qquad g^{\mathsf{T}}Kg = K,$$

where

$$H = \left(\begin{array}{cc} -I_n & 0 \\ 0 & I_n \end{array} \right), \qquad K = \left(\begin{array}{cc} 0 & I_n \\ I_n & 0 \end{array} \right).$$

Any matrix $g \in G$ has the form

$$g = \left(\begin{array}{cc} A & -\overline{B} \\ B & \overline{A} \end{array} \right),$$

where A and B are two $n \times n$ matrices such that

$$A^\mathsf{T} \overline{A} - B^\mathsf{T} \overline{B} = \mathrm{I}, \qquad A^\mathsf{T} B = -B^\mathsf{T} A.$$

The Lie algebra of the group G has the form

$$\mathfrak{G} = \left\{ \begin{pmatrix} \alpha & -\overline{\beta} \\ \beta & \overline{\alpha} \end{pmatrix} \;\middle|\; \alpha^\mathsf{T} = -\overline{\alpha},\ \beta^\mathsf{T} = -\beta \right\}.$$

We consider the linear system of differential equations

$$\dot{x} = \alpha x - \overline{\beta} y, \qquad \dot{y} = \beta x + \overline{\alpha} y.$$

As in the case of the Cartan–Siegel homogeneity domain of type I, we can use Theorem 6.4 to easily obtain the Riccati-type equation for describing the evolution of n-dimensional planes in the space $\mathbb{C}^{2n}$, i.e., for the complex $(n \times n)$-dimensional matrix that defines the coordinates of points of $G_n(\mathbb{C}^{2n})$:

$$\dot{W} = \beta + \overline{\alpha} W - W \alpha + W \overline{\beta} W, \qquad (6.63)$$

where $\alpha^\mathsf{T} = -\overline{\alpha}$ and $\beta^\mathsf{T} = -\beta$.

Theorem 6.6. *Equation* (6.63) *defines a flow on the domain described by conditions* (6.61). *Each stratum of the boundary of this domain is an integral manifold of Eq.* (6.63).

Proof. First, we verify that a solution to Eq. (6.63) is a skew-symmetrical matrix. Indeed,

$$\dot{W}^\mathsf{T} = -\beta - W^\mathsf{T} \alpha + \overline{\alpha} W^\mathsf{T} - W^\mathsf{T} \overline{\beta} W^\mathsf{T}.$$

Adding the obtained equation to (6.63), we obtain

$$\begin{aligned} (\dot{W} + \dot{W}^\mathsf{T}) = {} & \overline{\alpha}(W + W^\mathsf{T}) - (W + W^\mathsf{T})\alpha \\ & - (W + W^\mathsf{T})\overline{\beta} W^\mathsf{T} + W \overline{\beta}(W + W^\mathsf{T}). \end{aligned} \qquad (6.64)$$

The uniqueness theorem implies that if $(W + W^\mathsf{T})(t_0) = 0$, then $(W + W^\mathsf{T})(t) \equiv 0$. Further, we have

$$\begin{aligned} \frac{d}{dt}(W^\mathsf{T} \overline{W}) &= \dot{W}^\mathsf{T} \overline{W} + W^\mathsf{T} \dot{\overline{W}} \\ &= -\beta \overline{W} - W^\mathsf{T} \alpha \overline{W} + \overline{\alpha} W^\mathsf{T} \overline{W} - W^\mathsf{T} \overline{\beta} W^\mathsf{T} \overline{W} \\ &\quad + W^\mathsf{T} \overline{\beta} + W^\mathsf{T} \alpha \overline{W} - W^\mathsf{T} \overline{W} \overline{\alpha} + W^\mathsf{T} \overline{W} \beta \overline{W}, \end{aligned}$$

$$\begin{aligned} \frac{d}{dt}(W^\mathsf{T} \overline{W} - \mathrm{I}) = {} & (W^\mathsf{T} \overline{W} - \mathrm{I})\beta \overline{W} - W^\mathsf{T} \overline{\beta}(W^\mathsf{T} \overline{W} - \mathrm{I}) \\ & + \overline{\alpha}(W^\mathsf{T} \overline{W} - \mathrm{I}) - (W^\mathsf{T} \overline{W} - \mathrm{I})\overline{\alpha}. \end{aligned} \qquad (6.65)$$

Equation (6.65) directly implies that the Shilov boundary of domain (6.61),

$$s_0 = \{W \mid W^\top \overline{W} = I, \ W^\top = -W\},$$

is an integral manifold of Eq. (6.63).

Further, let $Y = W^\top \overline{W} - I$. Then Eq. (6.65) is written in the form

$$\dot{Y} = (\overline{\alpha} - W^\top \overline{\beta})Y - Y(\overline{\alpha} - \beta \overline{W}). \tag{6.66}$$

Applying Lemma 6.7, we conclude that if $\operatorname{rk} Y(t_0) = k$, then $\operatorname{rk} Y(t) \equiv k$.

In the case where the trajectory passes through an infinitely distant point, the rank of the matrix $Y(t)$ is preserved. The proof of this is similar to that in the proof of Theorem 6.5.

Remark. Let the matrices α and β be real and skew symmetrical. Then relation (6.64) for real matrix equation (6.63) shows that the manifold S of skew-symmetrical matrices, which was studied in Chap. 5, is an integral manifold of Eq. (6.63). Another proof of this is implied by Proposition 5.10, which yields a description of the manifold of skew-symmetrical matrices in the space with an indefinite metric of zero signature as the set of isotropic planes. Indeed, the group of analytic automorphisms of the Siegel domain of type II preserves the nondegenerate quadratic form of zero signature. Specifically because of this, the manifold of isotropic planes is an integral manifold of the Riccati equation that corresponds to this group.

4.4. Flow on the Siegel Homogeneity Domain of Type IV. We recall that the Siegel symmetrical homogeneity domain $\mathfrak{T}$ of type IV is the domain in the space $\mathbb{C}^p$ that is defined by the inequalities

$$|z_1|^2 + \ldots + |z_p|^2 < (1 + |z_1^2 + \ldots + z_p^2|^2)/2 < 1. \tag{6.67}$$

On the domain $\mathfrak{T}$, the subgroup G of index 2 of the group $O(p,2)$ acts and leaves the canonical quadratic form with two positive and p negative squares invariant. Matrices g belonging to this group satisfy the equations

$$g = \overline{g}, \qquad g^\top H g = H, \qquad H = \begin{pmatrix} I_2 & 0 \\ 0 & -I_p \end{pmatrix}. \tag{6.68}$$

We deduce the Riccati-type equation that describes a flow on the domain $\mathfrak{T}$. Differentiating (6.68) at the point e, we obtain

$$(h^\top)H + Hh = 0.$$

This means that the matrix Hh is skew symmetrical. Because postmultiplication by the matrix $H^{-1} = H$ changes the signs of the last p rows, the Lie algebra of the group G consists of matrices of the form

$$\begin{pmatrix} 0 & -c & b \\ c & 0 & d \\ b^\top & d^\top & A \end{pmatrix}, \tag{6.69}$$

where c is a scalar, b and d are p-dimensional rows, and A is a skew-symmetrical $p \times p$ matrix. Under the change of variables $(t_1, t_2) \mapsto (t_1 - it_2, t_1 + it_2)$, matrix (6.69) is premultiplied by the matrix

$$\begin{pmatrix} 1 & -i & 0 \\ 1 & i & 0 \\ 0 & 0 & I_p \end{pmatrix}$$

and is postmultiplied by the inverse matrix

$$\begin{pmatrix} 1/2 & 1/2 & 0 \\ i/2 & -i/2 & 0 \\ 0 & 0 & I_p \end{pmatrix}.$$

As a result, we obtain matrices of the form

$$\begin{pmatrix} -ic & 0 & b - id \\ 0 & ic & b + id \\ (b^\mathsf{T} + id^\mathsf{T})/2 & (b^\mathsf{T} - id^\mathsf{T})/2 & A \end{pmatrix}.$$

Let $U = b + id$. In this notation, matrices belonging to the Lie algebra of the group acting on the domain $\mathfrak{T}$ have the form

$$\begin{pmatrix} -ic & 0 & \overline{U} \\ 0 & ic & U \\ U^\mathsf{T}/2 & \overline{U}^\mathsf{T}/2 & A \end{pmatrix}, \tag{6.70}$$

where $U = (U^1, \ldots, U^p)$ is a complex p-dimensional vector row and $A = (a_j^i)$ is a real skew-symmetrical $p \times p$ matrix. Taking the entry $-ic$ of matrix (6.70) as B_{11} (the remaining part of the partition of the matrix B is uniquely defined by this entry), we apply Theorem 6.4. For such a choice, W is a $(p+1)$-dimensional vector that coincides with the coordinates $(z_1, \ldots, z_p, z_{p+1})$ of the projective space $\mathbb{CP}^{p+1}$. Equation (6.51) in the coordinates z_i has the form

$$\begin{aligned}
\dot{z}_k &= \frac{1}{2}U^k + \sum_{j=1}^{p} a_k^j z_j + \frac{1}{2}\overline{U}^k z_{p+1} + icz_k - \sum_{j=1}^{p} z_k \overline{U}^j z_j, \\
\dot{z}_{p+1} &= icz_{p+1} + \sum_{j=1}^{p} U^j z_j + icz_{p+1} - \sum_{j=1}^{p} \overline{U}^j z_j z_{p+1},
\end{aligned} \tag{6.71}$$

$$k = 1, \ldots, p.$$

Theorem 6.4 implies that the manifold

$$\mathfrak{W} = \left\{ z \in \mathbb{CP}^{p+1} \,\middle|\, z_{p+1} = \sum_{j=1}^{p} z_j^2 \right\} \tag{6.72}$$

is an integral manifold of system (6.71).

Exercise. Verify this fact by direct differentiation.

Equation (6.72) implies that for initial values $z \in \mathfrak{W}$, the last equation of system (6.71) is a consequence of the first p equations. Therefore, the first p equations of system (6.71) after $z_{p+1} = \sum_{j=1}^{p} z_j^2$ is substituted in them define the Riccati-type equation that acts on $\mathfrak{T}$, the Siegel domain of type IV:

$$\dot{z}_k = \frac{1}{2}U^k + \sum_{j=1}^{p} a_k^j z_j + \frac{1}{2}\overline{U}^k \sum_{j=1}^{p} z_j^2 + icz_k - \sum_{j=1}^{p} z_k \overline{U}^j z_j, \qquad (6.73)$$
$$k = 1, \ldots, p.$$

§5. Matrix Analog of the Schwarz Differential Operator

In the preceding sections, we discussed the complexification of a solution W to the matrix Riccati equation. Here, we consider the Riccati-type equation with the complex independent variable t. We consider the linear $2n$-dimensional system of differential equations

$$\dot{\xi} = \left(\begin{array}{cc} A(t) & B(t) \\ C(t) & D(t) \end{array} \right) \xi. \qquad (6.74)$$

We assume that the matrix of coefficients of system (6.74) is composed of analytic functions and they are therefore defined on a certain Riemannian surface $\mathfrak{X}$. We consider the fundamental system of solutions to (6.74). Let γ be an arbitrary closed path on the surface $\mathfrak{X}$. Under a continuous change of solutions of the considered fundamental system along the path γ, new values of these solutions returning to the initial point are different in general from the initial values. However, they remain solutions of system (6.74). Because the initial values form a basis in the solution space, their new values are linear combinations of the old values. The matrix of coefficients of these linear combinations is called the *monodromy matrix* corresponding to the path γ. The corresponding linear transformation is called the *monodromy transformation*.

We recall that the set of classes of homotopic closed paths on the manifold $\mathfrak{X}$ emanating from a given point x forms a group whose group operation corresponds to sequentially passing first along the first path and then along the second one. This group is called the *fundamental group* of the manifold with the distinguished point x and is denoted by $\pi_1(\mathfrak{X}, x)$. For a connected manifold, all the groups $\pi_1(\mathfrak{X}, x)$ are isomorphic. Therefore, the argument x can be omitted, and we can write merely $\pi_1(\mathfrak{X})$.

Obviously, the monodromy transformation defines a representation of the fundamental group $\pi_1(\mathfrak{X})$ of the Riemannian surface $\mathfrak{X}$ on which the coefficients of system (6.74) are defined:

$$\rho: \pi_1(\mathfrak{X}) \to \mathrm{GL}(2n, \mathbb{C}). \qquad (6.75)$$

When passing from system (6.74) to the matrix Riccati-type equation

$$\dot{W} = C + DW - WA - WBW, \qquad (6.76)$$

which describes the change of coordinates of n-dimensional planes of the space $\mathbb{C}^{2n}$, representation (6.75) induces a representation

$$\tilde{\rho}\colon \pi_1(\mathfrak{X}) \to \mathrm{SL}(2n, \mathbb{C}) \tag{6.77}$$

of the group $\pi_1(\mathfrak{X})$ in the group $\mathrm{SL}(2n, \mathbb{C})$, which is considered as the group of generalized linear-fractional transformations of the Grassmann manifold.

The representation $\tilde{\rho}$ uniquely defines the *principal bundle* $\mathfrak{U}$ with the base $\mathfrak{X}$ and fiber $\mathrm{SL}(2n, \mathbb{C})$ and the flat bundle $\mathfrak{O}$ with the fiber $G_n(\mathbb{C}^{2n})$ associated with it.

Remark. We describe the standard method for constructing $\mathfrak{O}$. Let $U(\mathfrak{X})$ be the universal covering of the surface $\mathfrak{X}$. We define the diagonal action of $\pi_1(\mathfrak{X})$ on $U(\mathfrak{X}) \times G_n(\mathbb{C}^{2n})$ as follows: for $g \in \pi_1(\mathfrak{X})$ and $u \in U(\mathfrak{X})$, we set $g(u, W) = (g(u), \tilde{\rho}(g)W)$. The desired bundle is the quotient manifold of $U(\mathfrak{X}) \times G_n(\mathbb{C}^{2n})$ by this action,

$$\mathfrak{O} = U(\mathfrak{X}) \times G_n(\mathbb{C}^{2n})/\pi_1(\mathfrak{X}).$$

To introduce coordinates in the fibers of $\mathfrak{O}$, we must choose a section $W_0(t)$ of this bundle to define a vertical plane specifying a coordinate system on $G_n(\mathbb{C}^{2n})$.

A good choice of this section can essentially simplify the equation. For example, we take a certain particular solution to Eq. (6.76) as $W_0(t)$. Then, under the change of variables

$$V = (W - W_0(t))^{-1}, \qquad W = V^{-1} + W_0(t),$$

the quadratic term disappears in the Riccati-type equation. Indeed, Eq. (6.76) becomes

$$\begin{aligned}
\dot{W} = C + DW - WA - WBW &= -V^{-1}\dot{V}V^{-1} + \dot{W}_0(t) \\
&= -V^{-1}\dot{V}V^{-1} + C + DW_0 - W_0 A - W_0 B W_0 \\
&= C + D(V^{-1} + W_0) - (V^{-1} + W_0)A - (V^{-1} + W_0)B(V^{-1} + W_0).
\end{aligned}$$

Because $W_0(t)$ is a solution to Eq. (6.76), we obtain

$$\begin{aligned}
-V^{-1}\dot{V}V^{-1} &= DV^{-1} - V^{-1}A - V^{-1}BW_0 - W_0 BV^{-1} - V^{-1}BV^{-1}, \\
\dot{V} &= -VD + AV + (BW_0)V + V(W_0 B) + B.
\end{aligned}$$

Introducing a new notation for brevity, we have

$$\dot{V} = K(t)V + VL(t) + M(t). \tag{6.78}$$

We can construct a solution to Eq. (6.78) as follows. Let $\Phi(t)$ be a solution to the matrix equation

$$\dot{\Phi} = K(t)\Phi, \qquad \Phi(t_0) = I,$$

and let $\Psi(t)$ be a solution to the matrix equation

$$\dot{\Psi} = \Psi L(t), \qquad \Psi(t_0) = I.$$

Then
$$X = \Phi(t)C\Psi(t) \tag{6.79}$$
is a solution to the equation $\dot{X} = K(t)X + XL(t)$ for any constant matrix C.

Further, a solution to (6.78) is obtained from $X(t)$ by varying constants. That a key role in all these constructions is played by generalized linear-fractional transformations gives an idea of the necessity of introducing a matrix analog of the Schwarz derivative.

5.1. Classical Schwarz Differential Operator. The *Schwarz derivative* of an analytic function $f: \mathbb{C}^1 \to \mathbb{C}^1$ of the complex argument z is the expression

$$[S(f)]_z = \frac{d}{dz}\left(\frac{d^2 f/dz^2}{df/dz}\right) - \frac{1}{2}\left(\frac{d^2 f/dz^2}{df/dz}\right)^2. \tag{6.80}$$

The Schwarz derivative arose in the studies by Lagrange, Schwarz, Klein, and others (see, e.g., the classic papers [56, 61, 94] as well as the more modern ones [22, 106]). This derivative is related to conformal mapping theory and uniformization problems. As is known, for simply connected domains G of one complex variable z, an important role is played by conformal mappings $f: G \to K$ onto the interior of the unit disk $K = \{z \in \mathbb{C} \mid |z| < 1\}$ centered at the origin. For multiconnected domains G, a similar role is played by conformal mappings $f: G \to \widetilde{K}$ onto the interior $\widetilde{K}$ of the unit disk K with cuts along arcs of concentric circles centered at the origin. The functions that assign this mapping are solutions to the Schwarz equation whose right-hand side is explicitly written using the Green's function of the domain G (see [98], Sec. 2.12). The operator $[S(f)]_z$ has the following two important properties: it is invariant with respect to linear-fractional transformations of the function f, and it transforms as the quadratic differential with respect to linear-fractional changes of the argument z.

Along with connections to conformal mapping theory, the operator $[S(f)]_z$ is directly related to the moduli problem of algebraic curves. Namely, with the Schwarz operator, the set of flat structures on the Riemannian surface $\mathfrak{X}$ can be related to the space of regular quadratic differentials on $\mathfrak{X}$ [106].

Following the F. Klein presentation, we demonstrate one way the Schwarz differential operator appears. Let z be a complex variable assuming values on the Riemann sphere (the complex plane together with an infinitely distant point). The group of linear-fractional transformations of the Riemann sphere is the group of linear-fractional transformations

$$\tilde{z} = \frac{\alpha z + \beta}{\gamma z + \delta}.$$

After Abel proved the unsolvability in radicals of equations of the fifth degree, a number of authors (Hermite, Briosci, Klein, and others [9, 47, 56]) found a method for expressing roots of a fifth-degree equation in terms of elliptic and hypergeometric functions [80]. For this, it is necessary to consider a finite subgroup Γ of the group of linear-fractional transformations that transform a certain right polyhedron Π inscribed in the Riemann sphere into itself. Further, we consider algebraic functions $f(z)$ that are invariant with respect to the action of the group Γ whose degree is equal to the number of vertices of the polyhedron Π. If we equate such a function to a constant Z and consider the equation $f(z) = Z$, then the points of the Riemann sphere corresponding to vertices of the polyhedron Π are solutions to this equation. Then we consider the inverse function $Z(z)$ to $f(z)$, which is multivalued. To compute the function $Z(z)$, we find a differential operator that is invariant under any linear-fractional substitution.

Let η be an arbitrary function of the argument Z. Applying a linear-fractional transformation to it, we obtain the function

$$\frac{\alpha\eta + \beta}{\gamma\eta + \delta}.$$

From this function and its first, second, and third derivatives, we can compose an invariant of the group of linear-fractional transformations (i.e., an expression that does not depend on the constants α, β, γ, and δ). Therefore, we obtain a third-order differential operator S.

If we take the function Z as the function η, then the result of applying this operator is the same for all branches of the function $Z(z)$. Therefore, the value of the function $S(z)$ is the same for all branches of the function Z. Therefore, $S(z)$ is a single-valued function; moreover, this function is a rational function because Z is an algebraic function of the argument z. Equating $S(z)$ to this rational function, we obtain a third-order differential equation with rational coefficients, which has the particular solution $\eta = Z$.

We find an explicit expression of the operator S. Let

$$\zeta = \frac{\alpha\eta + \beta}{\gamma\eta + \delta}.$$

Differentiating the identity

$$\gamma\eta\zeta - \alpha\eta + \delta\zeta - \beta = 0$$

with respect to z, we obtain

$$\gamma(\eta'\zeta + \eta\zeta') - \alpha\eta' + \delta\zeta' = 0,$$
$$\gamma(\eta''\zeta + 2\eta'\zeta' + \eta\zeta'') - \alpha\eta'' + \delta\zeta'' = 0,$$
$$\gamma(\eta'''\zeta + 3\eta''\zeta' + 3\eta'\zeta'' + \eta\zeta''') - \alpha\eta''' + \delta\zeta''' = 0.$$

Excluding α, γ, and δ from the obtained system, which is linear in these variables, we obtain the equation

$$\begin{vmatrix} 0 & \zeta' & \eta' \\ 2\eta'\zeta' & \zeta'' & \eta'' \\ 3\eta''\zeta' + 3\eta'\zeta'' & \zeta''' & \eta''' \end{vmatrix} = 0$$

after simple transformations. Separating variables, we obtain

$$\frac{\zeta'''}{\zeta'} - \frac{3}{2}\left(\frac{\zeta''}{\zeta'}\right)^2 = \frac{\eta'''}{\eta'} - \frac{3}{2}\left(\frac{\eta''}{\eta'}\right)^2.$$

Therefore, the desired differential operator has the form

$$S(\eta) = \frac{\eta'''}{\eta'} - \frac{3}{2}\left(\frac{\eta''}{\eta'}\right)^2. \tag{6.81}$$

The differential operator $S(\eta)$ is called the *Schwarz operator* (or the *Schwarz derivative*).

The above arguments prove the following statement.

Proposition 6.7. *The Schwarz operator* (6.81) *is invariant with respect to linear-fractional substitutions of a differential function.*

5.2. Schwarz Operator and a Linear Second-Order Differential Equation.

We consider the linear second-order differential equation[4]

$$y''(t) + p(t)y'(t) + q(t)y(t) = 0, \tag{6.82}$$

whose coefficients $p(t)$ and $q(t)$ are given rational functions. Further, let $y_1(t)$ and $y_2(t)$ be two linearly independent solutions to Eq. (6.82). We consider the ratio

$$\eta = \frac{y_1}{y_2} \tag{6.83}$$

of these solutions.

Proposition 6.8. *The Schwarz differential operator applied to the function η is a rational function of t.*

Proof. In general, zeros of denominators of the coefficients of Eq. (6.82) are branch points of the solutions $y_1(t)$ and $y_2(t)$. Therefore, if the variable t circumscribes a closed curve l on the complex plane that contains at least one of these zeros, then the functions $y_1(t)$ and $y_2(t)$ being continuously varied assume new values that are linear combinations of the old values.

When going along the closed contour, the ratio of solutions $\eta = y_1/y_2$ is transformed into a linear-fractional function of its old value. Therefore, after going around, $S(\eta)$ is not changed and is hence a rational function of t.

We let $r(t)$ denote the expression $S(\eta)$. The function $r(t)$ depends on only the coefficients p and q of differential equation (6.82). We find this dependence in an explicit form. We have

$$y_1'' + py_1' + qy_1 = 0, \qquad y_2'' + py_2' + qy_2 = 0.$$

---

[4] We note that the argument t in this equation can be real as well as complex.

From this, we obtain

$$(y_1'' y_2 - y_2'' y_1) + p(y_1' y_2 - y_2' y_1) = 0. \tag{6.84}$$

Differentiating Eq. (6.83), we obtain

$$\frac{\eta''}{\eta'} = -p - 2\frac{y_2'}{y_2}. \tag{6.85}$$

Differentiating Eq. (6.85) once more, we have

$$\frac{\eta'''}{\eta'} - \left(\frac{\eta''}{\eta'}\right)^2 = -p' - 2\frac{y_2''}{y_2} + 2\left(\frac{y_2'}{y_2}\right)^2.$$

Substituting (6.85) in this relation, we obtain

$$S(\eta) = -\frac{1}{2}p^2 - p' - 2\frac{y_2''}{y_2} - 2p\frac{y_2'}{y_2}.$$

The terms containing y_2 in the right-hand side yield $2q$ by Eq. (6.82). Finally, we have

$$S(\eta) = 2q - \frac{1}{2}p^2 - p'. \tag{6.86}$$

Equation (6.86) is called the *Schwarz equation* constructing according to Eq. (6.82).

5.3. Schwarz Operator and the Riccati Equation. Summarizing, we can say that the Schwarz equation is closely related to the Riccati equation because both these equations are obtained by projectivization (by considering the ratio of coordinates) from the linear second-order equation

$$y''(t) + p(t)y'(t) + q(t)y(t) = 0. \tag{6.87}$$

Under the projectivization that leads to the Riccati equation, we consider the ratio $W = y'/y$, and under the projectivization that leads to the Schwarz equation, we consider the ratio $\eta = y_1/y_2$, where y_1 and y_2 are two particular solutions to Eq. (6.87).

The Riccati equation as well as the Schwarz equation are closely related to the group of linear-fractional transformations. The relation of the Riccati equation to this group was revealed sufficiently completely in the preceding chapters. The Schwarz equation has a number of properties that are similar to those of the Riccati equation. For example, the following statement holds (compare with Theorem 0.1 in the introduction).

Proposition 6.9. *The general solution to Schwarz equation* (6.86) *is a linear-fractional function (with constant coefficients) of a certain particular solution to this equation.*

Proof. Let $\eta_1(t)$ be a particular solution to Eq. (6.86). We consider the function

$$\eta = \frac{\alpha\eta_1 + \beta}{\gamma\eta_1 + \delta}$$

with four arbitrary constants α, β, γ, and δ. By Proposition 6.7, we have

$$S(\eta) = S(\eta_1) = 2q - \frac{1}{2}p^2 - p'.$$

We prove that the obtained solution is the general solution to the Schwarz equation. Indeed, let $\eta_1(0) = a$, $\eta_1'(0) = b$, and $\eta_1''(0) = c$ be initial data for a given particular solution, and let $\eta(0)$, $\eta'(0)$, and $\eta''(0)$ be initial data for an arbitrary solution.

The choice of the constants α, β, γ, and δ gives a possibility to obtain any initial data of the Cauchy problem $\eta(0)$, $\eta'(0)$, and $\eta''(0)$ that belong to an open set of the three-dimensional space whose closure coincides with the whole space. Namely, to determine the coefficients α, β, γ, and δ, we obtain the following set of equations:

$$\eta = \frac{\alpha a + \beta}{\gamma a + \delta},$$

$$\eta' = \frac{b}{(\gamma a + \delta)^2},$$

$$\eta'' = \frac{c}{(\gamma a + \delta)^2} - \frac{2\gamma b^2}{(\gamma a + \delta)^3},$$

$$\alpha\delta - \beta\gamma = 1.$$

It is easy to see that from this set of equations, we can sequentially find α, β, γ, and δ for all initial values, except for those manifolds on which the denominators of the obtained equations vanish.

We state a direct interconnection between the Riccati equation

$$w' + w^2 + pw + q = 0 \tag{6.88}$$

and the Schwarz equation

$$\frac{\eta'''}{\eta'} - \frac{3}{2}\left(\frac{\eta''}{\eta'}\right)^2 = 2q - \frac{1}{2}p^2 - p', \tag{6.89}$$

which are obtained from the same equation (6.82). From relation (6.86), we express the ratio y_2'/y_2 and let $w(t)$ denote it:

$$w(t) = -\frac{1}{2}\frac{\eta''}{\eta'} - \frac{1}{2}p. \tag{6.90}$$

Proposition 6.10. *Formula* (6.90) *states a connection between solutions to Eqs.* (6.88) *and* (6.89).

Proof. Let η be an arbitrary solution to Schwarz equation (6.89). Substituting (6.90) in (6.88), we can easily verify that $w(t)$ is a solution to the Riccati equation.

Conversely, let $w(t)$ be a solution to Riccati equation (6.88). Substituting the expression $w(t)$ for y_2'/y_2 in Eq. (6.85), we solve the obtained equation in η':

$$\eta' = \exp\left(-\int (p + 2w)dt\right).\tag{6.91}$$

Substituting (6.91) in (6.89) and using Eq. (6.88), we can verify that the function η is a solution to Schwarz equation (6.89).

5.4. Matrix Analog of the Schwarz Operator. To study the matrix Riccati equation, it is important to construct a generalization of the scalar Schwarz operator to the case of matrix-valued functions.

Definition 6.1. (See [121].) The expression

$$S_t W = \{W'^{-1}W''\}' - \frac{1}{2}\{W'^{-1}W''\}^2,$$

where W is an $n \times n$ matrix depending on the one-dimensional complex argument t, is called the *matrix Schwarz operator*.

Proposition 6.11. *Any solution of the third-order differential equation*

$$S_t W = 0 \tag{6.92}$$

is a generalized linear-fractional function (i.e., a function that has the form $(A + Bt)^{-1}(C + Dt))$, and any such function is a solution of (6.92).

Proof. In Eq. (6.92), we make the change of variables $\phi = W'$ and $\phi^{-1}\phi' = V$. Then

$$V' = V^2/2, \qquad -V^{-1}V'V^{-1} = -I/2.$$

Integrating, we obtain

$$2V^{-1} = (a - It), \qquad V = 2(a - It)^{-1}, \qquad \phi^{-1}\phi' = 2(a - It)^{-1},$$

$$\phi = c(a - It)^{-2}, \qquad W = c(a - It)^{-1} + b.$$

Changing the notation, we obtain

$$W = (A + Bt)^{-1} + C, \tag{6.93}$$

where A, B, and C are arbitrary constant matrices.

Proposition 6.12. *Curves with the scalar double ratio[5] and only they are solutions to differential equation (6.92) on $G_n(\mathbb{C}^{2n})$.*

---

[5] For the corresponding definition, see Sec. 4 of Chap. 5.

Proof. Proposition 6.12 means that the matrix double ratio of any four points lying on a solution $W(t)$ to Eq. (6.92) is scalar. To prove this, it suffices to compare (6.93) and (5.26) and apply Lemma 5.7.

Proposition 6.13. *Under a change of the independent variable $t = t(\tau)$, the Schwarz operator is transformed as*

$$(S_\tau W)(t(\tau)) = (S_t W)\left(\left(\frac{dt}{d\tau}\right)^2 + (S_\tau t)(t(\tau))\right)^2. \qquad (6.94)$$

Proof. We have

$$\frac{dW}{d\tau} = W' \frac{dt}{d\tau},$$

$$\frac{d^2 W}{d\tau^2} = W''\left(\frac{dt}{d\tau}\right)^2 + W'\left(\frac{d^2 t}{d\tau^2}\right),$$

$$\frac{d\tau}{dt}(W')^{-1} = \left(\frac{dW}{d\tau}\right)^{-1},$$

$$\left\{\left(\frac{dW}{d\tau}\right)^{-1}\left(\frac{d^2 W}{d\tau^2}\right)\right\} = (W')^{-1} W'' \frac{dt}{d\tau} + I\left(\frac{dt}{d\tau}\right)^{-1}\left(\frac{d^2 t}{d\tau^2}\right),$$

$$\left\{\left(\frac{dW}{d\tau}\right)^{-1}\left(\frac{d^2 W}{d\tau^2}\right)\right\}' = \left\{(W')^{-1} W''\right\}'\left(\frac{dt}{d\tau}\right)^2$$

$$+ \left\{(W')^{-1} W''\right\}\left(\frac{d^2 t}{d\tau^2}\right) + I\left\{\left(\frac{dt}{d\tau}\right)^{-1} \frac{d^2 t}{d\tau^2}\right\}',$$

$$\left\{\left(\frac{dW}{d\tau}\right)^{-1}\left(\frac{d^2 W}{d\tau^2}\right)\right\}^2 = \left\{(W')^{-1} W''\right\}^2\left(\frac{dt}{d\tau}\right)^2$$

$$+ 2\left\{(W')^{-1} W''\right\}\frac{d^2 t}{d\tau^2} + I\left[\left(\frac{dt}{d\tau}\right)^{-1} \frac{d^2 t}{d\tau^2}\right]^2.$$

Using the last two relations, we obtain (6.94).

Corollary 6.2. *Under linear-fractional changes of the argument, we have*

$$(S_\tau W(t(\tau)))(d\tau)^2 = (S_t W)(dt)^2.$$

Indeed, for

$$t = \frac{\alpha \tau + \beta}{\gamma \tau + \delta},$$

we have $S_\tau t = 0$ by Proposition 6.11.

Corollary 6.3. *We have*

$$\left(S_{\frac{\alpha \tau + \beta}{\gamma \tau + \delta}} W\right)\frac{(\alpha\delta - \beta\gamma)^2}{(\gamma\tau + \delta)^4} = S_\tau W.$$

The image of the matrix Schwarz differential operator is the space of tensors corresponding to matrix quadratic differentials.

Under generalized linear-fractional changes of the function W, the matrix Schwarz differential operator, in contrast to the scalar one, is not invariant. But the class of similar matrices corresponding to the image of $S_t W$ is invariant. Namely, the following proposition holds.

Proposition 6.14. *Let*

$$V = (A + BW)(C + DW)^{-1}.$$

Then there exists a matrix Γ such that

$$S_t V = \Gamma \left(S_t W \right) \Gamma^{-1}.$$

Proof. The group of linear-fractional transformations is generated by the following transformations:

1. $W \mapsto W + A$;   3. $W \mapsto AW$;

2. $W \mapsto WA$;   4. $W \mapsto W^{-1}$.

Obviously, transformations of type 1 do not change W' and, moreover, do not change $S_t W$. Under premultiplication by the matrix A (transformations of type 2), the matrix $(W')^{-1}W''$ is not changed because $(AW')^{-1}(AW'') = (W')^{-1}W''$.

Under postmultiplication by the matrix (transformations of type 3), we have $(W'A)^{-1}W''A = A^{-1}(W')^{-1}W''A$. Therefore,

$$S_t(WA) = A^{-1}\{(W')^{-1}W''\}'A$$
$$A^{-1}(W')^{-1}W''AA^{-1}(W')^{-1}W''A/2 = A(S_t W)A^{-1}.$$

In the case of transformations of type 4, for $Y = W^{-1}$, we have

$$Y' = -W^{-1}W'W^{-1},$$
$$Y'' = 2W^{-1}W'W^{-1}W'W^{-1} - W^{-1}W''W^{-1},$$
$$(Y')^{-1}Y'' = -2W'W^{-1} + W\{(W')^{-1}W''\}W^{-1},$$
$$\{(Y')^{-1}Y''\}' = -2W''W^{-1} + W'W^{-1}W'W^{-1} + W''W^{-1}$$
$$\qquad + W\{(W')^{-1}W''\}'W^{-1} - W\{(W')^{-1}W''\}W^{-1}W'W^{-1},$$
$$\{(Y')^{-1}Y''\}^2 = 4W'W^{-1}W'W^{-1} - 2W''W^{-1}$$
$$\qquad - 2W\{(W')^{-1}W''\}W^{-1}W'W^{-1} + W\{(W')^{-1}W''\}^2W^{-1}.$$

From the last two formulas, we obtain the relation $S_t Y = W(S_t W)W^{-1}$.

Corollary 6.4. *The scalar differential operators that are coefficients of the characteristic polynomial of the operator S_t are invariant with respect to the group of generalized linear-fractional transformations.*

Chapter 7
Higher-Dimensional Calculus of Variations

In this chapter, we discuss the main ideas of the higher-dimensional calculus of variations, the theory of minimization of a multiple integral. The minimization problem for a multiple integral arises in various fields of the natural sciences. For example, various problems in continuous medium theory (arising in hydrodynamics and gas dynamics, elasticity and plasticity theory, shell theory, etc.) lead to the minimization problem for a multiple integral when applying variational principles. Variational principles are especially important in quantum field theory because this approach is practically the only way to obtain equations describing the evolution of a quantum system. We begin with one of the main problems of the higher-dimensional calculus of variations, the problem of minimal surfaces.

§1. Minimal Surfaces

Minimal surface theory, initiated by Euler, Lagrange, Monge, Riemann, and Weierstrass, was elaborated by a brilliant assemblage of great mathematicians. In the simplest statement (in studying a two-dimensional surface in the three-dimensional Euclidean space), the problem is as follows: given a smooth closed curve $l \in \mathbb{R}^3$, find a two-dimensional surface $\mathfrak{X}$ with the boundary l that has the minimum area.

Lebesgue called this problem the *Plateau problem*, naming it for the blind physician who demonstrated very beautiful experiments with soap films spanned by a wire frame. Plateau extracted nontrivial consequences from his experiments that refer to the self-intersection properties of minimal surfaces [32]. One of the natural generalizations of the Plateau problem is the following problem (which is also called the Plateau problem).

Let $L^{\nu-1} \subset M^m$ be a smooth closed $(\nu-1)$-dimensional submanifold of an m-dimensional Riemannian manifold M^m. It is required to find a ν-dimensional submanifold $\mathfrak{N} \in M^m$ with the boundary L that has the minimum volume.

Remark. Not every closed submanifold L can serve as the boundary of a certain manifold $\mathfrak{N}$. Necessary and sufficient conditions for this are formulated on the language of (co)bordism theory. We do not discuss them here. If a contour L is not the boundary, then the Plateau problem obviously has no solution.

We assume that the contour l as well as the minimal surface $\mathfrak{X}$ in $\mathbb{R}^3$ itself are projected to the plane (x, y) in a one-to-one way. Then the surface $\mathfrak{X}$ is given by a certain (smooth) function $z = z(x, y)$ that is defined on a certain domain D of the plane (x, y). The area of this surface is equal to

$$\int\int_D \sqrt{1 + z_x^2 + z_y^2}\, dx \wedge dy. \tag{7.1}$$

Let a function

$$z = \phi(x, y) \tag{7.2}$$

defining the contour l in $\mathbb{R}^3$ be given on the boundary of the domain D. Then (we prove this in Sec. 2) the desired minimal surface is a solution to the nonlinear second-order equation

$$z_{xx}(1 + z_y^2) - 2z_{xy}z_x z_y + z_{yy}(1 + z_x^2) = 0 \tag{7.3}$$

with boundary conditions (7.2); this equation is the Euler equation for functional (7.1).

We note that the condition that the contour l is projected in a one-to-one way does not imply that the desired minimal surface also has a one-to-one projection on (x, y). Therefore, it is preferable to study minimal surfaces given in the parametric form $x = x(t)$, where $x = (x_1, x_2, x_3)$ are the coordinates of a point in $\mathbb{R}^3$ and $t = (t_1, t_2)$ is a parameter varying in an auxiliary two-dimensional domain B bounded by a curve β. In this case, the area of the surface $\mathfrak{X}$ is given by the integral

$$A(x(\cdot)) = \int\int_B \sqrt{g_{11}g_{22} - g_{12}^2}\, dt_1 \wedge dt_2, \tag{7.4}$$

where

$$g_{ij} = \left\langle \frac{\partial x}{\partial t_i}, \frac{\partial x}{\partial t_j} \right\rangle$$

is the metric tensor on $\mathfrak{X}$ induced by the metric of the ambient space $\mathbb{R}^3$.

It is easy to verify that functional (7.4) is invariant with respect to transformations of the coordinates (t). Using this, we introduce those local coordinates (t_1, t_2) on the surface $\mathfrak{X}$ for which the conditions

$$g_{11} = g_{22}, \qquad g_{12} = 0 \tag{7.5}$$

hold. In these coordinates, the metric on the surface $\mathfrak{X}$ becomes

$$ds^2 = g_{11}(dt_1^2 + dt_2^2).$$

The coordinates (t_1, t_2) are called *conformal* in this case because they assign a conformal mapping of the domain B onto $\mathfrak{X}$ (the angles made by curves are preserved at each point).

Remark. If the surface metric is real analytic, then each point has a neighborhood in which local conformal coordinates can be introduced, and on a surface that is a solution to an elliptic equation, it is possible to introduce an analytic metric.

We can write functional (7.4) in the conformal coordinates as

$$d(x(\cdot)) = \frac{1}{2} \int \int_B \left[\left(\frac{\partial x}{\partial t_1} \right)^2 + \left(\frac{\partial x}{\partial t_2} \right)^2 \right] dt_1 \wedge dt_2. \tag{7.6}$$

The Euler equation becomes the vector Laplace equation

$$\Delta x_i = 0, \quad i = 1, 2, 3.$$

A solution to the vector Laplace equation is called a *harmonic surface*. The nonlinearity of the problem is that we seek a harmonic surface satisfying conformality condition (7.5). By the obvious inequality

$$\frac{g_{11} + g_{22}}{2} \geq \sqrt{g_{11}g_{22} - g_{12}^2},$$

in which the equality holds only for $g_{11} = g_{22}$ and $g_{12} = 0$, we have the inequality $D(x(\cdot)) \geq A(x(\cdot))$ for an arbitrary harmonic surface $x(\cdot)$; this inequality becomes an equality only if conditions (7.5) hold. Therefore, a surface that minimizes the functional $D(x(\cdot))$ for given boundary conditions also solves the problem for the minimum of the area $A(x(\cdot))$.

Integral (7.6) is called the *Dirichlet integral*. The Dirichlet integral is positive definite; therefore, the existence of its minimum was long accepted. Riemann even used this as a basis of his geometric function theory and called it the *Dirichlet principle*. In 1869, Weierstrass published a paper in which he criticized the Dirichlet principle, arguing that a continuous function given on a closed set of an infinite-dimensional space (which is not compact) need not attain its greatest lower bound.

Since then, many studies have been devoted to the question of the existence of a solution to the Plateau problem. Only in the 1930s, Douglas [27] and Rado [84] independently and almost simultaneously proved the existence theorem for the Plateau problem in the particular case of two-dimensional surfaces in the n-dimensional Euclidean space. More than 30 years after that, the existence theorem for the Plateau problem (and its far-reaching generalizations) was proved for Riemannian manifolds of an arbitrary dimension through the efforts of many mathematicians (see, e.g., [24, 32, 53], where detailed references are given). The term *Dirichlet integral* is now used in a very broad sense and is applied to any quadratic integral functional in the classical calculus of variations.

A minimal surface can be characterized by the following geometric property: its mean curvature vanishes at all points. We recall that the *mean curvature* of a surface is the sum of its principal curvatures; by the Vieta theorem, we can find it as the trace of the quadratic form $\text{Tr}(g^{-1}f)$, where g is the matrix of the first fundamental form defining the metric of the surface and f is the matrix of the second fundamental form that is defined by the embedding of the surface into the Euclidean space, i.e.,

$$f_{ij} = \left\langle \frac{\partial^2 x}{\partial t_i \partial t_j}, n \right\rangle,$$

where n is a unit vector normal to the surface.

Proposition 7.1. *A two-dimensional surface in the three-dimensional Euclidean space is a minimal surface iff its mean curvature vanishes.*

Proof. In the conformal coordinates, the mean curvature of a surface equals

$$\begin{aligned} H &= \mathrm{Tr}(g^{-1}f) \\ &= (g_{11}g_{22} - g_{12}^2)^{-1}(g_{22}f_{11} - 2g_{12}f_{12} + g_{11}f_{22}) \\ &= g_{11}^{-1}\langle \Delta x, n \rangle. \end{aligned} \tag{7.7}$$

If this surface is minimal, then $\Delta x = 0$ in the conformal coordinates, and therefore $H = 0$.

Conversely, (7.7) implies that if $H = 0$, then $\langle \Delta x, n \rangle = 0$. In the conformal coordinate system, the vector Δx is orthogonal not only to the vector n but also to the vectors $\partial x / \partial t_i$, $i = 1, 2$. To verify this, it suffices to differentiate relations (7.5) with respect to t_1 and t_2. At any regular point of the surface, the vectors n, $\partial x / \partial t_1$, and $\partial x / \partial t_2$ are linearly independent. Therefore, $\Delta x = 0$, and the surface is minimal.

We present examples of minimal surfaces.

Example 7.1 (catenoid). One of the simplest examples of a minimal surface is the catenoid, which we considered in Chap. 1. Because the catenoid is a surface of revolution of a certain curve, to solve the minimization problem for the Dirichlet integral in the class of surfaces of revolution, we can apply the technique of the one-dimensional calculus of variations. We recall that in Chap. 1, we found conditions under which the catenoid yields the minimum area in this class of surfaces.

Example 7.2 (helicoid). The helicoid is obtained as a result of superposing two motions of a straight line, a translation and a rotation with a constant angular velocity in the plane orthogonal to the velocity vector of the translation. The equations of the helicoid (with accuracy up to a factor) have the form

$$z = \arctan(x/y).$$

Example 7.3 (Scherk surface). The simplest function of two variables is the function of the form

$$z = f(x) + g(y). \tag{7.8}$$

We find solutions to minimal surface equation (7.3) that have form (7.8). Substituting (7.8) in (7.3), we obtain

$$f''(x)(1 + g'(y)^2) + g''(y)(1 + f'(x)^2) = 0.$$

Separating variables, we obtain

$$\frac{f''(x)}{1 + (f'(x))^2} = c, \qquad \frac{g''(y)}{1 + (g'(y))^2} = -c.$$

Integration of these equations yields

$$f' = \tan(cx + a), \qquad g' = -\tan(cy + d),$$
$$f = -\frac{1}{c} \ln \cos(cx + a) + b, \qquad g = \frac{1}{c} \ln \cos(cy + d) + e.$$

Therefore, all surfaces of form (7.8) are obtained from the surface

$$z = \ln \frac{\cos y}{\cos x} \qquad (7.9)$$

by translations along the coordinate axes and the homothety (defined by the parameter c). Surface (7.9) is called the *Scherk surface*.

Example 7.4. Totally geodesic submanifolds of a Riemannian space are minimal manifolds. Indeed, the second fundamental form of a totally geodesic submanifold (and hence the mean curvature) vanishes.

Example 7.5. Compact complex submanifolds of Kähler manifolds are minimal.

Any complex submanifold $\mathfrak{X} \subset \mathbb{C}^n$ is minimal with respect to perturbations whose support belongs to a boundary domain of the space $\mathfrak{X}$ [31]. In particular, the submanifolds $\mathbb{CP}(k) \subset \mathbb{CP}(n)$ are minimal (their minimality is also implied by Example 7.4).

§2. Necessary Optimality Conditions for a Multiple Integral

Let M^m be a Riemannian manifold, and let $L^{\nu-1} \subset M^m$ be a closed bounded smoothly embedded submanifold such that it is the boundary of at least one ν-dimensional manifold $\mathfrak{N}^\nu \subset M^m$. For brevity, we omit the superscript denoting the dimension of a manifold in what follows.

To assign a functional of the multiple integral type on the set of smooth ν-dimensional manifolds $\mathfrak{N} \subset M$ with the boundary

$$\partial \mathfrak{N} = L, \qquad (7.10)$$

we introduce a generalized parametric assignment of surfaces $\mathfrak{N}$ in which a certain smooth ν-dimensional manifold W with boundary L now plays the role of the range of parameters. The interior of this manifold does not necessarily contain a global coordinate system but, as does any manifold, contains an atlas consisting of local coordinates. Local coordinates on W are denoted by $t = (t^1, \ldots, t^\nu)$, and local coordinates on M are denoted by $y = (y^1, \ldots, y^m)$.

Let $f: W \to M$ be a smooth embedding that is identical on L. The mapping f in local coordinates is given by smooth functions $y = y(t)$. The integral over the manifold W can be written in local coordinates using a partition of unity, i.e., in essence, dividing the domain of integration such that each part lies entirely in one of the charts. In what follows, bearing this procedure in mind, we use the following convention: admitting some freedom in the notation, we write the integrand in a local form as the integration in one of the charts but assume that the integration is in fact performed over the whole manifold W.

We define the functional being minimized by

$$J = \int_W F\left(t, y(t), \frac{\partial y}{\partial t}\right) dt,$$

where F is a given smooth function of three groups of variables: the vector $t = (t^1, \ldots, t^\nu)$, the vector $y = (y^1, \ldots, y^m)$, and the matrix $\partial y / \partial t = (\partial y^i / \partial t^\alpha)$.

2.1. Euler Equation. We consider first the case where the manifold M is the Euclidean space $\mathbb{R}^m$. As in Chap. 1 for the one-dimensional calculus of variations, we compute the first variation of the functional J by differentiating it with respect to the direction $h(\cdot) \in C_0^2(W, \mathbb{R}^m)$. Here, $C_0^2(W, \mathbb{R}^m)$ denotes the space of twice continuously differentiable functions on the manifold W that assume values in the space $\mathbb{R}^m$ and vanish on the boundary of L (this is needed in order to not violate boundary conditions (7.10)):

$$h|_L = 0. \tag{7.11}$$

In what follows, we assume that the point $\hat{y}(\cdot)$ at which we differentiate belongs to the space $C^2(W, \mathbb{R}^m)$.

Remark. When deducing the Euler equation in Chap. 1, we restrict ourselves to the assumption that the desired extremal belongs to the space C^1. Of course, this assumption is less restrictive as compared with the assumption of belonging to the space C^2; nevertheless, it remains an a priori constraint on the character of the smoothness of the unknown object. Although this assumption is, in fact, valid for most real problems, the Euler equation is not a necessary optimality condition in general. Therefore, we now, without any doubt, make a slightly stronger a priori assumption in order to simplify the presentation.

We therefore differentiate the function

$$\phi(\lambda) = \int_W F\left(t, \hat{y}(t) + \lambda h, \frac{\partial(\hat{y} + \lambda h)}{\partial t}\right) dt \tag{7.12}$$

with respect to λ for $\lambda = 0$.

We use the following abbreviations. The "hat" over a function means that instead of the variable y and its derivatives, we consider $\hat{y}$ as its arguments.

Further, we use the convention on summation over repeating sub- and superscripts. Latin indices refer to coordinates of the manifold M and are varied from 1 to m; Greek indices refer to coordinates on W and are varied from 1 to ν.

If $\hat{y}(\cdot)$ gives the minimum value to the functional J, then

$$\phi'(0) = \int_W \left[\frac{\partial \widehat{F}}{\partial y^i} h^i + \frac{\partial \widehat{F}}{\partial(\partial y^i/\partial t^\alpha)} \frac{\partial h^i}{\partial t^\alpha} \right] dt = 0. \tag{7.13}$$

We integrate the second group of summands by parts taking into account that boundary conditions (7.11) lead to terms outside the integral sign vanishing:

$$\phi'(0) = \int_W \left[\frac{\partial \widehat{F}}{\partial y^i} - \frac{\partial}{\partial t^\alpha} \left(\frac{\partial \widehat{F}}{\partial(\partial y^i/\partial t^\alpha)} \right) \right] h^i \, dt = 0 \tag{7.14}$$

for any functions $h(\cdot) \in C_0^2(W, \mathbb{R}^m)$.

The coefficients the functions h^i in the parentheses in (7.14) are continuous. We therefore obtain the following necessary condition for (7.14) to be satisfied:

$$\frac{\partial \widehat{F}}{\partial y^i} - \frac{\partial}{\partial t^\alpha} \left(\frac{\partial \widehat{F}}{\partial(\partial y^i/\partial t^\alpha)} \right) = 0, \quad i = 1, \dots, m, \tag{7.15}$$

which is called the *Euler equation*.

In the case where M is an arbitrary Riemannian manifold, a slightly more fundamental approach is required, which we do not carry out in detail, however: we only indicate the main steps of the corresponding technique. The difficulty arises because of the nonlinearity of the manifold M: to define a variation of the functional, it is not possible to multiply the variation $h(\cdot)$ by the number λ, and it is necessary to use the differential calculus on infinite-dimensional manifolds in essence. The infinite-dimensional manifold we deal with here is the manifold of embeddings $f : W \to M$ that are identical on the manifold L. It is called the *film manifold* for brevity and is denoted by $\mathfrak{W}$.

Let $\hat{f}$ be a certain point of the manifold $\mathfrak{W}$. A curve passing through a point $f \in \mathfrak{W}$ is a one-parameter family of embeddings f_α, where $f_0 = f$. The tangent to this curve is the derivative of the family f_α with respect to the parameter α for $\alpha = 0$, which is a vector field on the set $f_0(W)$ and is tangent to the manifold M. Because each of the embeddings f_α satisfies boundary condition (7.10), this vector field should vanish on the boundary L. Therefore, the tangent space to the manifold $\mathfrak{W}$ at the point $\hat{f}$ is the set of smooth vector fields on the image $\hat{f}(W)$ of the mapping $\hat{f}$ that have the following properties: they are tangent to the manifold M and vanish on L. We preserve the symbol $h(\cdot)$ for these vector fields.

The functional J assigns the mapping $J : \mathfrak{W} \to \mathbb{R}$. To find the derivative of this mapping in the space $\mathfrak{W}$ with respect to a direction given by a specific vector field h, we extend the vector field h from the manifold $\hat{f}(W)$ to its neighborhood in an arbitrary way as a smooth vector field. From each point of the manifold $\hat{f}(W)$, we move at time λ along the integral trajectories of the obtained field. For a sufficiently small λ, the result of such a movement is a smoothly embedded manifold $y(t, \lambda)$ that is diffeomorphic to W and passes through L. Substituting it in the functional J, we obtain a function $\phi(\lambda)$. Using the standard results of the theory of ordinary differential equations and calculus, it is easy to prove the following statement.

Proposition 7.2.
1. *The function $\phi(\lambda)$ is differentiable.*
2. *Its derivative for $\lambda = 0$ depends on only the values of the field h on the manifold $\hat{f}(W)$ itself and does not depend on the extension of this field.*
3. *The formula for the derivative $\phi'(0)$ coincides with formula (7.13) (with the interpretation of the field h changed in the corresponding way).*
4. *As before, the Euler equation is given by (7.15).*

We can obtain the formula for the second variation of the functional J by computing the second derivative of the function ϕ with respect to λ for $\lambda = 0$.

2.2. Second Variation. Again, we begin with the case where $M = \mathbb{R}^m$. We compute the second derivative of the function $\phi(\lambda)$ defined above by (7.12):

$$\phi''(0) = \delta^2 J$$

$$= \int_W \left[\frac{\partial^2 \hat{f}}{\partial(\partial y^i/\partial t^\alpha)\partial(\partial y^j/\partial t^\beta)} \frac{\partial h^i}{\partial t^\alpha} \frac{\partial h^j}{\partial t^\beta} \right.$$

$$\left. + 2\frac{\partial^2 \hat{f}}{\partial(\partial y^i/\partial t^\alpha)\partial y^j} \frac{\partial h^i}{\partial t^\alpha} h^j + \frac{\partial^2 \hat{f}}{\partial y^i \partial y^j} h^i h^j \right] dt. \tag{7.16}$$

Expression (7.16) is called the *second variation* of the functional J and is denoted by $\delta^2 J$.

To extend the formula of the second variation to an arbitrary manifold M, it suffices, as in the case of the first variation, for $h(\cdot)$ to be a vector field defined on $\hat{f}(W)$ and tangent to the manifold M.

Proposition 7.3. *Let $\hat{f}(W)$ satisfy Euler equation (7.15). Then a necessary minimality condition for the functional J is the nonnegativity of the second variation $(\delta^2 J \geq 0)$ for any smooth vector field $h(\cdot)$ that vanishes on L.*

The proof is obvious.

To obtain conditions for the nonnegativity of the second variation it is usual, as in the one-dimensional case, to consider the *associated problem*, i.e., the minimization problem of the quadratic functional $\delta^2 J$. The Euler equation for the functional $\delta^2 J$ is called the *Euler–Jacobi equation* for the initial functional J. It has the form

$$\frac{\partial}{\partial t^\alpha}\left\{\frac{\partial^2 \hat{f}}{\partial(\partial y^i/\partial t^\alpha)\partial(\partial y^j/\partial t^\beta)}\frac{\partial h^j}{\partial t^\beta} + \frac{\partial^2 \hat{f}}{\partial(\partial y^i/\partial t^\alpha)\partial y^j}h^j\right\}$$

$$+ \frac{\partial^2 \hat{f}}{\partial(\partial y^j/\partial t^\alpha)\partial y^i}\frac{\partial h^j}{\partial t^\alpha} + \frac{\partial^2 \hat{f}}{\partial y^i\partial y^j}h^j = 0, \quad i = 1,\dots,m. \quad (7.17)$$

2.3. Variational Equation. Let

$$\Phi\left(t, y, \frac{\partial y}{\partial t},\dots\right) = 0 \qquad (7.18)$$

be a partial differential equation with respect to the unknown functions $y \in M$ of the independent variables $t = (t^1,\dots,t^\nu) \in W$. Here, $\partial y/\partial t$ is the $m\times\nu$ matrix of partial derivatives; the ellipsis dots stand for higher-order derivatives.

The *variational equation* of Eq. (7.18) in a neighborhood of its solution $\hat{y}(\cdot)$ (or merely the variational equation if it is clear what solution we speak about) is the following linear partial differential equation obtained by linearization of Eq. (7.18) at the points $\hat{y}(\cdot)$:

$$\frac{\partial\widehat{\Phi}}{\partial y}h(t) + \frac{\partial\widehat{\Phi}}{\partial(\partial y/\partial t)}\frac{\partial h(t)}{\partial t} + \cdots = 0. \qquad (7.19)$$

The unknown function in this equation is $h(\cdot)$.

Proposition 7.4. *Let $y(t,\sigma)$ be a one-parameter family of solutions to Eq. (7.18) that contains the solution $\hat{y}(t)$ for $\sigma = 0$ and is smooth in all its arguments. Then*

$$h(t) = \left.\frac{\partial y}{\partial\sigma}\right|_{\sigma=0}$$

is a solution to variational equation (7.19).

Proof. It suffices to substitute the family of solutions $y(t,\sigma)$ in (7.18), to differentiate the resulting identity with respect to σ (for $\sigma = 0$), and to use the fact that mixed partial derivatives are independent of the order of differentiation.

Proposition 7.5. *Euler–Jacobi equation (7.17) is the variational equation for Euler equation (7.15).*

Proof. We substitute the family of solutions $y(t,\sigma)$ to the Euler equation (7.15) in (7.15) and differentiate the resulting identity with respect to σ for $\sigma = 0$. It is easy to verify that the result of differentiation is Euler–Jacobi equation (7.17).

Definition 7.1. Solutions to the Euler–Jacobi equation are called the *Jacobi field*.

We can describe Jacobi fields using geodesic variations.

Definition 7.2. A family of surfaces $y(t, \sigma)$ $(t \in W, |\sigma| \leq \epsilon)$ is called a *geodesic variation* if for each fixed σ, the manifold $y(t, \sigma)$ is a solution to the Euler equation.

Proposition 7.6. *Let $y(t, \sigma)$ be a geodesic variation. Then*

$$\frac{\partial y(t, \sigma)}{\partial \sigma}\bigg|_{\sigma=0}$$

is a Jacobi field on $f(W)$.

To prove this statement, it suffices to apply Propositions 7.4 and 7.5.

For an arbitrary manifold M on which the functional J is defined, the second variation $\delta^2 J$ is a quadratic form on the linear space of vector fields on M restricted to the extremal $\hat{x}(\cdot)$. The set of such fields forms a vector bundle.

§3. Vector Bundles

Definition 7.3. A smooth real vector (linear) bundle over a smooth manifold B is a smooth manifold E on which a smooth mapping $\pi \colon E \to B$ having the following properties is defined:

1. The inverse image of any point $b \in B$ is a submanifold $F_b \subset E$ that is a linear space isomorphic to $\mathbb{R}^n$ (moreover, the dimension n does not depend on the point $b \in B$).
2. Any point $b \in B$ has a neighborhood U such that its inverse image $\pi^{-1} U$ is diffeomorphic to the direct product $\mathbb{R}^n \times U$; this diffeomorphism is denoted by $h \colon \mathbb{R}^n \times U \to \pi^{-1} U$.
3. For any point $b \in U$, the restriction $h|_{\mathbb{R}^n \times b}$ is a linear one-to-one correspondence between $\mathbb{R}^n$ and $\pi^{-1} b$.

The manifold E is called the *bundle space*, B is called the *base*, the mapping π is called the *projection*, and the vector space $\pi^{-1} b$ is called the *fiber* over the point b. The pairs $(\pi^{-1} U, h^{-1})$ form an atlas of local coordinates on the bundle space E. Property 2 is called the *local triviality* property (if it is possible to take $U = B$ in this property, then the bundle is called *trivial*). Property 3 means that the *structure groups* of a vector bundle are subgroups of the group $\mathrm{GL}(n)$. On the intersection of charts $(\pi^{-1} U, h^{-1})$, we have the mapping $\mathbb{R}^n \to \mathbb{R}^n$. The group generated by the mappings in property 3 is called the *structure group* of the bundle.

If we replace property 3 by the following property 3', then the structural group of this bundle is a subgroup of the group $\mathrm{O}(n)$:

$3'$. For any point $b \in U$, the restriction $h|_{\mathbb{R}^n \times b}$ is an orthogonal mapping of $\mathbb{R}^n$ onto $\pi^{-1}b$.

Example 7.6. The well-known example of a vector bundle is the *Möbius band*, which is obtained as follows. In the plane $\mathbb{R}^2$ with Cartesian coordinates (x, y), we consider the direct product of the segment $[0, 1]$ of the axis Ox and the axis Oy. We identify the points $(0, y)$ of the line $x = 0$ with the points $(1, -y)$ of the line $x = 1$. We obtain the vector bundle whose base is a circle and whose fibers are the lines $x = \text{const}$. The structural group of this bundle is the cyclic group $\mathbb{Z}_2$ of the second order.

Example 7.7 (canonical linear bundle). We represent the real projective space $\mathbb{RP}^n$ as the set of nonordered pairs $(x, -x)$, where x ranges over the unit sphere $S^n \subset \mathbb{R}^{n+1}$. Let E be a subset of $\mathbb{RP}^n \times \mathbb{R}^{n+1}$ consisting of all pairs $(\{\pm x\}, v)$ in which the vector v is a multiple of the vector x. We define the projection $\pi(\{\pm x\}, v) = \{\pm x\}$. Then the fiber $\pi^{-1}(\{\pm x\})$ can be identified with the line passing through x and $-x$ in $\mathbb{R}^{n+1}$. The obtained linear bundle is called the *canonical linear bundle* over $\mathbb{RP}^n$ and is denoted by γ_n^1.

We prove that γ_n^1 is locally trivial. Let $U \subset S^n$ be an open set so small that it contains no pair of diametrically opposite points. Let U_1 be the image of U in $\mathbb{RP}^n$. Then the homeomorphism $h \colon U_1 \times \mathbb{R} \to \pi^{-1}(U_1)$ is given by

$$h(\{\pm x\}, t) = (\{\pm x\}, tx).$$

Obviously, (U_1, h^{-1}) is a local coordinate system.

In the canonical linear bundle, the base is the set of lines (each of these line is considered as a point of the projective space) passing through the origin of the space $\mathbb{R}^{n+1}$; fibers are these lines themselves, which are now considered not as a whole but as the set of their points. Therefore, γ_n^1 is sometimes called the *tautological bundle*.

Example 7.8. The *canonical vector bundle* γ_n^k over the Grassmann manifold $G_k(\mathbb{R}^n)$ consists of all pairs $(V, x) \in G_k(\mathbb{R}^n) \times \mathbb{R}^n$, where V is a k-dimensional linear subspace and $x \in V$. The base of γ_n^k is the Grassmann manifold; in correspondence to each pair (V, x), the projection sets its first element, the subspace V.

Exercise. Prove that γ_n^k is locally trivial.

Example 7.9. The *tangent bundle* τ_M of a smooth manifold M consists of all pairs (t, x), where $t \in M$ and x is a vector tangent to M at the point t. The projection is given by $\pi(t, x) = t$. The local triviality property is obvious.

Example 7.10. The *normal bundle* ν_M of a smooth manifold M that is embedded in a Riemannian manifold $\mathfrak{M}$ consists of the set of all pairs (t, v), where $t \in M$ and v is a vector tangent to the manifold $\mathfrak{M}$ at the point t and orthogonal to the tangent space to M at the point t.

Definition 7.4. A *section* of a vector bundle with the base B and bundle space E is any continuous mapping

$$s \colon B \to E$$

that transforms any point $b \in B$ into a certain point of the fiber $\pi^{-1}b$. A section is said to be *everywhere nonzero* if for all points b, the vector $s(b)$ never vanishes.

A vector field on a manifold M is a section of the tangent bundle τM. Obviously, the trivial bundle has an everywhere nonzero section. We show that the bundle γ_n^1 in Example 7.7 has no such sections. The superposition of the standard mapping $S^n \to \mathbb{RP}^n$ and a section s transforms a point $x \in S^n$ into a certain pair $(\{\pm x\}, r(x)x)$. The function $r(x)$ is a continuous real function of x that satisfies the relation $r(-x) = -r(x)$. By the mean value theorem, because the space S^n is connected, there exists a point x_0 such that $r(x_0) = 0$. Therefore, γ_n^1 is a nontrivial bundle.

Exercise. Show that the bundle γ_1^1 coincides with the Möbius band (see Example 7.6).

Definition 7.5. Sections $s_1, s_2, \ldots, s_n$ are said to be *everywhere independent* if for each point $b \in B$, the vectors $s_1(b), \ldots, s_n(b)$ are linearly independent.

Theorem 7.1. *A vector bundle ξ of dimension n is trivial iff it admits n everywhere independent sections.*

Proof. If ξ is a trivial bundle, then in order to construct everywhere independent sections, it suffices to take a basis of one of its fibers and extend it as constant functions to all other fibers.

Conversely, let $s_1, \ldots, s_n$ be everywhere independent sections. Let $x = (x_1, \ldots, x_n)$ be coordinates in $\mathbb{R}^n$. We assign the mapping

$$f \colon B \times \mathbb{R}^n \to E$$

by

$$f(b, x) = x_1 s_1(b) + \ldots + x_n s_n(b).$$

Obviously, f is a continuous mapping that isomorphically maps each fiber of the trivial bundle to the corresponding fiber of the bundle ξ. Therefore, ξ is trivial.

§4. Distributions and the Frobenius Theorem

Let M be a smooth ν-dimensional manifold.

Definition 7.6. A rule that sets a k-dimensional vector subspace of the tangent space to the manifold M at the point t in correspondence to each point $t \in M$ is called a k-*dimensional distribution* (or a distribution of k-dimensional planes) on M.

We can assign distributions by two different methods: using a basis or using a set of linear equations. In the first case, a distribution W is given by a tuple of k vector fields $X_1(t), \ldots, X_\nu(t)$ linearly independent at each point; a vector of the tangent space $X \in \tau_{t_0} M$ belongs to the distribution W if X is a linear combination of the vectors $X_1(t_0), \ldots, X_\nu(t_0)$. In the second case, a distribution W is given by a tuple of $\nu - k$ linearly independent differential 1-forms (for brevity, we say 1-forms) $\omega_1, \ldots, \omega_{(\nu - k)}$; a vector of the tangent space X belongs to the distribution W if all these differential forms vanish at the vector X. We explain that at each fixed point $t \in M$, the set of conditions $\omega_i(X) = 0$ defines a set of $\nu - k$ linear equations for the coordinates X of tangent vectors at the point t. Because this system has linearly independent equations, it defines a k-dimensional linear subspace of the tangent space at the point t. When t varies, we obtain the distribution W. The differential forms $\omega_1, \ldots, \omega_{(\nu - k)}$ are called *basis forms* of the distribution W.

A distribution is called *smooth* if it can be given using a basis consisting of smooth vector fields (or smooth differential forms).

Lemma 7.1 (Cartan formula). *Let A and B be two vector fields, and let ω be a differential 1-form on a manifold W. Then*

$$d\omega(A, B) = A(\omega(B)) - B(\omega(A)) - \omega([A, B]). \qquad (7.20)$$

Proof. We write our values in local coordinates. Let

$$A = A^i(t)\frac{\partial}{\partial t^i}, \qquad B = B^i(t)\frac{\partial}{\partial t^i}, \qquad \omega = c_j(t)dt^j.$$

Then

$$d\omega = \sum_{i<j}\left(\frac{\partial c_j}{\partial t^i} - \frac{\partial c_i}{\partial t^j}\right)dt^i \wedge dt^j, \qquad [A, B] = \left(A^i\frac{\partial B^j}{\partial t^i} - B^i\frac{\partial A^j}{\partial t^i}\right)\frac{\partial}{\partial t^j}.$$

Substituting these expressions in (7.20), we obtain

$$d\omega(A, B) = \sum_{i<j}\left(\frac{\partial c_j}{\partial t^i} - \frac{\partial c_i}{\partial t^j}\right)(A^i B^j - A^j B^i). \qquad (7.21)$$

We find the action of the vector field A on the function $\omega(B) = c_j B^j$:

$$A(\omega(B)) = A^i\frac{\partial c_j}{\partial t^i}B^j + A^i c_j\frac{\partial B^j}{\partial t^i}. \qquad (7.22)$$

Similarly,

$$B(\omega(A)) = B^i \frac{\partial c_j}{\partial t^i} A^j + B^i c_j \frac{\partial A^j}{\partial t^i}. \tag{7.23}$$

Finally,

$$\omega([A, B]) = c_j \left(A^i \frac{\partial B^j}{\partial t^i} - B^i \frac{\partial A^j}{\partial t^i} \right). \tag{7.24}$$

Subtracting (7.23) and (7.24) from (7.22), we obtain the required result.

Definition 7.7. We say that a differential 2-form θ belongs to an ideal generated by 1-forms $\omega_1, \ldots, \omega_{(\nu-k)}$ if it is representable in the form

$$\theta = \sum_{i=1}^{\nu-k} \theta_i \wedge \omega_i,$$

where θ_i are 1-forms.

Proposition 7.7 (test for belonging to an ideal). *Let an ideal I be generated by forms $\omega_1, \ldots, \omega_{(\nu-k)}$. A differential 2-form θ belongs to the ideal I iff $\theta(X, Y) = 0$ for any two vectors X and Y belonging to the distribution W generated by the forms $\omega_1, \ldots, \omega_{(\nu-k)}$.*

Proof. In one direction, the implication is obvious: if $\theta \in I$ and $X, Y \in W$, then $\theta(X, Y) = 0$ because

$$\theta_i \wedge \omega_i(X, Y) = \det \begin{pmatrix} \theta_i(X) & \theta_i(Y) \\ \omega_i(X) & \omega_i(Y) \end{pmatrix}.$$

To prove the implication in the other direction, we let $\theta(X, Y) = 0$ for any $X, Y \in W$. We add forms $\alpha_1, \ldots, \alpha_k$ to the forms ω_i, $i = 1, \ldots, \nu - k$ such that at each point $x \in M$, the obtained tuple forms a basis in the cotangent space $\tau^* M$. We use this basis to expand the form θ:

$$\theta = f^{ij}\omega_i \wedge \omega_j + g^{is}\omega_i \wedge \alpha_s + h^{sl}\alpha_s \wedge \alpha_l. \tag{7.25}$$

In the tangent space τM, we choose a basis $A_1, \ldots, A_k, B_1, \ldots, B_{\nu-k}$ that is dual to the basis α_i, ω_j. This means that

$$
\begin{aligned}
\alpha_i(A_i) &= 1, & \alpha_i(A_j) &= 0 \text{ for } i \neq j, & \omega_s(A_i) &= 0 \text{ for all } s, \\
\omega_j(B_j) &= 1, & \omega_i(B_j) &= 0 \text{ for } i \neq j, & \alpha_s(B_j) &= 0 \text{ for all } s.
\end{aligned}
$$

The form θ belongs to the ideal I iff all the coefficients h^{sl} are identically equal to zero. We suppose that one of the coefficients $h^{sl} \neq 0$ at some point. Then, by (7.25), the value of the form θ on the pair of vectors A_s, A_l is different from zero. But $A_s, A_l \in W$. The obtained contradiction proves the assertion.

Let W be a smooth k-dimensional distribution.

Definition 7.8. The distribution W is said to be *involutive* if one of the following two conditions holds:

1. The commutator of any two vector fields that belong to the distribution W also belong to W.
2. If $\omega_1, \ldots, \omega_{(\nu-k)}$ are basis forms of the distribution W, then $d\omega_i$ belongs to the ideal generated by the forms ω_i, $i = 1, \ldots, \nu - k$.

Proposition 7.8. *Conditions 1 and 2 are equivalent.*

Proof. We consider two vector fields $A, B \in W$.

Let condition 2 hold. Then, according to Proposition 7.7, $d\omega(A, B) = 0$. By (7.20), we have $\omega([A, B]) = 0$; this means that condition 1 holds.

Let condition 1 hold, and let $\omega_1, \ldots, \omega_{\nu-k}$ be basis forms of the distribution W. Then we obtain

$$d\omega_i(A, B) = 0 \tag{7.26}$$

for all vector fields $A, B \in W$. By Proposition 7.7, this means that $d\omega_i \in I$.

The concept of *foliation* is closely related to the concept of *distribution*. We begin with the simplest example, which, however, describes the general case from the local structure point of view.

Let $\mathbb{R}^n$ be the n-dimensional Euclidean space, $0 \leq k \leq n$. We consider the space $\mathbb{R}^n$ as the union of k-dimensional planes $\mathbb{R}^k$ of the form

$$x^1 = c_1, \qquad \ldots, \qquad x^{n-k} = c_{n-k}, \tag{7.27}$$

where c_i are arbitrary constants. In this case, we say that the subspaces given by Eqs. (7.27) are *leaves* and the space $\mathbb{R}^n$ is *foliated* into k-dimensional leaves or a k-dimensional foliation is given on $\mathbb{R}^n$.

We now give the general definition of a foliation. Let A be a certain set of parameters α, and let M be a smooth n-dimensional manifold.

Definition 7.9. A family of smooth k-dimensional submanifolds $L_\alpha \subset M$ is called a *k-dimensional foliation* on the manifold M if one and only one of the submanifolds L_α passes through each point $x \in M$ and, moreover, the following continuity condition holds: each point $x \in M$ has a neighborhood U such that the restriction of leaves L_α to the neighborhood U is described by the equations

$$f_1(x) = c_1, \qquad \ldots, \qquad f_{n-k}(x) = c_{n-k},$$

where the functions f_i are smooth and the Jacobi matrix $(\partial f_i / \partial x^j)$ has the maximum rank (equal to $n-k$) at all points $x \in U$.

We present an example of a distribution that does not define a foliation. Let x^1, x^2, and x^3 be Cartesian coordinates in $\mathbb{R}^3$. In each of the horizontal planes $x^3 = c$, we choose its own family of parallel lines. We assume that the directions of these lines are discontinuously varied when the constant c varies. Then the constructed family of lines does not satisfy the continuity condition and therefore does not form a foliation on $\mathbb{R}^3$.

Let $M = \mathbb{R}^n$. We consider a k-dimensional foliation. The functions f_i can be taken as new variables in $\mathbb{R}^n$. Then the foliation turns out to be locally equivalent to the foliation given by formulas (7.27). However, the global structure of a foliation can be sufficiently complicated.

Example 7.11. The family of integral curves of arbitrary smooth fields of directions on a manifold M can serve as an example of a foliation. This is a typical example of a one-dimension foliation.

Example 7.12 (Reeb foliation). Let S^1 be the circle $S^1 = \exp(2\pi i \beta)$, $\beta \in \mathbb{R}$, and D^2 be a two-dimensional disk given by the relation $x^2 + y^2 \leq 1$ in the Cartesian coordinates. The direct product $P = D^2 \times S^1$ is called the *solid torus*.

We consider the function $\exp[r^2/(1 - r^2)]$. Let r be equal to the distance to the central axis of the solid torus P, and let β be equal to the angle of turn on the circle S^1. We consider a certain two-dimensional surface L_0 in P parameterized by the points (x, y) of the disk D^2:

$$L_0 = (x, y, \beta(x, y)),$$

where

$$\beta(x, y) = \exp\left[2\pi i \exp\left(\frac{x^2 + y^2}{1 - x^2 - y^2}\right)\right].$$

The leaves of the Reeb foliation are obtained from the surface L_0 by a turn of an arbitrary angle α:

$$L_\alpha = ((x, y), \beta(x, y) + \alpha)).$$

Obviously, one and only one leaf passes through each point of the solid torus. The constructed family of leaves obviously depends smoothly on a point $(x, y, \beta) \in P$.

We present an image to help the reader imagine a snake that is inside a toric tube and is swallowing its tail. In this case, the tail remains fixed, and the snake lengthens. The diameter of the head of the snake tends to the diameter of the tube. After infinitely many turns, the skin of the snake forms the leaf L_0 that is a two-dimensional surface homeomorphic to $\mathbb{R}^2$. Other leaves have the same structure; it is easy to see that distinct leaves either never intersect or are identically equal. A unique compact leaf of the Reeb foliation is the surface of the torus to which each of the leaves asymptotically tends (Fig. 7.1).

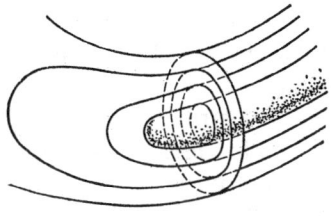

Fig. 7.1

Example 7.13 (Reeb foliation on the three-dimensional sphere).
We assign the sphere S^3 by the equation

$$x^2 + y^2 + z^2 + w^2 = 1,$$

where x, y, z, and w are Cartesian coordinates on $\mathbb{R}^4$. Obviously, the two-dimensional torus T described by the equations

$$T = \{x^2 + y^2 = z^2 + w^2 = 1/2\}$$

is a subset of this sphere.

The torus T divides the sphere S^3 into two congruent solid tori

$$x^2 + y^2 \leq z^2 + w^2,$$
$$z^2 + w^2 \leq x^2 + y^2.$$

Considering the Reed foliation from Example 7.12 inside each of these tori, we obtain the Reeb foliation on the sphere S^3 with a unique compact leaf T.

A submanifold $N \subset M$ is called an *integral manifold* of a distribution W if all tangent planes to the manifold N belong to the distribution W.

Definition 7.10. A distribution W of dimension k on a manifold M is said to be *completely integrable* if there exists a k-dimensional foliation on M and all its leaves are integral manifolds of the distribution W.

Theorem 7.2 (Frobenius). *A distribution is completely integrable iff it is involutive.*

Proof. We suppose that the distribution W is completely integrable. Let two vector fields A and B belong to the distribution W. We consider a certain k-dimensional integral manifold L of the distribution W. Then the fields A and B are tangent to L at all points $x \in L$. Therefore, its commutator is also tangent to L (see Chap. 3); therefore, it belongs to the distribution W. This means that the distribution W is involutive.

We now prove the inverse statement. We suppose that the distribution W is involutive. We proceed by induction on the dimension of the distribution.

For $k = 1$, the assertion on the existence of a foliation that is tangent to this distribution is implied by the theorem of existence and uniqueness for ordinary differential equations.

Let the theorem hold for distributions of dimensions not exceeding $k - 1$. We then prove that it holds for distributions of dimension k.

It suffices to carry out the proof in a certain sufficiently small neighborhood U of a fixed point. Without loss of generality, we can assume that the coordinates of this point are equal to zero.

Let $X_1, \ldots, X_k$ be smooth vector fields that form a basis of the involutive distribution W. Because this is a basis, $X_1(0) \neq 0$. By the theorem on the strengthening of a vector field ([4], Sec. 32) in the neighborhood U, we can choose a coordinate system such that the functions

$$x_1^1 = 1, \qquad x_1^2 = 0, \qquad \ldots, \qquad x_1^\nu = 0$$

are local coordinates of the vector field X_1. The differential operator corresponding to this field is

$$\frac{\partial}{\partial t^1}.$$

Further, instead of the vector fields $X_2, \ldots, X_k$, we construct the fields $\widetilde{X}_2, \ldots, \widetilde{X}_k$ defined by

$$\widetilde{X}_i = X_i - \phi_i X_1.$$

We choose scalar-valued functions $\phi_i(t)$ such that the first coordinate of the fields $\widetilde{X}_i(t)$ is equal to zero for all $t \in U$. In other words, the vector X_i is "turned" in the plane (X_1, X_i) such that it has a zero projection onto the vector X_1. It is easy to see that the set of vectors $X_1, \widetilde{X}_2, \ldots, \widetilde{X}_k$ remains a basis of the distribution W.

We consider a section V of the neighborhood U defined by the equation $t^1 = 0$. The fields $\widetilde{X}_2, \ldots, \widetilde{X}_k$ are tangent to the section V because their first coordinate vanishes. The vectors $\widetilde{X}_2, \ldots, \widetilde{X}_k$ forms a basis of a $(k-1)$-dimensional involutive distribution $\widetilde{W}$ on the section V (the involutiveness of $\widetilde{W}$ is implied by the involutiveness of W). Therefore, by the induction hypothesis, the distribution $\widetilde{W}$ is completely integrable and defines a certain $(k-1)$-dimensional foliation on V.

Let the leaves l_c of the constructed foliation be given by the equations

$$f_i(t^2, \ldots, t^\nu) = c_i, \quad i = 1, \ldots, \nu - k + 1, \tag{7.28}$$

where the constants $c = (c_1, \ldots, c_{\nu-k+1})$ parameterize the set of leaves. That the leaves l_c are tangent to the distribution $\widetilde{W}$ is analytically expressed by the derivatives of the functions f_i in the direction of the vectors $\widetilde{X}_j$ vanishing at points in V:

$$\widetilde{X}_j(f_i) = 0, \quad t \in V. \tag{7.29}$$

We draw a trajectory of the vector field X_1 through each point of the section V. The trajectories emanating from points of the leaf l_c form a k-manifold, which is denoted by L_c. In the neighborhood U, each leaf L_c is given by the same system of equations (7.28) as the leaf l_c. That these equations do not depend on t^1 means that the leaves L_c are "cylinders" in the coordinate t^1. Therefore, we have constructed a k-dimensional foliation defined on U.

We prove that the constructed foliation is an integral manifold of the distribution W. To do this, it suffices to show that the derivatives of the functions f_i in the directions $X_1, \widetilde{X}_j$ vanish at all points of the neighborhood U. First, it is obvious that the vector field X_1 is tangent to the leaves L_c. Indeed, by construction, it is tangent to the trajectories by which the leaves L_c are "woven." Formally, this is expressed by the relations

$$X_1(f_i) = \frac{\partial f_i}{\partial t^1} = 0, \qquad (7.30)$$

which are obvious because the functions f_i do not depend on t^1. To compute the derivative of f_i in other directions, we find a system of differential equations that these derivative satisfy. For any $t \in U$, we transform the expression

$$\frac{\partial}{\partial t^1} \widetilde{X}_j(f_i) = X_1(\widetilde{X}_j(f_i)),$$

subtracting the zero summand $\widetilde{X}_j(X_1(f_i))$ (by (7.30)) from it. Then

$$\frac{\partial}{\partial t^1}(\widetilde{X}_j(f_i)) = [X_1, \widetilde{X}_j](f_i). \qquad (7.31)$$

We use the involutiveness of the distribution W. The commutators of the fields $X_1, \widetilde{X}_2, \ldots, \widetilde{X}_k$ belong to the distribution W because the fields belong to W. Because these fields form a basis of W, there exist functions $\kappa_j^s(t)$ such that

$$[X_1, \widetilde{X}_j] = \kappa_j^1 X_1 + \sum_{s=2}^{k} \kappa_j^s \widetilde{X}_s. \qquad (7.32)$$

Substituting (7.32) in (7.31) and taking (7.30) into account, we obtain

$$\frac{\partial}{\partial t^1}(\widetilde{X}_j(f_i)) = \sum_{s=2}^{k} \kappa_j^s(\widetilde{X}_s(f_i)). \qquad (7.33)$$

We fix the subscript i and the independent variables $t^2, \ldots, t^\nu$ in (7.33). Then we can consider these relations as a system of ordinary differential equations with respect to the unknown functions $(\widetilde{X}_s(f_i))$ of the independent variable t^1. The initial conditions for these functions for $t^1 = 0$ are zero by (7.29). Because system (7.33) is linear and homogeneous, by the uniqueness theorem, we have

$$\widetilde{X}_s(f_i) \equiv 0; \qquad (7.34)$$

this proves the complete integrability of the distribution W.

We consider the case of a distribution W of codimension one. In this case, the distribution W is given by only one differential 1-form ω. It is easy to see that the involutiveness condition of the distribution W (i.e., the relation $d\omega = \theta \wedge \omega$) is equivalent to the easily verified condition

$$d\omega \wedge \omega = 0. \qquad (7.35)$$

A simple example of a not completely integrable 1-form in $\mathbb{R}^3$ is the form $x\,dy + dz$. It is easy to verify that condition (7.35) does not hold for this form. Therefore, there is no two-dimensional foliation on $\mathbb{R}^3$ whose leaves are tangent to the distribution given by the form $x\,dy + dz$. This assertion means that there is no family of two-dimensional surfaces that are orthogonal to the family of integral curves of the vector field $(0, x, 1)$.

§5. Connection in a Linear Bundle

Let $\pi\colon E \to M$ be a linear bundle with an n-dimensional fiber over a smooth m-dimensional manifold M. The differential of the mapping π at a point $y \in E$ assigns the mapping of the space tangent to E at the point y into the space tangent to M at the point $\pi(y)$. The kernel of this differential (i.e., the set of vectors that are tangent to the fiber passing through the point y) is called the *vertical subspace* at the point y. This subspace is denoted by $V(y)$. The family of subspaces $V(y)$ forms a smooth n-dimensional distribution on E.

Exercise. Explain why the distribution $V(y)$ is smooth and integrable.

Definition 7.11. If a smooth distribution of m-dimensional planes on the manifold E is transversal to $V(y)$ at each point $y \in E$, it is called a *connection of the general form* ∇ on the bundle $\pi\colon E \to M$ with the m-dimensional base M. The planes of this distribution are called *horizontal subspaces* and are denoted by $H(y)$. A smooth curve $y = \gamma(\theta)$ in E is said to be *horizontal* if any of its tangent vectors is horizontal, i.e., $\gamma'(\theta) \in H(\gamma(\theta))$.

This definition implies that the space $T_y E$ tangent to the manifold E at any point y is the direct sum of the vertical and horizontal planes $T_y E = H(y) \oplus V(y)$; therefore, any vector tangent to the manifold E is uniquely decomposed into horizontal and vertical components.

Lemma 7.2. *Let a general connection be given in a bundle $\pi\colon E \to M$, and let $t(\theta)$ be a certain curve on the base M. Then for any point $y_0 \in E$ such that $\pi y_0 = t(0)$ and for any sufficiently small part of the curve $t(\theta)$, $0 \leq \theta \leq \delta$, there is only one horizontal curve $\gamma(\theta) \in E$ emanating at the point $y_0 = \gamma(0)$ and projected onto $t(\theta)$:*

$$\pi\gamma(\theta) = t(\theta).$$

Proof. We can assume that a sufficiently small segment of a smooth curve has no self-intersections. We restrict the bundle to the curve $t(\theta)$. Then the base of the obtained bundle is the curve $t(\theta)$, and its fibers are the fibers of the initial bundle $\pi^{-1}t(\theta)$ lying over points of the curve $t(\theta)$. Because the base of the bundle is one-dimensional, horizontal directions of the connection in the bundle E generate (or "cut out") one-dimensional horizontal directions in the bundle $E_{t(\theta)}$ over the curve $t(\theta)$. Integral trajectories of this field of horizontal directions and only they are horizontal curves of the bundle $E_{t(\theta)}$. The application of standard theorems on the existence and uniqueness of a

solution to the Cauchy problem for ordinary differential equations completes
the proof of the lemma.

The defined curve γ is called the *covering path* of the curve $t(\theta)$.

Remark. In Lemma 7.2, we consider a small part of the curve $t(\theta)$ because
the horizontal curves of the general connection can go to infinity in a finite
time. In this case, the covering path is not defined. Below, we consider a
natural condition that excludes the possibility of a horizontal curve going to
infinity in a finite time.

For now, we assume that horizontal curves do not go to infinity in a finite
time. Then for any smooth path $t(\theta)$, $\theta_0 \leq \theta \leq \theta_1$), we can use Lemma 7.2 to
define the mapping $\phi_{t(\theta)}$ of the fiber $\pi^{-1}t(\theta_0)$ into the fiber $\pi^{-1}t(\theta_1)$, which
is called the *parallel translation* of a fiber along the path $t(\theta)$.

By the smooth dependence of the solution to differential equations on the
initial data, the mapping ϕ is smooth. Further, the mapping ϕ does not depend
on the parameterization of the path $t(\theta)$. Indeed, the change of parameterization can only change the length of a tangent vector while preserving its
direction; this has no influence on the mapping $\phi_{t(\theta)}$.

Definition 7.12. A connection in a vector bundle is said to be *linear* if
the parallel translation along any curve is a linear fiber mapping.

Remark. In this definition, the linear fiber mapping is compatible with the
structural group of the vector bundle.

We describe the operation of parallel translation in local coordinates of
a bundle E. Let $t = (t^1, \ldots, t^m)$ be local coordinates on M, and let $x =
(x^1, \ldots, x^n)$ be local fiber coordinates. The coordinates on E are denote by $y =
(t, x)$ Coordinates of vectors of the tangent space are denoted by corresponding
capital letters (T, X). Because the horizontal plane is transversal to a fiber, its
equation in the tangent space $T_{tx}E$ can be written in the form $X = \mathfrak{A}(t, x)T$,
where $\mathfrak{A}$ is a certain $n \times m$ matrix.

We decompose an arbitrary vector (T, X) into the horizontal and vertical components. The first coordinates of the horizontal component should be
equal to T and the first coordinate of the vertical component should vanish.
Let (T, W) be coordinates of the horizontal component; the coordinates of
the vertical components are $(0, V)$. We find V and W. By the equation of the
horizontal plane, we have $W = \mathfrak{A}T$. The sum of the horizontal and vertical
components is equal to the vector (T, X) if $V = X - \mathfrak{A}T$. Therefore, the horizontal component of the vector (T, X) is the vector $h(T, X) = (T, \mathfrak{A}T)$, and
the vertical component is the vector $v(T, X) = (0, X - \mathfrak{A}T)$.

We consider a curve $t = t(\theta)$. Then the parallel translation of a fiber along
this curve is described in local coordinates by the differential equation

$$\dot{x} = \mathfrak{A}(t(\theta), x)\dot{t}(\theta). \tag{7.36}$$

Proposition 7.9. *A connection is linear iff the function $\mathfrak{A}(t,x)$ is linear in the variables x.*

Indeed, the mapping of shifting along trajectories of a system of differential equations is linear for any θ iff the system is linear.

By Proposition 7.9, we have $\mathfrak{A}(t,x) = -\Gamma(t)x$. Then the horizontal coordinate of the vector (T, X) is

$$h(T, X) = (T, -\Gamma(t)xT). \tag{7.37}$$

The functions $\Gamma(t)$ are called the *coefficients of a linear connection*.

Remark. We note that the coefficients of a linear connection are not a tensor. Indeed, the expression $\Gamma(t)x$ is an $n \times m$ matrix, and $\Gamma(t)\dot{t}(\theta)$ is an $n \times n$ matrix. Both these matrices are not tensors because these matrices are coefficients of the system of linear equations defining the translation of vectors. Therefore, in a fiber with a matrix $K(t)$ depending on a point $t \in M$, the coefficients of the linear system are transformed under a change of coordinates by using derivatives of the transformation matrix K.

Therefore, the system of linear differential equations (7.36), which defines the parallel translation of a fiber for the linear connection, is linear:

$$\dot{x} + (\Gamma(t(\theta))\dot{t}(\theta))x = 0. \tag{7.38}$$

Specifically because of this, by the theorem on the continuation of solutions to a linear system of differential equations, the horizontal curves of the linear connection cannot go to infinity in a finite time, and the parallel translation is defined for any smooth curve.

A connection can be characterized by a certain vector-valued differential form ω on the manifold E with values in the fiber $\mathbb{R}^n$. Namely, for each vector (T, X), we define the value $\omega(T, X)$ as the vector of the fiber that is the vertical component of the vector (T, X), i.e., $\omega(T, X) = X - \mathfrak{A}T$. The form ω is called the *connection differential form*.

Obviously, $\omega(T, X) = 0$ iff the vector (T, X) is horizontal. Therefore, the assignment of the form ω uniquely defines the distribution of horizontal planes and therefore the connection itself.

Proposition 7.10. *A connection is linear iff the coefficients of the form ω depend linearly on the coordinates of the fiber,*

$$\omega(T, X) = X + (\Gamma(t)x)T, \tag{7.39}$$

i.e., $\mathfrak{A}(t,x) = -\Gamma(t)x$.

The proof is a direct consequence of Proposition 7.9 and the definition of the form ω.

The parallel translation of fibers gives us a possibility to define the operation of covariant differentiation of sections of the bundle E. As the connection itself, the operator of covariant differentiation is denoted by ∇.

Let $T(t)$ be a vector field on the manifold M, $t(\theta)$ be an arbitrary integral curve of the field T in the base M, and s be a section of the bundle E.

Definition 7.13. The *covariant derivative* of the section s in the direction of the field T is the section $\nabla_T s$ defined by

$$\nabla_T s = \lim_{h \to 0} \frac{1}{h} [\tau_\theta^{\theta+h}(s(t(\theta+h))) - s(t(\theta))]. \tag{7.40}$$

Here, $\tau_\theta^{\theta+h} \colon \pi^{-1}(t(\theta+h)) \to \pi^{-1}(t(\theta))$ denotes the parallel translation of the fiber $\pi^{-1}(t(\theta+h))$ along the curve $t(\theta)$ from the point $t(\theta+h)$ to the point $t(\theta)$, which is realized along solutions to differential equation (7.38). Expanding this solution in h, we obtain

$$\lim_{h \to 0} \frac{1}{h} [s(t(\theta+h)) + \Gamma s(t(\theta+h))\dot{t}h + o(h) - s(t(\theta))] = \frac{\partial s}{\partial \theta} + \Gamma s \dot{t}.$$

Therefore,

$$\nabla_T s = \frac{\partial s}{\partial \theta} + \Gamma s \dot{t}. \tag{7.41}$$

Formula (7.41) shows that the covariant derivative in direction T at a point $t \in M$ depends only on the vector $T(t)$.

The following assertion is obvious.

Proposition 7.11. *A vector field is translated along a curve $t(\theta)$ parallel iff its covariant derivative with respect to $\dot{t}(\theta)$ is identically equal to zero.*

Proposition 7.11 implies that the operator of covariant differentiation and the connection uniquely define one another.

§6. Levi-Civita Connection

The most important (and the first) example of a connection is a connection in the tangent bundle of a Riemannian manifold. Let M be a Riemannian manifold, and let τ_M be its tangent bundle. Coordinates on M are denoted by t, and coordinates on tangent planes by x. The entries of the matrix $\Gamma(t)x$ are

$$\Gamma_{ij}^k(t)x^i, \quad 1 \le i, j, k \le n.$$

A connection in the bundle τ_M is called *symmetrical* if the relations

$$\Gamma_{ij}^k(t) = \Gamma_{ji}^k(t)$$

hold at all points t. In this case, Eq. (7.38) becomes

$$\dot{x}^k + \Gamma_{ij}^k \dot{t}^j x^i = 0. \tag{7.42}$$

We consider an arbitrary coordinate system $t^1, \ldots, t^m$ on the manifold M. The vector fields tangent to the coordinate lines of this system are denoted by $\partial/\partial t_i$, $i = 1, \ldots, m$.

Proposition 7.12. *The formula*

$$\nabla_{\partial/\partial t_i}\left(\frac{\partial}{\partial t_j}\right) = \Gamma_{ij}^k \frac{\partial}{\partial t_k}. \tag{7.43}$$

holds.

Proof. To prove the statement, it suffices to note that the partial derivatives of the section $\partial/\partial t_j$ vanish because the coordinates of these vector fields in the considered coordinate system are constant. Further, it remains to apply (7.41). $\square$

The parallel translation of vectors allows us to define the parallel translation of any tensor on the tangent bundle to the manifold M. First, the covariant derivative of ordinary (scalar-valued) functions on a manifold coincides with the usual directional derivative because these functions can be considered as sections of a trivial bundle with the fiber $\mathbb{R}$ equipped with the trivial connection. The translation of vectors of the tangent plane (contravariant vectors) is defined by Eq. (7.42)

The translation of covariant vectors (linear functionals on the space of vectors of the tangent plane) is defined by the condition for preservation of the value of the translated linear functional on the translated vector field. Namely, if $Z \in T_{t(0)}^*$, then the value of the functional $\tau_\theta^0 Z$ on the vector T should be equal to the value of the functional Z on $\tau_0^\theta X$ for all θ. If this condition holds, then, obviously, $Z(\theta)(T(\theta)) = \text{const}$. This immediately implies that the parallel translation of linear functionals is described by the differential equation that is adjoint to Eq. (7.42):

$$\dot{Z}_i - Z_k \Gamma_{ij}^k \dot{t}^j (\theta) = 0.$$

In exactly the same way, for an arbitrary tensor field Q of type (r, s), the parallel translation is defined such that the convolution of Q with a tuple of r covariant vector fields and s contravariant vector fields (which is a scalar-valued function) remains constant under the parallel translation. For example, the covariant derivative of a tensor Q of type $(0, 2)$ in the direction $\partial/\partial t^l$ equals

$$\nabla_{\partial/\partial t^l} Q_{ij} = \frac{\partial Q_{ij}}{\partial t^l} - Q_{ik}\Gamma_{lj}^k - Q_{kj}\Gamma_{il}^k. \tag{7.44}$$

Definition 7.14. A connection is said to be *Riemannian* or is called the *Levi-Civita connection* if it is symmetrical and if the parallel translation preserves the inner product of tangent vectors.

We note that the structural group of the tangent bundle of a Riemannian manifold can be considered the orthogonal group, i.e., the group preserving the metric that the planes tangent to the manifold M are equipped with. Because of this, the parallel translation that preserves the inner product is compatible with the structure group of the bundle.

Definition 7.15. A tensor field $Q(t)$ on a manifold M is said to be *covariantly constant* if the result of the parallel translation of the tensor $Q(t(0))$ along any curve $t = t(\theta)$, $0 \le \theta \le 1$, coincides with $Q(t(1))$.

This definition is equivalent to the following: a field $Q(t)$ is covariantly constant if the covariant derivative of the tensor Q in all directions is identically zero at all points of the manifold.

Exercise. Prove that these definitions are equivalent.

Theorem 7.3. *The metric tensor is covariantly constant with respect to the Levi-Civita connection.*

Proof. Let $t(\theta)$ be an arbitrary path on the manifold M. At the point $t(0)$, we consider two arbitrary vectors X_0 and Y_0. We translate these vectors along the curve $t(\theta)$. The fields constructed on the curve $t(\theta)$ are denoted by X and Y. Let $\tilde{g}_{ij}^{t(\theta)}$ be the result of the parallel translation of the tensor g_{ij} along the curve $t(\theta)$. By the definition of the parallel translation of tensors, the value of the metric tensor $\tilde{g}_{ij}^{t(\theta)}$ on the pair of vectors X and Y is constant along the curve $t(\theta)$ and coincides with its initial value at the point $t(0)$.

On the other hand, by the definition of the Levi-Civita connection, the inner product of the vectors X and Y with respect to the metric on the manifold M (i.e., the action of the initial metric tensor g_{ij} on the fields X and Y) is also constant along $t(\theta)$ and coincides with the same initial value. By the arbitrariness of the fields X and Y on the curve $t(\theta)$, the relation $\tilde{g}_{ij}^{t(\theta)}(t(\theta)) = g_{ij}(t(\theta))$ holds. By the arbitrariness of $t(\theta)$, we find that the tensor g_{ij} is covariantly constant.

Corollary 7.1. *The covariant derivative of the metric tensor with respect to the Levi-Civita connection is identically equal to zero:*

$$\frac{\partial g_{ij}}{\partial t^l} - \Gamma_{lj}^k g_{ik} - \Gamma_{il}^k g_{kj} = 0. \tag{7.45}$$

Proof. It suffices to use Theorem 7.3 and (7.44).

Further, the matrix inverse to the matrix (g_{ij}) is as usual denoted by (g^{ij}).

Theorem 7.4. *The Levi-Civita connection on a Riemannian manifold exists and is unique. In local coordinates it is given by*

$$\Gamma_{ij}^l = \frac{1}{2} g^{lk} \left(\frac{\partial g_{ik}}{\partial t^j} + \frac{\partial g_{kj}}{\partial t^i} - \frac{\partial g_{ij}}{\partial t^k} \right).$$

Proof. Let $\Gamma_{s,ij}$ denote $\Gamma_{ij}^k g_{ks}$. Then relation (7.45) becomes

$$\Gamma_{i,jk} + \Gamma_{j,ki} = \frac{\partial g_{ij}}{\partial t^k}.$$

The conditions for symmetry of the connection mean that $\Gamma_{i,jk} = \Gamma_{i,kj}$. Permuting the subscripts i, j, and k cyclically, we obtain

$$\Gamma_{i,jk} + \Gamma_{j,ik} = \frac{\partial g_{ij}}{\partial t^k},$$

$$\Gamma_{j,ki} + \Gamma_{k,ji} = \frac{\partial g_{kj}}{\partial t^i}, \qquad (7.46)$$

$$\Gamma_{i,kj} + \Gamma_{k,ij} = \frac{\partial g_{ik}}{\partial t^j}.$$

We add the second and third equations of system (7.46) and subtract the first from the obtained sum. Using the condition for symmetry of the connection, we obtain

$$\Gamma_{k,ij} = \frac{1}{2}\left(\frac{\partial g_{kj}}{\partial t^i} + \frac{\partial g_{ik}}{\partial t^j} - \frac{\partial g_{ij}}{\partial t^k} \right) = g_{kl}\Gamma_{ij}^l;$$

this proves the theorem.

The values Γ_{ij}^k are called the *Christoffel symbols*. We met them in Chap. 1 when writing the geodesic equations on a Riemannian manifold. We recall that these equations have the form

$$\frac{d^2 t^k}{d\theta^2} + \Gamma_{ij}^k \frac{dt^i}{d\theta} \frac{dt^j}{d\theta} = 0, \quad k = 1, \ldots, n.$$

Comparing the geodesic equation with the equation for the parallel translation of vectors (7.42), we conclude that the curve $t(\cdot)$ is a geodesic iff the field of its tangent vectors $dt/d\theta$ forms a vector field parallel translated (in the sense of the Levi-Civita connection) along the curve $t(\cdot)$. This property is often taken as the definition of a geodesic on a manifold with a connection.

6.1. Torsion and Curvature of a Connection of a Vector Bundle.
In what follows, we consider an arbitrary vector bundle $\pi: E \to M$. Let ∇ be a connection on an arbitrary vector bundle E, and let X and Y be two arbitrary vector fields on E. Using the operator of covariant differentiation, we construct the following two tensor fields with values in the bundle E [59].

Definition 7.16. The *torsion of the connection* ∇ is the tensor field T whose value on a pair of fields X and Y is equal to the vector field defined by

$$T(X,Y) = \nabla_X(Y) - \nabla_Y(X) - [X,Y],$$

where $[X,Y]$ is the commutator of the vector fields X and Y.

Definition 7.17. The *curvature of the connection* ∇ is the tensor field R whose value on a pair of fields X and Y is equal to the field of linear operators defined by

$$R(X,Y) = \nabla_X \nabla_Y - \nabla_Y \nabla_X - \nabla_{[X,Y]}.$$

In the case where the bundle E is the tangent bundle, the following theorem holds.

Theorem 7.5. *A connection on the tangent bundle τ_M is symmetrical iff it is torsion free.*

Proof. Indeed,

$$T\left(\frac{\partial}{\partial t_i}, \frac{\partial}{\partial t_j}\right) = \nabla_{\partial/\partial t_i}\left(\frac{\partial}{\partial t_j}\right) - \nabla_{\partial/\partial t_j}\left(\frac{\partial}{\partial t_i}\right) - \left[\frac{\partial}{\partial t_i}, \frac{\partial}{\partial t_j}\right].$$

We use Proposition 7.12 and the vanishing of the commutator of the coordinate (and therefore constant in this coordinate system) vector fields. Then it is easy to complete the proof by verifying that the obtained expression becomes

$$(\Gamma_{ij}^k - \Gamma_{ji}^k)\frac{\partial}{\partial t_k}.$$

We can now find the expression of the curvature tensor through the coefficients of the linear connection of the bundle E. Let ∇_i denote the operator $\nabla_{\partial/\partial t^i}$. In this notation, (7.43) becomes

$$\nabla_i\left(\frac{\partial}{\partial t^k}\right) = \Gamma_{ik}^l \frac{\partial}{\partial t^l}.$$

This formula gives a possibility to compute the result of the action of the operator $R(\partial/\partial t^i, \partial/\partial t^j)$ on the basis vector $\partial/\partial t^l$. We have

$$\nabla_j \nabla_i\left(\frac{\partial}{\partial t^k}\right) = \nabla_j\left(\Gamma_{ik}^l \frac{\partial}{\partial t^l}\right) = \frac{\partial \Gamma_{ik}^l}{\partial t^j}\frac{\partial}{\partial t^l} + \Gamma_{ik}^s \Gamma_{js}^l \frac{\partial}{\partial t^l}.$$

The commutator of the vector fields $\partial/\partial t^i$ and $\partial/\partial t^j$ vanishes. Therefore,

$$R\left(\frac{\partial}{\partial t^j}, \frac{\partial}{\partial t^i}\right)\frac{\partial}{\partial t^k} = \left(\frac{\partial \Gamma_{ik}^l}{\partial t^j} - \frac{\partial \Gamma_{jk}^l}{\partial t^i} + \Gamma_{ik}^s \Gamma_{js}^l - \Gamma_{jk}^s \Gamma_{is}^l\right)\frac{\partial}{\partial t^l}.$$

Let R_{kij}^l denote the expression

$$\left(\frac{\partial \Gamma_{ik}^l}{\partial t^j} - \frac{\partial \Gamma_{jk}^l}{\partial t^i} + \Gamma_{ik}^s \Gamma_{js}^l - \Gamma_{jk}^s \Gamma_{is}^l\right).$$

Then we have

$$R\left(\frac{\partial}{\partial t^j}, \frac{\partial}{\partial t^i}\right)\frac{\partial}{\partial t^k} = R_{kij}^l \frac{\partial}{\partial t^l}. \tag{7.47}$$

We can rewrite (7.47) in the invariant terms using the differential connection form

$$\omega = dx + \Gamma(t)x\,dt.$$

Definition 7.18. The *exterior covariant differential* of a vector-valued differential form ω is the differential form $D\omega$ whose value on a pair of fields Y and Z tangent to E is equal to the value of the exterior differential $d\omega$ on the horizontal components of the vectors Y and Z:

$$D\omega(Y, Z) = d\omega(h(Y), h(Z)).$$

The form $D\omega$, where ω is the connection form, is called the *curvature form* of the connection ∇.

Proposition 7.13. *The coordinates $R^l_{kij}x^k$ of the curvature tensor of the vector bundle E at a point (t, x) are equal to the coordinates of the curvature form of this bundle:*

$$(D\omega)^l = R^l_{kij}x^k\,dt^i \wedge dt^j.$$

Proof. The exterior differential of the form ω is equal to

$$d\omega = d(\Gamma^l_{kj}x^k\,dt^j) = \frac{1}{2}\left(\frac{\partial \Gamma^l_{kj}}{\partial x^i} - \frac{\partial \Gamma^l_{ki}}{\partial x^j}\right)x^k\,dt^i \wedge dt^j + \Gamma^l_{kj}dx^k \wedge dt^j. \quad (7.48)$$

To obtain the value of the form $d\omega$ on horizontal components of the vectors Y and Z, it is necessary to use (7.37) and substitute the expressions $dx^s = -\Gamma^s_{ki}x^k\,dt^i$ for dx^s (preparatorily changing the summation subscript k to s in the second group of summands in (7.48), which contains dx). As a result, we obtain the differential form

$$\frac{1}{2}\left(\frac{\partial \Gamma^l_{kj}}{\partial x^i} - \frac{\partial \Gamma^l_{ki}}{\partial x^j} - \Gamma^s_{ik}\Gamma^l_{js} + \Gamma^s_{jk}\Gamma^l_{is}\right)x^k\,dt^i \wedge dt^j.$$

A connection is called *flat* if its curvature tensor vanishes. We fix the first subscript k of the coefficients of the connection Γ^l_{ki} (corresponding to summation over coordinates of the vector of the fiber) and let Γ_k denote the obtained matrix with the entries Γ^l_{ki}. Then the condition that the connection is flat can be written in the matrix form as

$$\frac{\partial \Gamma_k}{\partial t^s} - \frac{\partial \Gamma_s}{\partial t^k} + [\Gamma_k, \Gamma_s] = 0,$$

where $[\Gamma_k, \Gamma_s]$ is the commutator of the matrices Γ_k and Γ_s.

Remark. All statements in this section are valid for a pseudo-Riemannian manifold, i.e., a manifold with a nondegenerate, but not necessarily positive-definite, metric.

§7. Nonnegativity Conditions of the Second Variation

We return to the higher-dimensional calculus of variations. The study of conditions for the nonnegativity of the second variation of a multiple integral began with Clebsch [19], who studied the *Dirichlet functional*

$$\phi''(0) = \delta^2 J$$

$$= \int_W \left[\frac{\partial^2 \hat{f}}{\partial(\partial y^i/\partial t^\alpha)\partial(\partial y^j/\partial t^\beta)} \frac{\partial h^i}{\partial t^\alpha} \frac{\partial h^j}{\partial t^\beta} \right.$$

$$\left. + 2\frac{\partial^2 \hat{f}}{\partial(\partial y^i/\partial t^\alpha)\partial y^j} \frac{\partial h^i}{\partial t^\alpha} h^j + \frac{\partial^2 \hat{f}}{\partial y^i \partial y^j} h^i h^j \right] dt.$$

Clebsch's idea was to reduce this functional to the integral of the leading terms of the integrand, i.e., to terms containing products of derivatives of the desired vector-valued function. Probably, Clebsch started from the intuitive idea that a direct analogue of the Legendre condition (the second variation is nonnegative) holds for a multiple integral if the quadratic form of the leading terms of the integrand defined on $n \times \nu$ matrices,

$$y^i_\alpha = \frac{\partial x^i}{\partial t^\alpha},$$

is positive semidefinite for all t. But approximately half a century after the Clebsch paper, J. Hadamard revealed that this is not true [41]. Hadamard found the following condition for the nonnegativity of the second variation.

Theorem 7.6 (Hadamard necessary condition). *Let the functional*

$$\delta^2 J = \int_W \left(a^{\alpha\beta}_{ij}(t) \frac{\partial x^i}{\partial t^\alpha} \frac{\partial x^j}{\partial t^\beta} + 2c^\alpha_{ij}(t)\frac{\partial x^i}{\partial t^\alpha} x^j + b_{ij}(t)x^i x^j \right) dt \qquad (7.49)$$

be nonnegative for $x(\cdot)$ satisfying the boundary conditions

$$x|_{\partial W} = 0.$$

Then for all $t \in W$, the quadratic form $a^{\alpha\beta}_{ij}(t)y^i_\alpha y^j_\beta$ assumes nonnegative values on all $n \times \nu$ matrices of the form $y^i_\alpha = \xi^i \eta_\alpha$ (i.e., on all matrices of rank one).

The conclusion of Theorem 7.6 can be reformulated as follows: *for all $t \in W$, the biquadratic form $a^{\alpha\beta}_{ij}(t)\xi^i\xi^j\eta_\alpha\eta_\beta$ is nonnegative for all $\xi \in \mathbb{R}^n$ and $\eta \in \mathbb{R}^\nu$.*

Proof (of Theorem 7.6). We suppose the contrary: for some $t_0 \in W$, there exists a matrix $\xi^i\eta_\alpha$ such that

$$a^{\alpha\beta}_{ij}(t)\xi^i\xi^j\eta_\alpha\eta_\beta < 0.$$

Using this inequality, we construct a variation of the section $x = 0$ on which functional (7.49) assumes a negative value.

Without loss of generality, we can assume that $t_0 = 0$, the module of the vectors ξ and η equals 1, and, moreover,

$$\eta = \frac{\partial}{\partial t_1}$$

and

$$\xi = \frac{\partial}{\partial x_1}.$$

Because we construct the variation of the section $x = 0$ in an arbitrarily small neighborhood of the point $t = 0$, we can assume, for simplicity of the calculations, that the manifold W is a plane. The changes that appear in this case lead to the addition of an infinitely small value of a higher order as compared with the leading terms that define the sign of the second variation.

In the plane $x = 0$, we consider a rectangular parallelepiped q centered at the point $t = 0$. One of the edges of this parallelepiped is directed along the axis η and is of length $2\epsilon^2$ (where ϵ is a sufficiently small positive constant); the length of other edges is 2ϵ. They are also directed along the coordinate axes. The face of the parallelepiped q lying in the plane $t^1 = \epsilon^2$ is denoted by λ_+; the face lying in the plane $t^1 = -\epsilon^2$ is denoted by λ_-. Both faces are cubes with the edge 2ϵ.

Lemma 7.3. *The intersection of the ν-dimensional plane given by the equation $x = Pt + K$ with the plane $x = 0$ (in the case where this intersection is not empty) is a $(\nu-1)$-dimensional plane iff the rank of the matrix P is one.*

Proof. The desired intersection is defined by solutions to the linear system of equations $Pt + K = 0$, which has $\nu-1$ independent solutions iff the rank of the matrix P is one.

On the parallelepiped q, as on the base, we construct a polyhedron Q whose lateral surface replaces the base of q in constructing the variation. We construct the variation in the plane $x^2 = x^3 = \ldots = x^n = 0$.

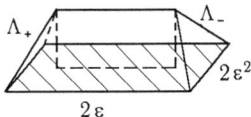

We choose an ν-dimensional plane Λ_- inclined by $\xi^i \eta_\alpha$, i.e., with the equation $x^1 = t^1 + K$, and choose the free term K such that the plane Λ_- intersects the plane $x = 0$ along the face λ_-. Then the plane Λ_- is given by the equation $x^1 = t^1 + \epsilon^2$. Similarly, we consider the plane Λ_+ inclined by $-\xi^i \eta_\alpha$ and having the equation $x^1 = -t^1 + \epsilon^2$, which intersects the plane $x = 0$ along the face Λ_+. The intersection of these planes $\Lambda_+ \cap \Lambda_-$ is described by the equations $t^1 = 0$ and $x^1 = \epsilon^2$.

In the $(\nu-1)$-dimensional plane $\Lambda_+ \cap \Lambda_-$, we choose vertices of the polyhedron Q as follows. Let $H_{1-\epsilon} : \lambda_+ \to \lambda_+$ be the homothety with the coefficient $1 - \epsilon$, which is described by the relation $H_{1-\epsilon} : t^i \mapsto (1 - \epsilon)t^i$, $i = 2, \ldots, n$. Under the homothety $H_{1-\epsilon}$, the cube with the edge 2ϵ lying in the plane λ_+ passes to the cube with the edge $2\epsilon - 2\epsilon^2$. As the vertices of the polyhedron Q, we take a tuple of points with the coordinates

$$(0, \pm(\epsilon - \epsilon^2), \ldots, \pm(\epsilon - \epsilon^2), \epsilon^2),$$

i.e., the first coordinate is always zero; the coordinate along the axis x^1 is ϵ^2 for all vertices; and the coordinates with the indices $2, \ldots, \nu$ are coordinates of vertices of the cube centered at the origin with the edge $\epsilon - \epsilon^2$.

Therefore, the upper base of the polyhedron Q is the cube $H_{1-\epsilon}\lambda_+$, which is translated to the plane $t^1 = 0$ and is raised along the axis x^1 to the height ϵ^2. The upper base has dimension $\nu-1$ and has no direct influence on the value of functional (7.49).

Other faces of the polyhedron Q are called *lateral* faces. They are higher-dimensional truncated pyramids. We write the equations of ν-dimensional lateral faces of the polyhedron Q in the form $x^1 = P_s t + K_s$, where P_s are certain vectors; their subscript s defines the number of the lateral face.

Lemma 7.4. *The coefficients P_s remain bounded as $\epsilon \to 0$. The volume of faces lying in the planes Λ_+ and Λ_- has the order $O(\epsilon^{\nu+1})$. The volume of other lateral faces of the polyhedron Q has the order $o(\epsilon^{\nu+1})$.*

Proof. Because $H_{1-\epsilon}$ is a homothety, the distance from projections of vertices of the polyhedron Q on the plane $x = 0$ to the nearest $(\nu-1)$-dimensional faces of q is equal to ϵ^2. Therefore, the heights of lateral faces are of order ϵ^2. Two ν-dimensional planes in the $(\nu+1)$-dimensional space of the variables $(t^1, \ldots, t^\nu, x^1)$ intersect along a $(\nu-1)$-dimensional plane. Their mutual location is determined by a unique dihedral angle that corresponds to the angle made by vectors normal to them. In the case considered, all these angles are equal to $\pi/4$; this proves the first assertion of the lemma.

The base of the face Λ_+ (as well as of Λ_-) is the cube of volume $\epsilon^{\nu-1}$; the height of this face is of order ϵ^2. The height of each of the remaining lateral faces also has the order ϵ^2; however, the volume of its bases equals ϵ^ν. Lemma 7.4 is proved.

Continuing the proof of the theorem, we compute functional (7.49) on the constructed variation. The volume of the face Λ_+ is denoted by V_+. Expanding the coefficients $a_{ij}^{\alpha\beta}(t)$ with the Taylor formula at the point $t = 0$ and taking into account that the functions $x(t)$ have the order of smallness ϵ^2 on the variation considered, we obtain

$$\delta^2 J = a_{ij}^{\alpha\beta}(0)\xi^i\xi^j\eta_\alpha\eta_\beta 2V_+\epsilon^{\nu+2} + o(\epsilon^{\nu+2}) < 0.$$

Theorem 7.6 is proved.

Lemma 7.3 explains why only matrices of rank one are used in the Hadamard–Legendre condition. At that time, this was a sensation in the higher-dimensional calculus of variations.

The paper by Hadamard stimulated the appearance of a number of papers whose authors tended to reduce the gap between necessary and sufficient optimality conditions for multiple integrals. One of the brilliant papers in this number was by Terpstra [105], who considered abstract algebraic questions induced by the above-mentioned problems.

Namely, let

$$a_{ij}^{\alpha\beta} y_\alpha^i y_\beta^j \tag{7.50}$$

be a quadratic form defined on $n\times\nu$ matrices y_α^i. We consider the cone of matrices of rank one (i.e., matrices of the form $y_\alpha^i = \xi^i\eta_\alpha$). On this cone, form (7.50) becomes the biquadratic form

$$a_{ij}^{\alpha\beta}\xi^i\xi^j\eta_\alpha\eta_\beta \tag{7.51}$$

defined on pairs of vectors $\xi \in \mathbb{R}^n$ and $\eta \in (\mathbb{R}^\nu)^*$.

We assume that form (7.50) is nonnegative on all matrices of rank one. Terpstra stated the following question: is it possible to transform form (7.50) into a positive-definite form on the whole space of matrices by adding the terms

$$r_{ij}^{\alpha\beta}(y_\alpha^i y_\beta^j - y_\beta^i y_\alpha^j)$$

to it, where the tensors $r_{ij}^{\alpha\beta}$ are skew symmetrical in the superscripts α and β as well as in the subscripts i and j?

We note that the added terms vanish on matrices of rank one and biquadratic form (7.51) therefore does not change but form (7.50) certainly changes essentially.

Terpstra proved that the answer is positive in the case where n or ν does not exceed two. Otherwise (when both these dimensions are greater than two), the answer is negative: Terpstra constructed an example of a positive-definite biquadratic form (7.51) on the space $\mathbb{R}^3 \times (\mathbb{R}^3)^*$ that cannot be transformed into a positive-definite form (7.50) using the above procedure. Using the Terpstra construction, D. V. Matyukhin recently proved [69] that for $\min(n,\nu) \geq 3$, forms having this property form an open subset in the space of all forms.

It seemed that the Terpstra results established an essential gap between the necessary Hadamard–Legendre condition and sufficient minimality conditions for a multiple integral. However, in the late 1940s, L. Van Hove proved [107] that a natural strengthening of the Hadamard condition,

$$\frac{\partial^2 \hat{f}}{\partial(\partial x^i/\partial t^\alpha)\partial(\partial x^j/\partial t^\beta)}\xi^i\xi^j\eta_\alpha\eta_\beta \geq \epsilon|\xi|^2|\eta|^2, \tag{7.52}$$

is a locally sufficient condition for a C_1 minimum. The term "locally sufficient" means that the domain over which the integration is performed is sufficiently small.

The idea of the van Hove proof is as follows. First, the coefficients are frozen, i.e., the argument $t = t_0$ in the coefficients of quadratic form (7.49) is fixed (this has no influence on the estimates arising in the proof of the positive definiteness because an arbitrary small neighborhood of the point t_0 can be chosen as the domain of integration). Second, the Fourier transform and the Parseval identity are applied to the obtained functional.

This construction reveals one more (different from Lemma 7.3) intrinsic reason why only matrices of rank one participate in the Hadamard–Legendre condition. Namely, when the Fourier transform is applied, the operation of differentiation passes to the operation of multiplication by the corresponding independent variable. If the function is the image of the function $x^i(t)$ under the Fourier transform, then the image of the derivative $\partial x^i/\partial t^\alpha$ is $\xi^i\eta^\alpha$. As a result, the biquadratic form, which appears in the left-hand side of (7.52), arises under the integral sign in the Parseval identity, and inequality (7.52) guarantees the positive definiteness of the considered functional.

To prove a sufficient condition for a strict minimum on "large" pieces of a surface, it is necessary to construct an analogue of the Hilbert field theory for the higher-dimensional calculus of variations. The first variant of such a theory was proposed by C. Caratheodory in the 1920s. Another (relatively simpler) variant, which is based on a slightly stronger condition on the minimized functional, was obtained by H. Weyl in the 1930s. We first present the H. Weyl construction.

§8. Field Theory in the Weyl Form

We consider the minimization problem for the multiple integral

$$J = \int_W f\left(t, x(t), \frac{\partial x}{\partial t}\right) dt. \tag{7.53}$$

Generalizing the field theory for the one-dimensional integral (see Chap. 1), H. Weyl [109] proposed to take the integral

$$I = \int_W \sum_{\alpha=1}^{\nu} (-1)^{\alpha+1}\, dS^\alpha(t, x) \wedge \{dt\}_\alpha$$

as the invariant integral of the field theory [35, 37, 64]; here, $\{dt\}_\alpha$ denotes the product of all differentials dt^γ in which the factor dt^α is omitted. It is easy to see that

$$I = \int_W \left(\frac{\partial S^\alpha}{\partial t^\alpha} + \frac{\partial S^\alpha}{\partial x^i}\frac{\partial x^i}{\partial t^\alpha}\right) dt. \tag{7.54}$$

We note that the expression under the integral sign is equal to the trace of the matrix composed of the total derivatives dS^α/dt^β. The integral I does not depend on the choice of the embedding of the surface W whenever this surface has a fixed boundary.

The main idea of the field theory consists in the construction of a ν-dimensional foliation Λ on a certain domain Ω of the space (t, x) that has the following properties:

1. The functional J coincides with the invariant integral I on leaves of the foliation Λ.
2. The leaves yield a solution to the minimization problem for the integral J; in other words, the difference between the functional J on a leaf L and on any other surface with the boundary ∂L in the domain Ω is not positive.

Following Weyl [109], we describe the construction of a field of extremals that has properties 1 and 2.

Let planes tangent to the leaves of Λ have the inclination

$$\frac{\partial x^i}{\partial t^\alpha} = \hat{p}^i_\alpha(t, x).$$

The obtained distribution of planes is called the *geodesic inclination* of the field. Therefore, we assume that $\hat{p}^i_\alpha$ is a ν-dimensional integrable distribution on the domain Ω.

Property 1 leads to the following equation for the functions S^α:

$$\frac{\partial S^\alpha}{\partial t^\alpha} = f(t, x, \hat{p}^i_\alpha) - \frac{\partial S^\alpha}{\partial x^i} \hat{p}^i_\alpha. \tag{7.55}$$

Property 2 yields the criteria for the distribution $\hat{p}^i_\alpha(t, x)$, which are obtained as follows. Let there be two ν-dimensional surfaces with a comman boundary: one is a leaf L of the foliation Λ, and the other is an arbitrary surface V lying in the domain Ω. Then

$$J(V) - J(L) = \int_V f\left(t, x, \frac{\partial x}{\partial t}\right) dt - \int_L f(t, x, \hat{p}) \, dt.$$

By property 1, the second summand is equal to the integral I, and by the invariance of the integral I, integration over the surface L can be replaced by integration over the surface V. Then

$$J(V) - J(L) = \int_V \left(f\left(t, x, \frac{\partial x}{\partial t}\right) - \frac{\partial S^\alpha}{\partial t^\alpha} - \frac{\partial S^\alpha}{\partial x^i} \frac{\partial x^i}{\partial t^\alpha} \right) dt.$$

Substituting (7.55) for $\partial S^\alpha/\partial t^\alpha$ in this expression, we obtain

$$J(V) - J(L) = \int_V \left[f\left(t, x, \frac{\partial x}{\partial t}\right) - f(t, x, \hat{p}) + \frac{\partial S^\alpha}{\partial x^i} \hat{p}^i_\alpha - \frac{\partial S^\alpha}{\partial x^i} \frac{\partial x^i}{\partial t^\alpha} \right] dt. \tag{7.56}$$

The expression under the integral sign should be nonnegative. For this, it suffices that the maximum of the function

$$-f\left(t, x, \frac{\partial x^i}{\partial t^\alpha}\right) + \frac{\partial S^\alpha}{\partial x^i}\frac{\partial x^i}{\partial t^\alpha} \tag{7.57}$$

in the variables $\partial x^i/\partial t^\alpha$ is attained for $\partial x^i/\partial t^\alpha = \hat{p}_\alpha^i$. The corresponding stationarity condition for function (7.57) has the form

$$\frac{\partial S^\alpha}{\partial x^i} = \frac{\partial f(t, x, p)}{\partial(\partial x^i/\partial t^\alpha)}. \tag{7.58}$$

Substituting (7.58) in the integrand of formula (7.56), we obtain the following function, which is similar to the Weierstrass function in the one-dimensional calculus of variations:

$$E = f\left(t, x, \frac{\partial x}{\partial t}\right) - f(t, x, \hat{p}) - \frac{\partial f(t, x, p)}{\partial(\partial x^i/\partial t^\alpha)}\left(\frac{\partial x^i}{\partial t^\alpha} - \hat{p}_\alpha^i\right). \tag{7.59}$$

To find the geodesic inclination of the field, we must solve Eq. (7.58) in the variables p. The equation for the functions S^α is obtained from (7.55) if we substitute the obtained expression in them. Then, only the functions $S^\alpha(t, x)$, $\alpha = 1, \ldots, \nu$, which completely determine the field of extremals, remain unknown in (7.55). The condition for the local solvability of (7.58) in the variables p_α^i is that the determinant of the $n\nu \times n\nu$ matrix

$$\frac{\partial^2 f(t, x, p)}{\partial(\partial x^i/\partial t^\alpha)\partial(\partial x^j/\partial t^\beta)}$$

is different from zero. The necessary maximality condition for function (7.57) is the positive semidefinitness of this matrix. This condition is called the *Legendre–Weyl condition*.

Therefore, the above construction leads to a complete analogue of the Legendre transform in which the role of momenta is played by the functions $\partial S^\alpha/\partial x^i$ and the role of the Hamiltonian is played by function (7.57).

An analogue of the strengthened Legendre condition in the one-dimensional calculus of variations is the following condition: the quadratic form

$$\frac{\partial^2 f(t, x, p)}{\partial p_\alpha^i \partial p_\beta^j}y_\alpha^i y_\beta^j \tag{7.60}$$

is positive definite on the set of $n \times \nu$ matrices y_α^i. This condition gives a possibility to (locally) solve Eq. (7.58) in the variables p.

The role of the Weierstrass function is played by (7.59), which is positive if the function $f(t, x, p)$ is convex in the variables p (in this case, the matrix p is considered as one "long" vector having $n\nu$ coordinates).

After the expression for the geodesic inclination is substituted in (7.55), this equation becomes a direct analogue of the *Hamilton–Jacobi equation*. In this case, the functions are naturally called the *actions*. As compared with the one-dimensional calculus of variations, the essential difference is that in the higher-dimensional case, we have only one equation that connects ν unknown functions $S^\alpha(t, x)$. Therefore, there arises a principal possibility to arbitrarily assign $\nu-1$ of these functions and solve the equation in the one remaining unknown function after their substitution in the equation. In this sense, the higher-dimensional problem in the calulus of variations has "essentially more" fields of extremals than the one-dimensional problem.

We see that the Legendre–Weyl transform lies in the base of constructing the field of extremals that corresponds to the closed differential form $dS^\alpha \wedge \{dt\}_\alpha$. Moreover, if the function f is convex in the variables $p_\alpha^i = \partial x^i/\partial t^\alpha$, the nonnegativity of the Weierstrass function is guaranteed; this implies a sufficient strong minimality condition in the form of the field theory.

Another approach to the construction of a field of extremals of integral (7.53) was proposed by C. Caratheodory. His idea was more complicated and subtle. First, he proposed to include the Legendre transform in a certain family of transformations, which are called *contact* transformations. Second, he introduced and studied a certain specific contact transformation that is related to the differential form

$$dS^1 \wedge \ldots \wedge dS^\nu \tag{7.61}$$

in the same way the Legendre–Weyl transform is related to the differential form $dS^\alpha \wedge \{dt\}_\alpha$.

Before describing the Caratheodory construction, we look at the Legendre–Weyl transform from the purely algebraic standpoint. Let

$$(f, \phi, p_\alpha^i, \pi_i^\alpha), \quad 1 \le i \le n, \quad 1 \le \alpha \le \nu, \tag{7.62}$$

be a tuple consisting of $2(n\nu+1)$ variables. We assume that the variables in (7.62) are related by

$$f + \phi = p_\alpha^i \pi_i^\alpha = \mathrm{Tr}(\pi p). \tag{7.63}$$

In what follows, f is called the initial function, ϕ the Hamiltonian, $p = (p_\alpha^i)$ the $n \times \nu$ matrix of variables (it corresponds to the matrix of partial derivatives $\partial x^i/\partial t^\alpha$ in the integrand), and $\pi = (\pi_i^\alpha)$ the $\nu \times n$ matrix of momenta (it corresponds to the partial derivatives of the integrand in the derivatives of the unknown functions).

We set a tuple of elements of the same form in correspondence to tuple (7.62). These elements are denoted by the corresponding capital letters:

$$(F, \Phi, P_i^\alpha, \Pi_\alpha^i), \quad 1 \le i \le n, \quad 1 \le \alpha \le \nu.$$

The correspondence is defined by

$$F = \phi, \qquad \Phi = f, \qquad P_i^\alpha = \pi_i^\alpha, \qquad \Pi_\alpha^i = p_\alpha^i. \tag{7.64}$$

Under mapping (7.64), the initial function becomes the Hamiltonian, the Hamiltonian the initial function, the variables the momenta, and the momenta the variables. Clearly, the new matrix of variables has ν rows and n columns, and the new matrix of momenta has n rows and ν columns. Also, relation (7.63) obviously remains valid for the new tuple (7.64):

$$F + \Phi = P_i^\alpha \Pi_\alpha^i. \tag{7.65}$$

Therefore, the considered mapping is birational and involutive. Computing the differential of relation (7.63) and using (7.64), we can easily verify that

$$dF - \Pi_\alpha^i dP_i^\alpha = -(df - \pi_i^\alpha dp_\alpha^i). \tag{7.66}$$

The differential form in the right-hand side of (7.66) vanishes iff the variables π_i^α are the derivatives of the function f with respect to the corresponding arguments p (momenta). The differential form like $df - \pi_i^\alpha dp_\alpha^i$ is called the *contact* form for the function f. Relation (7.66) means that under the Legendre–Weyl transformation, a contact form passes to a contact form. The transformations under which a contact form passes to a contact form are called *contact* transformations.

Caratheodory found a new contact transformation that takes the matrix nature of the variables and momenta into account. We describe the Caratheodory construction in more detail.

§9. Caratheodory Transformation

We consider the same tuple of variables (7.62) satisfying relation (7.63). Caratheodory proposed [13] a more complicated transformation that connects the variables and momenta via a certain matrix a, called the *conjugating matrix*.

Therefore, we consider

$$a = f I_\nu - \pi p, \tag{7.67}$$

where I_ν is the ν-dimensional identity matrix and πp is the product of matrices that is also a $\nu \times \nu$ matrix. Using the matrix a, we define the Caratheodory transformation by the following formulas:

$$P = a^{-1}\pi, \qquad \Pi = \frac{f^{\nu-2}}{\det a}pa, \qquad F = \frac{f^{\nu-2}}{\det a}f, \qquad \Phi = \frac{f^{\nu-2}}{\det a}\phi. \tag{7.68}$$

We verify that the new tuple (7.68) also satisfies relation (7.63). Indeed,

$$P\Pi = (a^{-1}\pi)\left(\frac{f^{\nu-2}}{\det a}pa\right);$$

the matrices $a = I_\nu f - \pi p$ and πp commute; therefore,

$$PII = \frac{f^{\nu-2}}{\det a}\pi p$$

or

$$\frac{PII}{F} = \frac{\pi p}{f}.$$ (7.69)

Therefore,

$$F + \Phi = \frac{f^{\nu-2}}{\det a}(f + \phi) = \frac{f^{\nu-2}}{\det a}\text{Tr}(\pi p).$$

Using (7.69), we finally obtain

$$F + \Phi = \text{Tr}(PII).$$

Proposition 7.14. *The Caratheodory transformation is involutive.*

Proof. To prove the involutiveness, we introduce the conjugating matrix

$$A = FI_\nu - PII$$

for tuple (7.68). We then obtain

$$\frac{A}{F} = \frac{a}{f}, \qquad \frac{\det A}{F^\nu} = \frac{\det a}{f^\nu}$$ (7.70)

from (7.69). To obtain the second relation in (7.70), we use the fact that when a matrix is multiplied by a scalar, its determinant is multiplied by the same multiplier raised to the power equal to the order of the matrix. From (7.70) and (7.68), we sequentially obtain

$$f = \frac{F^{\nu-2}}{\det A}F, \qquad \phi = \frac{F^{\nu-2}}{\det A}\Phi, \qquad p = IIA^{-1}, \qquad \pi = AP\frac{F^{\nu-2}}{\det A}.$$

This completes the proof of involutiveness.

Proposition 7.15. *The Caratheodory transformation is a contact transformation.*

Proof. By (7.68) and (7.67), we have

$$Pp = a^{-1}\pi p = a^{-1}(fI_\nu - a).$$

This implies

$$I_\nu + Pp = fI_\nu a^{-1}.$$ (7.71)

Therefore,

$$\det(I_\nu + Pp) = \frac{f^\nu}{\det a} = fF.$$ (7.72)

Let g be an arbitrary nonsingular matrix and G_α^β denote the algebraic complement to the entry g_β^α of this matrix.

Lemma 7.5. *For any nonsingular matrix $g = (g_\beta^\alpha)$, the formula*

$$d(\det g) = \det g \operatorname{Tr}(g^{-1} dg) \tag{7.73}$$

holds.

The proof of the lemma is implied by the formula for differentiation of the determinant of the matrix,

$$d(\det g) = G_\alpha^\beta dg_\beta^\alpha, \tag{7.74}$$

where G_α^β denotes the algebraic complement to the entry g_β^α.

Exercise. Obtain (7.73) from the Jacobi formula

$$\det \exp X = \exp(\operatorname{Tr} X).$$

Returning to the proof of Proposition 7.15, we obtain

$$(I + Pp)^{-1} = \frac{a}{f}$$

from (7.71). Using this relation and (7.73), we write the differential of (7.72):

$$f dF + F df = \det(I + Pp) \operatorname{Tr}\left(\frac{a}{f}((dP)p + P(dp))\right)$$

$$= Ff\left(\operatorname{Tr}\left(\frac{a(dP)p}{f}\right) + \operatorname{Tr}\left(\frac{aa^{-1}\pi(dp)}{f}\right)\right).$$

Because the trace of the product of matrices does not depend on the order of the factors, we can substitute $pa = (\det a / f^{\nu-2})\Pi$ for p in the first summand of the latter relation. To transform the second summand, we use the relation $F = f^{\nu-1} / \det a$. We then have

$$f dF + F df = f \operatorname{Tr}(\Pi(dP)) + F \operatorname{Tr}(\pi(dp)).$$

We finally obtain

$$f(dF - \operatorname{Tr}(\Pi(dP))) + F(df - \operatorname{Tr}(\pi(dp))) = 0,$$

but this means that the Caratheodory transformation is a contact transformation.

We describe the action of the Caratheodory transformation on functions $f(p)$ of the matrix argument p. First, using the function f, we compute the momenta

$$\pi_i^\alpha(p) = \frac{\partial f}{\partial p_\alpha^i};$$

we then construct the conjugating matrix

$$a(p) = f I_\nu - \pi p$$

and introduce new independent variables

$$P = a^{-1}\pi. \tag{7.75}$$

Finally, Eq. (7.75) should be solved in the variables p, i.e., we find $p = p(P)$ and then compose the function

$$F(P) = \frac{f^{\nu-1}}{\det a}(p(P))$$

of the new variable P.

For the Caratheodory transformation to be defined, it is necessary to require that Eq. (7.75) be solvable in p.

9.1. Condition for Realizability of the Caratheodory Transformation.

We rewrite (7.75) in the form

$$\pi - aP = 0 \tag{7.76}$$

and require that the Jacobian of the left-hand side of (7.76) with respect to the matrix variables p does not vanish for fixed P. We write this Jacobian in the coordinates:

$$J = \frac{\partial}{\partial p_\beta^j}(\pi_s^\alpha - (f\delta_\sigma^\alpha - \pi_k^\alpha p_\sigma^k)P_s^\sigma)$$

$$= \frac{\partial^2 f}{\partial p_\alpha^s \partial p_\beta^j} - \pi_j^\beta P_s^\alpha + \pi_j^\alpha P_s^\beta + \frac{\partial^2 f}{\partial p_\alpha^k \partial p_\beta^j}p_\sigma^k P_s^\sigma. \tag{7.77}$$

We introduce the matrix

$$b = fI_n - p\pi. \tag{7.78}$$

Lemma 7.6. *The matrix b is nonsingular, and the relation $Pb = \pi$ holds.*

Proof. It is easy to verify that the determinant of the matrix a is equal to the determinant of the $(n+\nu)\times(n+\nu)$ matrix

$$\begin{pmatrix} I_n & \pi \\ p & fI_\nu \end{pmatrix}.$$

Similarly, the determinant of the matrix b is equal to the determinant of the matrix

$$\begin{pmatrix} fI_n & \pi \\ p & I_\nu \end{pmatrix}.$$

Therefore, $f^n \det a = f^\nu \det b$; this implies the first assertion of the lemma.

To prove the second assertion, we write

$$Pb = fP - Pp\pi.$$

Substituting $Pp = fa^{-1} - I$ from (7.71) in this expression, we obtain

$$Pb = fP - fa^{-1}\pi + \pi.$$

The first and second summands are mutually annihilated by definition (7.68); we thus obtain $Pb = \pi$.

We postmultiply the Jacobi matrix given by (7.77) by the matrix b. We obtain

$$Jb = \frac{\partial^2 f}{\partial p_\alpha^k \partial p_\beta^j}(\delta_s^k b_i^s + p_\sigma^k \pi_i^{\sigma_i}) - \pi_i^\alpha \pi_j^\beta + \pi_j^\alpha \pi_i^\beta.$$

Using definition (7.78), we replace $(\delta_s^k b_i^s + p_\sigma^k \pi_i^{\sigma_i})$ with $f\delta_i^k$.

As a result, we have proved that the condition for realizability of the Caratheodory transformation is that the determinant of the matrix Jb/f (which is the matrix of the quadratic form defined on $n\times\nu$ matrices y_α^i) is not equal to zero, i.e.,

$$\det \frac{Jb}{f} = \det\left(\frac{\partial^2 f}{\partial p_\alpha^i \partial p_\beta^j} - \frac{1}{f}(\pi_i^\alpha \pi_j^\beta - \pi_j^\alpha \pi_i^\beta)\right) \neq 0. \qquad (7.79)$$

§10. Field Theory in the Caratheodory Form

We return to the minimization problem for the functional

$$J = \int_W f\left(t, x(t), \frac{\partial x}{\partial t}\right) dt. \qquad (7.80)$$

In what follows, the $n\times\nu$ matrix $\partial x/\partial t$ is denoted by p. As an invariant integral of the field theory [13, 35, 37, 64], we take the integral

$$\int_W dS^1(t,x) \wedge \ldots \wedge dS^\nu(t,x) = \int_W \det\left(\frac{\partial S^\alpha}{\partial t^\beta} + \frac{\partial S^\alpha}{\partial x^i}\frac{\partial x^i}{\partial t^\beta}\right) dt. \qquad (7.81)$$

The determinant under the integral sign in (7.81) is denoted by Δ. For given functions $S(t,x)$, we have $\Delta = \Delta(t,x,p)$.

In what follows, we assume that the function f is positive, which, however, is not an essential restriction because solutions of the problem are not changed when we add a constant to the function f. Therefore, if the function f is continuous on a certain compact set, then we can add a sufficiently large positive constant to it and transform it into a positive function.

We note that the determinant of the matrix composed of the derivatives dS^α/dt^β is under the integral sign in (7.81).

As in Sec. 9, we seek an ν-dimensional foliation Λ in a domain Ω that has the following properties:

1. On leaves L of the foliation Λ, the functional J coincides with the invariant integral I.
2. The leaves L yield solutions to the minimization problem for the invariant integral J: the difference between the functional J on a leaf L and on any other surface with the boundary ∂L in the domain Ω is not positive.

Let the tangent planes of the leaves of the foliation Λ have the inclination

$$\frac{\partial x^i}{\partial t^\alpha} = \hat{p}_\alpha^i(t, x).$$

We therefore assume that $\hat{p}_\alpha^i(t, x)$ is a ν-dimensional integrable distribution on the domain Ω.

Property 1 implies

$$f(t, x, \hat{p}(t, x)) = \Delta(t, x, \hat{p}) = \det\left(\frac{\partial S^\alpha}{\partial t^\beta} + \frac{\partial S^\alpha}{\partial x^i}\hat{p}_\beta^i(t, x)\right). \qquad (7.82)$$

Property 2 yields a criterion for the distribution $\hat{p}_\alpha^i(t, x)$, which is obtained as follows. Let there be two ν-dimensional surfaces with a common boundary: one is a leaf L of the foliation Λ, and the other is an arbitrary surface V in the domain Ω. Then

$$J(V) - J(L) = \int_V f(t, x, p)dt - \int_L f(t, x, \hat{p})dt.$$

By property 1, the second summand is equal to the integral I, and by the invariance of the integral I, integration over the surface L can be replaced by integration over the surface V. Then

$$J(V) - J(L) = \int_V (f(t, x, p) - \Delta(t, x, p))dt$$

$$= \int_V \Delta(t, x, p)\left(\frac{f(t, x, p)}{\Delta(t, x, p)} - 1\right)dt.$$

We substitute the expression $f(t, x, \hat{p}(t, x))/\Delta(t, x, \hat{p})$ (which is equal to one by (7.82)) in this formula. Because $f > 0$, we obtain $\Delta(t, x, \hat{p}) > 0$ by (7.82).

If p is slightly different from $\hat{p}$, then $\Delta(t, x, p)$ is slightly different from $\Delta(t, x, \hat{p})$, and $\Delta(t, x, p) > 0$. Then the difference

$$J(V) - J(L) = \int_V \Delta(t, x, p)\left(\frac{f(t, x, p)}{\Delta(t, x, p)} - \frac{f(t, x, \hat{p})}{\Delta(t, x, \hat{p})}\right)dt$$

is nonnegative whenever

$$\frac{f(t, x, p)}{\Delta(t, x, p)} \geq \frac{f(t, x, \hat{p})}{\Delta(t, x, \hat{p})}. \qquad (7.83)$$

Inequality (7.83) means that f/Δ as a function of the variables p for fixed (t, x) attains its minimum value on the planes of the distribution $\hat{p}(t, x)$. Of course, $\hat{p}(t, x)$ depends on the choice of the functions $S^\alpha(t, x)$.

If we find an integrable distribution R given in the domain Ω by the functions $\hat{p}(t, x)$ and the functions $S^\alpha(t, x)$, $\alpha = 1, \ldots, \nu$, that satisfy relations (7.82) and (7.83), then we can construct a foliation Λ corresponding to the distribution R.

We consider an arbitrary leaf L of the foliation Λ and any compact piece K of this leaf lying in Ω. We now take any other surface N with the boundary ∂K lying in Ω and assume that $\Delta(t, x, p) > 0$ on this surface.

The above argument can be summarized in the following statement.

Proposition 7.16. *Under the above assumption, the value of the functional J on the surface K is less than or equal to its value on the surface N.*

Therefore, we must find the functions S^α and $\hat{p}^i_\alpha(t, x)$ from relations (7.82) and (7.83). We show that these functions are found using the Caratheodory transformations.

The function f/Δ attains the minimum value at the point $\hat{p}$; therefore,

$$\frac{\partial}{\partial p}\left(\frac{f}{\Delta}\right) = 0.$$

Taking (7.82) into account, we obtain

$$\frac{\partial f}{\partial p^i_\alpha} = \frac{\partial \Delta}{\partial p^i_\alpha} \tag{7.84}$$

from the obtained relation. By Lemma 7.5 (formula (7.73)), this implies

$$\pi = c^{-1}S \det c, \tag{7.85}$$

where S denotes the matrix $(\partial S^\alpha/\partial x^i)$ and c denotes the matrix under the the determinant sign Δ.

In what follows, we need the following algebraic statement. Let (p, π, f) and (P, Π, F) be two tuples of variables obtained one from another under the Caratheodory transformation. Let the following set of matrices be given: a nonsingular $\nu \times \nu$ matrix $s = (s^\alpha_\beta)$, a nonsingular $\nu \times n$ matrix $S = (S^\alpha_i)$, and a nonsingular $\nu \times \nu$ matrix $c = (c^\alpha_\beta)$. Further, let these matrices satisfy the relation

$$c = s + Sp. \tag{7.86}$$

Lemma 7.7. *The system of equations*

$$S = sP, \tag{7.87}$$

$$\det s = 1/F \tag{7.88}$$

is equivalent to the system of equations

$$\pi = (\det c)c^{-1}S, \tag{7.89}$$

$$\det c = f. \tag{7.90}$$

Proof. We consider relations (7.87) and (7.88). Substituting (7.87) in (7.86), we obtain

$$c = s(I + Pp).\qquad(7.91)$$

Considering the determinant and using Eq. (7.88) and the relation

$$\det(I + Pp) = Ff,$$

we obtain (7.90). Using the relation

$$I + Pp = fa^{-1},$$

we transform (7.91) to the form

$$a = fc^{-1}s.$$

We premultiply the obtained relation by P and use condition (7.87), the relation $\pi = aP$, relations (7.68), and condition (7.90), which is already proved. We then obtain (7.89).

Conversely, let relations (7.89) and (7.90) hold. It is easy to see that these relations imply

$$c\pi = fS.\qquad(7.92)$$

We substitute (7.92) in (7.86) and multiply the obtained relation by f:

$$fc = fs + c\pi p.$$

By the relation

$$\pi p = fI - a,$$

we obtain

$$fs = ca\qquad(7.93)$$

from the above. Computing the determinant in both sides of Eq. (7.93) and using (7.90), we obtain

$$f^{\nu} \det s = f \det a.$$

By (7.86), we have $\det s = 1/F$; this proves (7.88). Further, we postmultiply (7.93) by P:

$$fsP = caP.\qquad(7.94)$$

By (7.68), the right-hand side of (7.94) is equal to $c\pi$; this, in turn, is equal to fS by (7.92). Therefore, $fsP = fS$; this proves (7.87).

Using the Caratheodory transformation, we can directly express the Weierstrass–Caratheodory function

$$E = f(t, x, \hat{p}) - \Delta(t, x, \hat{p})\qquad(7.95)$$

through the function f (without unknown functions of the action $S^{\alpha}(t, x)$).

Properties 1 and 2 of the Caratheodory field theory imply (7.89) and (7.90) if we take the matrices s, S, and c as

$$s = \left(\frac{\partial S^\alpha}{\partial t^\beta}\right), \qquad S = \left(\frac{\partial S^\alpha}{\partial x^i}\right), \qquad c = s + Sp.$$

By Lemma 7.7, we have

$$S = sP, \tag{7.96}$$

$$\det s = 1/F. \tag{7.97}$$

We transform the equation $c = s + Sp$, carrying out the matrix s from parentheses, applying the theorem on the determinant of the product of matrices, and using formulas (7.96) and (7.97):

$$\Delta(t, x, \hat{p}) = \frac{1}{F} \det(I + P\hat{p}). \tag{7.98}$$

We define the increment of the coordinates p by the formula

$$\hat{p}_\alpha^i - p_\alpha^i = f h_\alpha^i. \tag{7.99}$$

The matrix with the entries (h_α^i) is denoted by h. Substituting (7.99) in (7.98), we obtain

$$\Delta(t, x, \hat{p}) = \frac{1}{F} \det(I + Pp + fPh). \tag{7.100}$$

We substitute the expression

$$I + Pp = fa^{-1}$$

in (7.100) and carry out $\det a^{-1}$ from the parentheses and the determinant:

$$\Delta(t, x, \hat{p}) = \frac{1}{F \det a} \det(fI + faPh).$$

Applying

$$aP = \pi, \qquad F = \frac{f^{\nu-1}}{\det a},$$

we obtain

$$\Delta(t, x, \hat{p}) = \frac{1}{f^{\nu-1}} \det(fI + \pi(\hat{p} - p)).$$

Therefore, the Weierstrass–Caratheodory function is equal to

$$f(t, x, \hat{p}) - \frac{1}{f^{\nu-1}} \det\left[fI + \frac{\partial f}{\partial(\partial x/\partial t)}(\hat{p} - p)\right].$$

If we expand the Weierstrass–Caratheodory function $E = f - \Delta$ in powers of $\hat{p} - p$, then the free term and the linear terms vanish by (7.82) and (7.84). By (7.79), the quadratic terms of this expansion have the form

$$(\hat{p}_\alpha^i - p_\alpha^i)(\hat{p}_\beta^j - p_\beta^j)\left\{\frac{\partial^2 f}{\partial p_\alpha^i \partial p_\beta^j} - \frac{1}{f}(\pi_i^\alpha \pi_j^\beta - \pi_j^\alpha \pi_i^\beta)\right\}. \tag{7.101}$$

The nonnegativity condition of the Weierstrass–Caratheodory function (7.83) yields the condition for the positive semidefiniteness of quadratic form (7.101). This condition is naturally called the *Legendre–Caratheodory condition*.

Remark. Fulfillment of the strengthened Legendre–Weyl condition implies fulfillment of the strengthened Legandre–Caratheodory condition because we can add a sufficiently large positive constant to the function f; as a result, the two quadratic forms differ from each other by an arbitrarily small amount. On the other hand, we can give examples where the Legandre–Caratheodory condition holds but the Legendre–Weyl condition does not hold [114].

Remark. The Legendre–Weyl transformation originates from a differential form that is defined by the trace of the matrix of partial derivatives of the action functions; we can set the first characteristic Chern class in correspondence to this matrix. The Caratheodory transformation originates from a differential form that is defined by the determinant of the matrix of partial derivatives of action functions; we also can set the higher characteristic Chern class in correspondence to this matrix.

It is very desirable to find specific transformations that correspond to other characteristic classes (other invariant polynomials in eigenvalues of the matrix of partial derivatives of action functions). Probably, the totality of all such transformations gives a possibility to find sufficient optimality conditions (in the form of the field theory) that are close to the necessary optimality conditions.

Chapter 8
On the Quadratic System of Partial Differential Equations Related to the Minimization Problem for a Multiple Integral

In this chapter, we obtain and study the quadratic system of partial differential equations that serves as an analogue of the Riccati equation as applied to the minimization problem for a multiple integral [122]. In deducing this equation, we cannot assume that the matrix corresponding to the leading terms of the integrand is nonsingular: in contrast to the necessary Legendre condition for the one-dimensional integral, the necessary Hadamard–Legendre condition does not gives reasons for this assumption. Therefore, we temporarily return to the case of the one-dimensional integral and give one more deduction of the Riccati equation without assuming that the strengthened Legendre condition holds.

§1. Riccati Equation in the Case of the Degenerate Legendre Condition

We consider the Dirichlet functional

$$\int_{t_0}^{t_1} f(t, x, \dot{x})\, dt = \int_{t_0}^{t_1} \left[a_{ij}(t)\dot{x}^i \dot{x}^j + 2c_{ij}(t)x^i \dot{x}^j + b_{ij}x^i x^j \right] dt \qquad (8.1)$$

on the set of smooth functions $x(\cdot)$ defined on the closed interval $[t_0, t_1]$ such that

$$x(t_0) = x(t_1) = 0. \qquad (8.2)$$

This means that we take the direct product $[t_0, t_1] \times \mathbb{R}^n$ as the bundle E on which the Dirichlet integral is defined. The leading terms of the integrand are terms containing the products of derivatives of the functions $x(\cdot)$, i.e., $a_{ij}(t)\dot{x}^i \dot{x}^j$.

We study the question of a possible reduction of the Dirichlet functional to the functional of its leading terms

$$\int_{t_0}^{t_1} a_{ij}(t)(\dot{x}^i - w_k^i(t)x^k)(\dot{x}^j - w_s^j(t)x^s)\, dt. \qquad (8.3)$$

To reduce functional (8.1) to form (8.3), it is necessary to change the integrand without changing the value of the integral. This can be done by adding a closed differential form on the manifold $E = \mathbb{R}^n \times [t_0, t_1]$ under the integral sign. To preserve the quadratic character of the functional, this differential form should depend on x quadratically. This form is denoted by

$$\alpha = p_{ij}(t)x^i dx^j + q_{ij}(t)x^i x^j \, dt. \tag{8.4}$$

We prove the following auxiliary lemma.

Lemma 8.1. *The Euler equation of functional* (8.1) *holds identically for any function* $x(\cdot)$ *iff the integrand* $f \, dt$ *of functional* (8.1) *is represented as a closed differential form on the bundle* E.

Proof. We first suppose that the integrand $f \, dt$ of functional (8.1) can be written in form (8.4), where the differential form α is closed. We consider an arbitrary curve $x(\cdot)$ satisfying boundary conditions (8.2) and an arbitrary variation of this curve $x(\cdot) + \lambda h(\cdot)$, $\lambda \in \mathbb{R}$, that does not violate condition (8.2). The latter means that

$$h(t_0) = h(t_1) = 0. \tag{8.5}$$

Because the differential form α is closed, its integral over any contractible contour equals zero. Because the bundle E is a vector bundle, any closed contour on E is contractible. Because the sections $x(\cdot)$ and $x(\cdot) + \lambda h(\cdot)$ satisfy the same boundary conditions, their union forms a closed contour in E. Therefore, for any smooth function $h(\cdot)$ satisfying conditions (8.5) and for any λ, we have

$$\int_{x(\cdot)} \alpha = \int_{x(\cdot)+\lambda h(\cdot)} \alpha.$$

Consequently, the derivative of the function

$$\phi(\lambda) = \int_{x(\cdot)+\lambda h(\cdot)} \alpha$$

with respect to λ for $\lambda = 0$ vanishes identically. In Chap. 1, we showed that this derivative equals

$$\int (\hat{f}_x h + \hat{f}_{\dot{x}} \dot{h}) dt = 0;$$

this implies the fulfillment of the Euler equation for the function $x(\cdot)$. Because x is arbitrary, the Euler equation holds identically. We prove the converse statement. Let the Euler equation for functional (8.1) hold identically. We first prove that the integrand $f \, dt$ can be written as a differential form on the manifold E. Indeed, the terms

$$b_{ij}(t)x^i x^j \, dt, \qquad c_{ij}(t)x^i \dot{x}^j \, dt = c_{ij}(t)x^i dx^j$$

are in fact written as a differential form. As for the leading terms $a_{ij}(t)\dot{x}^i \dot{x}^j \, dt$, we see that for the Euler equation to hold identically, the matrix $\|a_{ij}\|$ must vanish identically. Otherwise, the summands containing the second derivatives of the functions x^i cannot annihilate any other summands of the Euler equation.

Therefore, functional (8.1) is the integral of the differential form α. We show that this form is closed. We consider a closed contour formed by two arbitrary sections $x_1(\cdot)$ and $x_2(\cdot)$ of the bundle E. Because the fibers of a vector bundle are simply connected, there exists a homotopy given by a continuous mapping

$$y: [t_0, t_1] \times [0, 1] \to \mathbb{R}^n$$

that depends on a scalar parameter $\lambda \in [0, 1]$ and connects these sections. That is,

$$y(t, 0) = x_1(t), \qquad y(t_0, \lambda) = x_1(t_0),$$
$$y(t, 1) = x_2(t), \qquad y(t_1, \lambda) = x_1(t_1).$$

Substituting $y(t, \lambda)$ in functional (8.1), we obtain the function

$$\psi(\lambda) = \int f(t, y(t, \lambda), \dot{y}(t, \lambda)) \, dt.$$

For all values of the parameter λ, the derivative of the function ψ is equal to the integral of the left-hand side of the Euler equation and hence vanishes on the entire closed interval $[0, 1]$. Therefore, the function ψ is constant and

$$\int_{x_1} \alpha = \int_{x_2} \alpha.$$

The sections x_1 and x_2 form a closed contour. Because the sections x_1 and x_2 are arbitrary, the Stokes formula implies that the form α is closed.

Corollary 8.1. *If the left-hand sides of the Euler equation for two integral functionals of form (8.1) are identically equal, then the integrands of these functionals differ from one another by a closed differential form defined on the bundle E. Therefore, these functionals coincide on all curves satisfying conditions (8.2).*

By Corollary 8.1, functional (8.1) is reduced to form (8.3) if the left-hand sides of the Euler equation for both functionals coincide:

$$-\frac{d}{dt}(a_{ik}\dot{x}^k + c_{ki}x^k) + (c_{ik}\dot{x}^k + b_{ik}x^k)$$
$$\equiv -\frac{d}{dt}(a_{ik}\dot{x}^k - a_{ij}w_k^j x^k) + (-a_{kj}w_i^j \dot{x}^k + a_{sj}w_k^s w_i^j x^k). \quad (8.6)$$

Equating the coefficients of $\dot{x}^k$ in this identity, we obtain the relation

$$-c_{ki} + c_{ik} = a_{ij}w_k^j - a_{kj}w_i^j,$$

which express the symmetry property of the matrix $a_{ij}w_k^j + c_{ki}$.

Let A be an $n \times n$ matrix with the entries a_{ij}, C be an $n \times n$ matrix with the entries c_{ij}, and W be an $n \times n$ matrix with the entries w_{ij}. We then have

$$(AW + C^\mathsf{T})^\mathsf{T} = (AW + C^\mathsf{T}). \quad (8.7)$$

Equating the coefficients of x^k in (8.6), we obtain the relations

$$\frac{d}{dt}(a_{ij}w_k^j + c_{ki}) + a_{sj}w_k^s w_i^j - b_{ik} = 0,$$

which in matrix form are

$$\frac{d}{dt}(AW + C^\top) + W^\top AW - B = 0. \tag{8.8}$$

In the case where the matrix A is nonsingular (this corresponds to fulfill-ment of the strengthened Legendre condition), we can obtain Riccati equation (2.8) from Eq. (8.8). Indeed, let $-Z$ denote the matrix $AW + C^\top$. Then

$$W = -A^{-1}Z - A^{-1}C^\top,$$

and Eq. (8.8) becomes Riccati equation (2.8):

$$\frac{dZ}{dt} - (Z + C)A^{-1}(Z + C^\top) + B = 0.$$

Symmetry condition (8.7) becomes the previous symmetry condition for the matrix Z (or, equivalently, the condition that the plane with the coordinates Z is Lagrangian).

We thus obtain another form of the Riccati equation, Eq. (8.8). This equa-tion is not resolved in the derivative. Moreover, it has a singularity at those points t where the matrix A degenerates. Its advantage over Eq. (2.8) is that it remains valid without any assumptions on the matrix A.

Theorem 8.1. *Let $A \geq 0$. Let there exist a solution $W(t)$ to Eq. (8.8) defined on the entire closed interval $[t_0, t_1]$ such that the matrix $AW + C^\top$ is symmetrical. Then functional (8.1) is nonnegative.*

Proof. The conditions of Theorem 8.1 make it possible to reduce functional (8.1) to functional (8.3) whose nonnegativity is implied by the nonnegativity of the matrix A.

§2. Reducing the Dirichlet Integral to the Integral of Its Principal Part

We consider a multiple integral as a functional given on a submanifold of a Riemannian space over which the integration is performed. The boundary of the submanifolds is assumed to be fixed.

Let $\mathfrak{M}$ be a ν-dimensional Riemannian manifold with the boundary $\partial\mathfrak{M}$, and let $\pi : \xi \to \mathfrak{M}$ be a vector bundle over the base $\mathfrak{M}$. A fiber $\pi^{-1}t$, i.e., the full inverse image of the point $t \in \mathfrak{M}$ under the mapping π, is an n-dimensional linear subspace, which is denoted by F_t. Local coordinates on $\mathfrak{M}$

are denoted by $t = (t^1, \ldots, t^\nu)$, and coordinates on the fibers are denoted by $x = (x^1, \ldots, x^n)$.

We consider the quadratic integral functional (Dirichlet functional)

$$\mathcal{J} = \int_{\mathfrak{N}} F \, dt, \qquad (8.9)$$

where the function F in local coordinates has the form

$$F = a_{ij}^{\alpha\beta}(t) \frac{\partial x^i}{\partial t^\alpha} \frac{\partial x^j}{\partial t^\beta} + 2c_{ij}^\alpha(t) \frac{\partial x^i}{\partial t^\alpha} x^j + b_{ij}(t) x^i x^j,$$

$dt = dt^1 \wedge \ldots \wedge dt^\nu$ is a volume element of the manifold $\mathfrak{N}$, and $a_{ij}^{\alpha\beta}$, c_{ij}^α, and b_{ij} are smooth functions given on the manifold $\mathfrak{N}$. Here and what follows, we use the convention on summation over repeating sub- and superscripts. Latin indices refer to coordinates on the fibers and assume values from 1 to n; Greek indices refer to coordinates on the base and assume values from 1 to ν. The functional $\mathcal{J}$ is considered on the space of $C^\infty(\xi)$-smooth sections $x(\cdot)$ of the bundle ξ that satisfy the boundary conditions

$$x\big|_{\partial\mathfrak{N}} = 0. \qquad (8.10)$$

The idea of studying the Dirichlet functional, which goes back to Clebsch [19], consists in reducing it (by choosing an appropriate representative in the corresponding cohomology class) to the integral of principal terms of the integrand

$$\int_{\mathfrak{N}} a_{ij}^{\alpha\beta}(t) \left(\frac{\partial x^i}{\partial t^\alpha} - w_{\alpha k}^i(t) x^k \right) \left(\frac{\partial x^j}{\partial t^\beta} - w_{\beta k}^j(t) x^k \right) dt = \int_{\mathfrak{N}} a_{ij}^{\alpha\beta} y_\alpha^i y_\beta^j \, dt, \quad (8.11)$$

where $y_\alpha^i = \partial x^i / \partial t^\alpha - w_{\alpha k}^i x^k$ and $w_{\alpha k}^i(t)$ are certain functions that should be chosen. We add a closed differential form under the integral sign in (8.9). It is easy to see that as a consequence of the contractibility of the fibers of the vector bundle and by boundary conditions (8.10), the value of the integral does not change. In order to retain the structure of the integrand, the added expression (as the initial integrand of the functional $\mathcal{J}$) should be quadratic in the variables x and $\partial x / \partial t$.

The Euler equation for functional (8.9) is

$$-\frac{\partial}{\partial t^\alpha} \left(a_{ij}^{\alpha\beta} \frac{\partial x^j}{\partial t^\beta} + c_{ij}^\alpha x^j \right) + \left(c_{ji}^\alpha \frac{\partial x^j}{\partial t^\alpha} + b_{ij} x^j \right) = 0. \qquad (8.12)$$

Lemma 8.2. *The Euler equation for functional (8.9) holds identically iff the integrand can be written as a closed differential form θ on the bundle ξ.*

Proof. We show first that identical fulfillment of the Euler equation implies that the integrand in (8.9) can be written as a differential form on ξ. Let $\{dt\}_\alpha$ be the exterior product of all differentials $dt^1 \wedge \ldots \wedge dt^\nu$ except for the factor dt^α, and let $\{dt\}_{\alpha\beta}$ be the exterior product of all differentials $dt^1 \wedge \ldots \wedge dt^\nu$ except for the factors dt^α and dt^β. The summands in the integrand of the form $b_{ij} x^i x^j \, dt$ and

$$c_{ij}^\alpha \frac{\partial x^i}{\partial t^\alpha} x^j \, dt = (-1)^{\alpha-1} c_{ij}^\alpha x^j \, dx^i \wedge \{dt\}_\alpha$$

are already differential forms on ξ. The coefficients of the principal terms of the integrand $a_{ij}^{\alpha\beta}$ should be skew symmetrical in the superscripts α and β because the leading terms of the Euler equation,

$$a_{ij}^{\alpha\beta} \frac{\partial^2 x^j}{\partial t^\alpha \partial t^\beta},$$

otherwise do not vanish identically. Because these coefficients are symmetrical with respect to permutations of pairs of indices (i, α) and (j, β), they should also be skew symmetrical with respect to the subscripts i and j. Therefore, the principal term in functional (8.9) can be written as the differential form on ξ

$$(-1)^{\alpha+\beta+1} a_{ij}^{\alpha\beta} \, dx^i \wedge dx^j \wedge \{dt\}_{\alpha\beta}.$$

To prove that the differential form θ appearing under the integral sign is closed, we consider an arbitrary section $x(\cdot)$ of the bundle ξ and its variation $h(\cdot)$ such that $h|_{\partial\mathfrak{N}} = 0$. Because ξ is a vector bundle, its fibers are contractible. Therefore, the manifolds defined by the sections $x(\cdot)$ and $x(\cdot) + h(\cdot)$ are homotopic. Identical fulfillment of the Euler equation means that the derivative of functional (8.9) computed on any section $x(\cdot)$ of the bundle ξ in any direction $h(\cdot)$ satisfying the boundary conditions vanishes. Therefore, under the homotopy, the value of the functional remains constant:

$$\int_{x(\cdot)} \theta = \int_{x(\cdot)+h(\cdot)} \theta.$$

The manifolds $x(\cdot)$ and $x(\cdot) + h(\cdot)$ satisfying the same boundary conditions (8.10) bound a certain $(\nu+1)$-dimensional manifold $\mathfrak{M}$; by the Stokes formula,

$$\int_{\mathfrak{M}} d\theta = 0.$$

Because the sections $x(\cdot)$ and $x(\cdot) + h(\cdot)$ are arbitrary, this implies that the form θ is closed.

Conversely, the closedness of the form θ implies

$$\int_{x(\cdot)} \theta = \int_{x(\cdot)+h(\cdot)} \theta.$$

From this, it is easy to show that the derivative of the functional $\mathcal{J}$ on any section $x(\cdot)$ of the bundle ξ in any direction $h(\cdot)$ vanishes; this means the identical fulfillment of the Euler equation.

According to Lemma 8.2, in order to reduce functional (8.9) to form (8.11), we must require that the results of applying the Euler operator (i.e., the left-hand side of Eq. (8.12)) to both integrands are identically equal, i.e.,

$$-\frac{\partial}{\partial t^\alpha}\left(a_{ij}^{\alpha\beta}\frac{\partial x^j}{\partial t^\beta} + c_{ij}^\alpha x^j\right) + \left(c_{ji}^\alpha\frac{\partial x^j}{\partial t^\alpha} + b_{ij}x^j\right)$$
$$\equiv -\frac{\partial}{\partial t^\alpha}\left(a_{ij}^{\alpha\beta}\frac{\partial x^j}{\partial t^\beta} - a_{ij}^{\alpha\beta}w_{\beta k}^j x^k\right) + \left(-a_{kj}^{\alpha\beta}\frac{\partial x^j}{\partial t^\beta}w_{\alpha i}^k + w_{\alpha i}^k a_{kj}^{\alpha\beta}w_{\beta s}^j x^s\right).$$

The terms with second derivatives are mutually annihilated. Equating the coefficients of $\partial x^j/\partial t^\alpha$, we obtain the relations

$$a_{il}^{\alpha\beta}w_{\beta j}^l + c_{ij}^\alpha = a_{kj}^{\beta\alpha}w_{\beta i}^k + c_{ji}^\alpha. \tag{8.13}$$

Equating the coefficients of x^j, we obtain the relations

$$\frac{\partial}{\partial t^\alpha}(a_{ik}^{\alpha\beta}w_{\beta j}^k + c_{ij}^\alpha) = b_{ij} - a_{ks}^{\alpha\beta}w_{\alpha i}^k w_{\beta j}^s. \tag{8.14}$$

We have thus proved the following theorem.

Theorem 8.2. *The functional $\mathcal{J}$ admits the reduction to form (8.11) iff there exists a global (i.e., defined on the whole manifold $\mathfrak{M}$) solution to system (8.13), (8.14).*

For tensors whose indices vary from 1 to n, we introduce the matrix notation using capital letters corresponding to the initial small ones. For example, for fixed α and β, the matrix with the entries $a_{ij}^{\alpha\beta}$ is denoted by $A^{\alpha\beta}$, and so on. Transposed matrices are denoted by the superscript T. Using the relations $a_{kj}^{\alpha\beta} = a_{jk}^{\beta\alpha}$, we rewrite Eq. (8.13) in the form

$$A^{\alpha\beta}W_\beta + C^\alpha = \left[A^{\alpha\beta}W_\beta + C^\alpha\right]^\mathsf{T} \tag{8.15}$$

and Eq. (8.14) in the form

$$\frac{\partial}{\partial t^\alpha}\left[A^{\alpha\beta}W_\beta + C^\alpha\right] = B - W_\alpha^\mathsf{T}A^{\alpha\beta}W_\beta. \tag{8.16}$$

Equation (8.16) is called the *Riccati partial differential equation*. Equation (8.16) with symmetry condition (8.15) plays the same role in the minimization problem for a multiple integral that the Riccati ordinary differential equation plays in the minimization problem for a one-dimensional integral. Namely, the following theorem holds.

Theorem 8.3. *Let the following conditions hold:*

1. *The quadratic form $a_{ij}^{\alpha\beta}y_\alpha^i y_\beta^j$ given on the space of $n\times\nu$ matrices y_α^i is positive semidefinite for all $t \in \mathfrak{M}$.*
2. *There exists a global (i.e., defined on the whole manifold $\mathfrak{M}$) solution to system (8.15), (8.16).*

Then functional (8.9), (8.10) *is positive semidefinite. The existence of a solution to system* (8.15), (8.16) *that is defined on a domain* $\mathfrak{O} \subset \mathfrak{N}$ *implies the local positive semidefiniteness on any subdomain* $\mathfrak{O}' \subset \mathfrak{O}$, *i.e., the condition*

$$\int_{\mathfrak{O}'} F\, dt \geq 0.$$

The proof is directly implied by Theorem 8.1 and formula (8.11).

Therefore, to study the question of the positive definiteness of the Dirichlet functional, we must first try to correct the principal terms

$$a_{ij}^{\alpha\beta} \frac{\partial x^i}{\partial t^\alpha} \frac{\partial x^j}{\partial t^\beta}$$

by adding a closed differential form $r_{ij}^{\alpha\beta}(t) dx^i \wedge dx^j \wedge \{dt\}_{\alpha\beta}$ such that they satisfy the conditions of Theorem 8.2.

We assume that the following condition holds: there exist functions $r_{ij}^{\alpha\beta}(t)$ corresponding to the closed differential form $r_{ij}^{\alpha\beta}(t) \{dt\}_{\alpha\beta}$ such that the quadratic form

$$a_{ij}^{\alpha\beta} y_\alpha^i y_\beta^j + r_{ij}^{\alpha\beta}(y_\alpha^i y_\beta^j - y_\beta^i y_\alpha^j)$$

assumes nonnegative values on any matrices y_α^i. For example, we can take $r_{ij}^{\alpha\beta}$ to be constant.

This condition is sufficiently strong because the Hadamard condition is a necessary condition for the positive semidefiniteness of functional (8.9), i.e., the quadratic form $\mathfrak{A} = a_{ij}^{\alpha\beta} y_\alpha^i y_\beta^j$ given on $n \times \nu$ matrices y_α^i is nonnegative on the cone of matrices of rank one or, in other words, the biquadratic form

$$a_{ij}^{\alpha\beta} \eta^i \eta^j \lambda_\alpha \lambda_\beta \geq 0 \tag{8.17}$$

is nonnegative on all pairs of vectors $\eta = (\eta^1, \ldots, \eta^n)$ and $\lambda = (\lambda_1, \ldots, \lambda_\nu)$.

Under fulfillment of the considered condition, we can seek a globally defined solution to Eqs. (8.15) and (8.16); if this solution is found, we can be sure that Theorem 8.2 is applicable. Otherwise, we must consider the problem taking differential constraints on the matrices y_α^i, which are obtained below in Theorem 8.7, into account.

§3. Relation of the Riccati Partial Differential Equation to the Euler Equation

As in the case of a one-dimensional integral (see Chap. 1), Eq. (8.16) is closely related to Euler equation (8.12), which in the matrix form is

$$-\frac{\partial}{\partial t^\alpha}\left[A^{\alpha\beta}\frac{\partial U}{\partial t^\beta} + C^\alpha U\right] + \left[(C^\alpha)^\mathsf{T}\frac{\partial U}{\partial t^\alpha} + BU\right] = 0, \tag{8.18}$$

where U is an $n \times \nu$ matrix corresponding to an n-parameter family of extremals or, in other words, to a field of extremals for functional (8.9).

Definition 8.1. A solution to Eq. (8.18) is said to be *nondegenerate on a manifold* $\mathfrak{N}$ if $\det U(t) \neq 0$ for all $t \in \mathfrak{N}$.

Definition 8.2. A solution $U(t)$ to Eq. (8.18) *satisfies the symmetry condition* if the matrix

$$A^{\alpha\beta} \frac{\partial U}{\partial t^\beta} U^{-1} + C^\alpha$$

is symmetrical.

Remark. In the case of a one-dimensional integral, the symmetry condition coincides with the Lagrange condition, i.e., with the condition for the symmetry of the matrix that assigns the coordinates of an n-dimensional linear subspace.

Theorem 8.4. *Let $U(t)$ be a solution to Eq. (8.18) that satisfies the symmetry condition and is nondegenerate in $\mathfrak{N}$. Then*

$$W_\alpha = \frac{\partial U}{\partial t^\alpha} U^{-1} \tag{8.19}$$

satisfies Eqs. (8.15) and (8.16).

Proof. The symmetry condition obviously coincides with condition (8.15). We verify the fulfillment of Eq. (8.16). Indeed,

$$\frac{\partial}{\partial t^\alpha} \left[A^{\alpha\beta} \frac{\partial U}{\partial t^\beta} U^{-1} + C^\alpha \right] = \frac{\partial}{\partial t^\alpha} \left\{ \left[A^{\alpha\beta} \frac{\partial U}{\partial t^\beta} + C^\alpha U \right] U^{-1} \right\}$$

$$= \left[(C^\alpha)^\top \frac{\partial U}{\partial t^\alpha} + BU \right] U^{-1}$$

$$- \left[A^{\alpha\beta} \frac{\partial U}{\partial t^\beta} + C^\alpha U \right] U^{-1} \frac{\partial U}{\partial t^\alpha} U^{-1}$$

$$= (C^\alpha)^\top W_\alpha + B - \left[A^{\alpha\beta} W_\beta + C^\alpha \right] W_\alpha.$$

Using condition (8.15), we obtain

$$\frac{\partial}{\partial t^\alpha} \left[A^{\alpha\beta} \frac{\partial U}{\partial t^\beta} U^{-1} + C^\alpha \right] = B - W_\beta^\top A^{\alpha\beta} W_\alpha.$$

This implies Eq. (8.16).

In contrast to the case corresponding to a one-dimensional integral, the converse theorem is not true. Not every solution to Eqs. (8.15) and (8.16) yields a solution to Euler equation (8.18). The converse theorem holds under an additional condition. We consider relation (8.19) as a system of equations with respect to the matrix $U(t)$ for given $W_\alpha(t)$:

$$\frac{\partial U}{\partial t^\alpha} = W_\alpha U. \tag{8.20}$$

The Cauchy problem for Eq. (8.20) consists in seeking a solution to this equation that satisfies the condition $U(t_0) = U_0$.

Lemma 8.3. *Let* $\det U_0 \neq 0$. *Then the Cauchy problem for Eq. (8.20) is locally solvable iff*

$$\frac{\partial}{\partial t^\alpha} W_\beta - \frac{\partial}{\partial t^\beta} W_\alpha - [W_\alpha, W_\beta] = 0. \tag{8.21}$$

Here, $[W_\alpha, W_\beta] = W_\alpha W_\beta - W_\beta W_\alpha$ *is the commutator of the two matrices* W_α *and* W_β.

Proof. The local solvability of system (8.20) is equivalent to the closedness of the matrix differential form

$$dU - W_\alpha(t)\, U\, dt^\alpha. \tag{8.22}$$

The coefficient of $dt^\alpha \wedge dt^\beta$ in the exterior differential of this form is equal to

$$\left(\frac{\partial W^\alpha}{\partial t^\beta} - \frac{\partial W^\beta}{\partial t^\alpha} + [W_\alpha, W_\beta] \right) U = 0. \tag{8.23}$$

Because $\det U(t) \neq 0$ in a sufficiently small neighborhood of the point t_0, condition (8.23) is equivalent to (8.21).

Proposition 8.1. *Let the functions* W_α, $\alpha = 1, \ldots, \nu$, *be continuous on a manifold* $\mathfrak{N}$ *and satisfy condition (8.21). Then a solution* $U(t)$ *to system (8.20) with a nondegenerate initial condition* $U(t_0) = U_0$ *(i.e.,* $\det U_0 \neq 0$) *exists, is unique, and can be continued (as a multivalued function) to the whole manifold* $\mathfrak{N}$. *Moreover,* $\det U(t) \neq 0$ *for all* $t \in \mathfrak{N}$.

Proof. On the manifold $\mathfrak{N}$, we consider an arbitrary curve $\gamma(s)$ passing through the point t_0. System (8.20) induces a linear system of ordinary differential equations with the independent variable s on the curve γ. The coefficients of this system are continuous, and the restriction $U(s) = U(t)|_\gamma$ is therefore defined and can be continued in s. Moreover, $\det U(s) \neq 0$.

The closedness of the differential form (8.22) guarantees that $U(t)$ is single-valued under a homotopic change of the path γ that connects the points t_0 and t. If the manifold $\mathfrak{N}$ is not simply connected, the function $U(t)$ can be multivalued, i.e., the value of $U(t)$ depends on the homotopy class of the path γ.

3.1. Compactification of the Space on Which the Riccati Partial Differential Equation is Defined.

To explain the geometric sense of Eq. (8.16), we consider the following auxiliary construction. On the Grassmann manifold $G_n(\mathbb{R}^{n(\nu+1)})$, i.e., on the manifold of n-dimensional subspaces of the space $\mathbb{R}^{n(\nu+1)}$, we introduce a specific coordinate system that is a matrix analogue of the projective coordinates on $\mathbb{RP}^k$. In $\mathbb{R}^{n(\nu+1)}$, we consider a $(\nu+1)$-tuple of n-dimensional planes that are in the general position, i.e.,

$$T_0 \oplus T_1 \oplus \ldots \oplus T_\nu = \mathbb{R}^{n(\nu+1)}.$$

Let π_α be the projection operator on the subspace T_α parallel to the subspace $\oplus_{\beta \neq \alpha} T_\beta$. Let W be an arbitrary n-dimensional plane. If we fix bases in the planes W and T_α, $\alpha = 0, \ldots, \nu$, then the $(\nu+1)$-tuple of $n \times n$ matrices of the mappings $\pi_\alpha : W \to T_\alpha$ (which are also denoted by π_α) is called the *matrix homogeneous coordinates* of the plane W.

Remark. The change of a basis in the plane W leads to the postmultiplication of all matrices π_α by the same nonsingular matrix.

We consider a chart M in $G_n(\mathbb{R}^{n(\nu+1)})$ consisting of planes W transversal to $\oplus_{\beta \neq 0} T_\beta$. Then the restriction of the projection π_0 to $W \in M$ is a one-to-one mapping. The *matrix affine coordinates* of a plane W in a given chart are matrices $W^\kappa = \pi_\kappa \pi_0^{-1}$, $\kappa \neq 0$.

The passage from one chart to the other is described by a linear transformation of the space $\mathbb{R}^{n(\nu+1)}$ that has the block form $Y^\sigma = \mathfrak{A}_\kappa^\sigma X^\kappa$ (the Greek subscripts number blocks that are $n \times n$ matrices). In this case, matrix affine coordinates are transformed as a vector composed of generalized linear-fractional transformations of the form

$$(\mathfrak{A}_0^\gamma + \mathfrak{A}_\alpha^\gamma W^\alpha)(\mathfrak{A}_0^0 + \mathfrak{A}_\alpha^0 W^\alpha)^{-1}, \quad \gamma = 1, \ldots, \nu.$$

The latter formula defines the action of $\mathrm{SL}(n(\nu + 1))$ on the manifold $G_n(\mathbb{R}^{n(\nu+1)})$. Considering this action as a change of variables in Eq. (8.16), we can define this equation on "infinitely distant" points of the chart M. We thus obtain a possibility to continue solutions to Eq. (8.16) that go to infinity by passing from the chart M_0 to another chart on $G_n(\mathbb{R}^{n(\nu+1)})$.

§4. Connection Defined by a Solution to the Riccati Partial Differential Equation

We recall that an *affine connection* ∇ on a vector bundle $\pi : \xi \to \mathfrak{N}^\nu$ is a smooth distribution of ν-dimensional planes (called horizontal) in the tangent planes $T_u \xi$. This distribution should have the following property: the projection operator of vectors tangent to the bundle on a fiber parallel to the horizontal plane depends linearly on a point of this fiber (see Chap. 7 and also [5, 25, 59, 71]).

To simplify formulas, instead of using differential operators to designate tangent vectors, we use differential forms dual to them, not specifying which we mean: the differential form (a functional on the set of tangent vectors) or the tangent vector (dual to it).

A connection ∇ is uniquely defined by a vector-valued differential form that defines the projection of tangent vectors (dt, dx) at a point (t, x) on a fiber. This form should depend linearly on x and yield the identity mapping on vertical vectors $(0, dx)$. Therefore, it should have the form

$$\omega = dx - W_\alpha(t)x\,dt^\alpha.$$

The corresponding operator of covariant differentiation in the direction of a vector v tangent to the manifold $\mathfrak{N}$ is

$$\nabla_v x(t) = \frac{\partial x}{\partial t^\alpha}v^\alpha - W_\alpha(t)xv^\alpha. \tag{8.24}$$

The projection operator of tangent vectors (dt, dx) at a point (t, x) on a fiber has the form

$$(dt, dx) \mapsto (0, dx - W_\alpha x\,dt^\alpha).$$

Horizontal vectors of the connection ∇ are those vectors (dt, dx) that lie in the kernel of this operator: $dx - W_\alpha x\,dt^\alpha = 0$. Therefore, the horizontal component of a vector (dt, dx) is the vector $(dt, W_\alpha x\,dt^\alpha)$.

We recall that a connection is called *flat* if its curvature tensor is equal to zero.

Proposition 8.2. *Condition* (8.21) *holds iff the connection* ∇ *is flat.*

Proof. We compute the differential form of curvature Ω of the connection ∇. We have

$$d\omega = -w^i_{\alpha k}dx^k \wedge dt^\alpha - \left(\frac{\partial}{\partial t^\beta}w^i_{\alpha k}\right)x^k dt^\beta \wedge dt^\alpha.$$

By definition (see Chap. 7), the exterior covariant derivative $D\omega$ of the form ω coincides with the value of the form $d\omega$ on the horizontal components of the vector (dt, dx). Therefore,

$$\Omega = D\omega = -w^i_{\alpha k}(w^k_{\beta j}x^j dt^\beta) \wedge dt^\alpha - \frac{\partial}{\partial t^\beta}w^i_{\alpha k}x^k dt^\beta \wedge dt^\alpha.$$

The coefficient of $x^k dt^\alpha \wedge dt^\beta$ in the exterior form $D\omega$ coincides with the left-hand side of (8.21).

Using Lemma 8.3 and Proposition 8.2, we obtain the following result.

Theorem 8.5 (on parameterization). *The affine connection* ∇ *defined by the differential form* ω *is flat iff there exists a matrix-valued function* $U(t)$ *such that* $\det U(t) \neq 0$ *and*

$$W_\alpha(t) = \frac{\partial U}{\partial t^\alpha}U^{-1}.$$

The function $U(t)$ *is defined with an accuracy up to postmultiplication by a constant matrix.*

Proof. Let

$$W_\alpha = \frac{\partial U}{\partial t^\alpha} U^{-1}.$$

Then, by Lemma 8.3, condition (8.21) holds; Proposition 8.2 implies that the connection ∇ is flat.

Now let the connection ∇ be flat. Then the system of equations

$$\frac{\partial U}{\partial t^\alpha} = W_\alpha U$$

satisfies the integrability conditions. By Proposition 8.1, for any U_0, $\det U_0 \neq 0$, there exists a unique solution $U(t)$ such that $U(t_0) = U_0$; moreover, $\det U(t) \neq 0$. Therefore,

$$W_\alpha = \frac{\partial U}{\partial t^\alpha} U^{-1}$$

and $U(t)$ is uniquely defined with an accuracy up to the matrix U_0.

Corollary 8.2. *The connection ∇ that is constructed using a solution to Eqs. (8.15) and (8.16) is generated by a nondegenerate solution to matrix Euler equation (8.18) iff it is flat.*

We note that condition (8.21) is simultaneously the integrability condition for the connection ∇ and, equivalently, the involutiveness condition for its horizontal planes. Namely, the following statement holds.

Proposition 8.3. *The affine connection ∇ defined by the form ω is integrable iff it is flat.*

Proof. By the Frobenius theorem, the vector-valued differential form ω is completely integrable iff its exterior derivative $d\omega$ lies in the ideal I generated by the components of the form ω. Bearing this in mind, we write $d\omega$ in the form

$$\begin{aligned}
\omega = {} & d - w^i_{\alpha k}(dx^k - w^k_{\beta s}x^s dt^\beta) \wedge dt^\alpha - \frac{\partial}{\partial t^\beta} w^i_{\alpha k} x^k dt^\beta \wedge dt^\alpha \\
& - w^i_{\alpha k} w^k_{\beta s} x^s dt^\beta \wedge dt^\alpha \\
= {} & -w^i_{\alpha k}(\omega^k \wedge dt^\alpha) \\
& - \frac{1}{2}\left[\frac{\partial}{\partial t^\alpha}w^i_{\beta k} - \frac{\partial}{\partial t^\beta}w^i_{\alpha k} + w^i_{\alpha j}w^j_{\beta k} - w^i_{\beta j}w^j_{\alpha k}\right]x^k dt^\alpha \wedge dt^\beta.
\end{aligned}$$

The first summand in the obtained expression for $d\omega$ lies in the ideal I. Therefore, the form $d\omega$ lies in the ideal I (and the connection ∇ is integrable) iff $\Omega = 0$.

Condition (8.21) is a consequence of the existence of n independent solutions to the Euler system of equations (8.12), which we consider as horizontal sections of the bundle ξ (see Theorem 7.1 and also [71]).

Condition (8.21) is also related to the concept of a field of extremals. A field of extremals is an n-parameter family of solutions to Euler equations such that the parameter space is diffeomorphically mapped onto fibers of the bundle ξ on which a certain invariant integral is defined.

A direct consequence of Propositions 8.1 and 8.2 is the following theorem.

Theorem 8.6. *A solution to Eqs. (8.15) and (8.16) generates a field of extremals if the affine connection generated by this solution is flat.*

Theorems 8.2 and 8.6 show that the approach to the question of the definiteness of functional (8.9) that uses Eqs. (8.15) and (8.16) is preferable to the approach using the concept of a field of extremals because Theorem 8.2 also yields a sufficient condition for definiteness when there is no field of extremals.

Theorem 8.1 determines conditions under which functional (8.9) is reduced to the form

$$\int_{\mathfrak{N}} a_{ij}^{\alpha\beta} y_\alpha^i y_\beta^j \, dt, \tag{8.25}$$

where y_α^i is the expression

$$y_\alpha^i = \frac{\partial x^i}{\partial t^\alpha} - w_{\alpha k}^i x^k. \tag{8.26}$$

Example 8.1. We consider the functional

$$\int\int_{\mathfrak{N}}\left[\left(\frac{\partial x^1}{\partial t^1}\right)^2 + \left(\frac{\partial x^1}{\partial t^2}\right)^2 + \left(\frac{\partial x^2}{\partial t^1}\right)^2 + \left(\frac{\partial x^2}{\partial t^2}\right)^2\right.$$
$$\left. +\left(1+\frac{1}{r^2}\right)((x^1)^2 + (x^2)^2) + \frac{2}{r}x^1 x^2\right] dt^1 \wedge dt^2,$$

where the domain of integration $\mathfrak{N}$ is the annulus $\rho_1 \leq r \leq \rho_2$, ϕ and r are polar coordinates, and the bundle ξ is trivial.

The positive definiteness of the considered functional is obvious. However, by examining this example, we show that a single-valued and globally defined solution to Eq. (8.16) can correspond to a multivalued solution of Euler equation (8.12). The system of Euler equations (8.12) has the form

$$\Delta x^1 - (1 + 1/r^2)x^1 - (1/r)x^2 = 0,$$
$$\Delta x^2 - (1 + 1/r^2)x^2 - (1/r)x^1 = 0.$$

Its matrix-valued solution is a multivalued function

$$U = e^\phi \begin{pmatrix} \cosh r & \sinh r \\ \sinh r & \cosh r \end{pmatrix}.$$

Equation (8.16) becomes

$$\frac{\partial W_1}{\partial t^1} + \frac{\partial W_2}{\partial t^2} + W_1^2 + W_2^2 - \begin{pmatrix} 1 + 1/r^2 & 1/r \\ 1/r & 1 + 1/r^2 \end{pmatrix} = 0.$$

We find a solution to this equation using a solution to the Euler equation

$$W_1 = \frac{\partial U}{\partial t^1} U^{-1} = \begin{pmatrix} -t^2/r^2 & t^1/r \\ t^1/r & -t^2/r^2 \end{pmatrix},$$

$$W_2 = \frac{\partial U}{\partial t^2} U^{-1} = \begin{pmatrix} t^1/r^2 & t^2/r \\ t^2/r & t^1/r^2 \end{pmatrix}.$$

The obtained solution is single valued.

Indeed, the matrices W_1 and W_2 commute, and $\partial W_1/\partial t^2 - \partial W_2/\partial t^1 = 0$; therefore, relation (8.21) holds.

Example 8.2. Let $\mathfrak{N}$ be a simply connected domain, and let $\xi = \mathfrak{N} \times \mathbb{R}^1$. In this case, W_α are scalar-valued functions, and integrability condition (8.21) becomes

$$\frac{\partial W_\beta}{\partial t^\alpha} = \frac{\partial W_\alpha}{\partial t^\beta}.$$

The latter means that the differential form $W_\alpha dt^\alpha$ is closed and is exact in the simply connected case, i.e., there exists a function $S(t)$ such that

$$\frac{\partial S}{\partial t^\alpha} = W_\alpha.$$

Then system of equations (8.16) is reduced to one quadratic equation with an elliptic principal part:

$$\frac{\partial}{\partial t^\alpha}\left(A^{\alpha\beta} \frac{\partial S}{\partial t^\beta} \right) + A^{\alpha\beta} \frac{\partial S}{\partial t^\alpha} \frac{\partial S}{\partial t^\beta} - B + \frac{\partial C^\alpha}{\partial t^\alpha} = 0. \tag{8.27}$$

Example 8.3. If we let $A^{\alpha\beta} = \delta^{\alpha\beta}$ be the identity matrix, $C^\alpha = 0$, and $B = -\omega^2$ in Example 8.2, then Eq. (8.27) becomes

$$\sum_\alpha \left[\frac{\partial^2 S}{(\partial t^\alpha)^2} + \left(\frac{\partial S}{\partial t^\alpha} \right)^2 \right] + \omega^2 = 0. \tag{8.28}$$

Equation (8.28) admits particular solutions of the form

$$S = k + \ln|\sin(\omega t^\gamma + L)|. \tag{8.29}$$

Therefore, if the domain $\mathfrak{N}$ lies between certain parallel planes Π_1 and Π_2 with the distance between them less than π/ω, then Eq. (8.28) has a solution defined on the whole domain $\mathfrak{N}$ independent of the size and shape of the domain $\mathfrak{N}$; this guarantees the positive definiteness of the functional

$$\int_{\mathfrak{N}} \left[\sum_\alpha \left(\frac{\partial x}{\partial t^\alpha} \right)^2 - \omega^2 x^2 \right] dt. \tag{8.30}$$

Indeed, Eq. (8.28) is invariant with respect to orthogonal transformations; this allows us to transform the planes Π_1 and Π_2 into planes parallel to one of the coordinate planes and use solution (8.29) that defines a field of extremals in $\mathfrak{N}$.

To formulate a more general result, we introduce the following definition. Let $\mathcal{L}_q$ be the line $t = ls + q$ with the directing vector l, where $s \in \mathbb{R}^1$ is a parameter on this line, and let $q \in \mathbb{R}^\nu$. Let $d_{ql}(\mathfrak{N})$ be the length of the minimum segment containing $\mathcal{L}_q \cap \mathfrak{N}$.

Definition 8.3. The *width* of the domain $\mathfrak{N}$ in the direction of a vector l is $d_l(\mathfrak{N}) = \sup_{q \in \mathbb{R}^\nu} d_{ql}(\mathfrak{N})$.

Proposition 8.4. *If $d_l(\mathfrak{N}) \leq \pi/\omega - \epsilon$, where $0 < \epsilon < \pi/\omega$, then functional* (8.30) *is positive semidefinite.*

Proof. We write Eq. (8.16) for functional (8.30):

$$\sum_\alpha \frac{\partial}{\partial t^\alpha} W_\alpha = -\omega^2 - \sum_\alpha W_\alpha^2. \tag{8.31}$$

In $\mathbb{R}^\nu$, we perform the orthogonal transformation that transforms l into the axis Ot^1. Equation (8.31) is preserved under this transformation. The left and right boundaries of the domain in the new coordinates can be written in the respective forms $t^1 = g^-(t^2, \ldots, t^\nu)$ and $t^1 = g^+(t^2, \ldots, t^\nu)$; moreover, $0 \leq g^+ - g^- \leq \pi/\omega - \epsilon$. The functions $W_1 = \omega \cot \omega(t^1 - g^-(t^2, \ldots, t^\nu) + \epsilon)$, $W_\gamma = 0$ for $\gamma > 1$ yield a solution to Eq. (8.31) that is defined for all $t \in \mathfrak{N}$. By Theorem 8.2, functional (8.30) is nonnegative.

We note that the obtained solution does not satisfy integrability condition (8.21) and the connection defined by it has a nonzero curvature. This means that it is not possible to construct a field of extremals generated by this connection (not even a multivalued field). Nevertheless, the sufficient condition is obtained in terms of a solution to Eq. (8.16).

Example 8.4. We consider the functional

$$\int_\mathfrak{N} \left[\left(\frac{\partial x^1}{\partial t^1} \right)^2 + \cdots + \left(\frac{\partial x^1}{\partial t^\nu} \right)^2 + \cdots + \left(\frac{\partial x^n}{\partial t^1} \right)^2 + \cdots \right.$$
$$\left. + \left(\frac{\partial x^n}{\partial t^\nu} \right)^2 + B_{ij} x^i x^j \right] dt^1 \wedge \cdots \wedge dt^\nu, \tag{8.32}$$

where B is a constant symmetrical matrix whose spectrum lies to the right of the point $-\lambda$ ($\lambda > 0$) and $\xi = \mathbb{R}^n \times \mathfrak{N}$.

Proposition 8.5. *Let $d_l(\mathfrak{N}) < \pi/\lambda - \epsilon$. Then functional* (8.32) *is nonnegative.*

Proof. Equation (8.16) for functional (8.32) has the form

$$\left(\frac{\partial W_1}{\partial t^1} \right) + \cdots + \left(\frac{\partial W_\nu}{\partial t^\nu} \right) = B - W_1^\top W_1 - \cdots - W_\nu^\top W_\nu. \tag{8.33}$$

As in Example 8.3, we transform the direction l into the axis Ot^1 by an orthogonal transformation. We now perform an orthogonal transformation in the fibers of $\mathbb{R}^n$ that diagonalizes the matrix B. We set $W_\gamma = 0$ for $\gamma > 1$ and seek diagonal solutions to Eq. (8.33): $W_1 = \text{diag}(v_1, \ldots, v_n)$. We have

$$\dot{v}_i = \lambda_i - v_i^2, \tag{8.34}$$

where λ_i are eigenvalues of the matrix B.

We consider the solution to system (8.34):

$$v_i = \sqrt{-\lambda_i} \cot \sqrt{-\lambda_i} \, (t^1 - g^-(t^2, \ldots, t^\nu) + \epsilon) \qquad \text{for } \lambda_i < 0,$$
$$v_j = (t^1 - g^-(t^2, \ldots, t^\nu) + \epsilon)^{-1} \qquad\qquad \text{for } \lambda_j = 0,$$
$$v_k = \sqrt{\lambda_k} \coth \sqrt{\lambda_k} \, (t^1 - g^-(t^2, \ldots, t^\nu) + \epsilon) \qquad \text{for } \lambda_k > 0.$$

It is easy to see that this solution is defined for all $t \in \mathfrak{N}$; this proves the negative definiteness of functional (8.32).

4.1. Potentiality Condition for Tensor Fields. In order to have a possibility to study functional (8.9) in the general case, it is necessary to reveal which functions y_α^i can be substituted under the integral sign in (8.25) if it is known that they are obtained by relation (8.26). Therefore, it is necessary to find the image of the differential operator in the right-hand side of relation (8.26). It is convenient to formulate an answer to this question in terms of the affine connection ∇. Namely, by (8.24), the right-hand side of relation (8.26) coincides with the values of the covariant derivative of the section $x(\cdot)$ in direction of the vector $\partial/\partial t^\alpha$.

Definition 8.4. A tensor field y_α^i is said to be *potential* if it can be written locally as the covariant derivative of a certain section.

By this definition, the image of the operator $\nabla_{\partial/\partial t^\alpha}$ consists of potential vector fields.

Theorem 8.7. *Let ∇ be a flat connection on the bundle ξ. Then the tensor field y_α^i is potential iff*

$$\nabla_{\partial/\partial t^\beta} y_\alpha^i = \nabla_{\partial/\partial t^\alpha} y_\beta^i. \tag{8.35}$$

Proof. We rewrite (8.26) in the form of the Pfaff system of equations

$$dx = (W_\alpha x + y_\alpha) dt^\alpha.$$

Using (8.21), it is easy to verify that the integrability condition of this Pfaff system,

$$\frac{\partial}{\partial t^\beta} (W_\alpha x + y_\alpha) = \frac{\partial}{\partial t^\alpha} (W_\beta x + y_\beta),$$

is equivalent to relation (8.35).

Remark. The proof of Theorem 8.7 can also be obtained from the formula for the curvature tensor R of the connection ∇,

$$R(X,Y) = \nabla_X \nabla_Y - \nabla_Y \nabla_X - \nabla_{[X,Y]}.$$

Because $R = 0$ and the commutator of the coordinate vector fields $\partial/\partial t^\alpha$ and $\partial/\partial t^\beta$ vanishes, we find that the operators $\nabla_{\partial/\partial t^\alpha}$ and $\nabla_{\partial/\partial t^\beta}$ commute, i.e., the second covariant derivatives of any section $x(\cdot)$ do not depend on the order of differentiation for a flat connection.

Again, we note that in the case where the manifold $\mathfrak{N}$ is not simply connected, the local section $x(\cdot)$ that is the potential of the tensor field y_α^i can yield a multivalued function in the continuation to $\mathfrak{N}$.

Epilogue

Equations with a quadratic right-hand side (Riccati equations) not only are related to the calculus of variations but also arise in the theory of infinite-dimensional integrable Hamiltonian systems [10, 73, 119], in the study of the moduli of algebraic curves and surfaces [106, 111], in the study of the uniformization problem in conformal mapping theory [98], in the application of the Bäcklund transformation in quantum field theory [43], and in many other classical and modern fields of mathematics. Many interesting topics related to these equations are not considered in this book. We only mention some of these topics.

Operator Riccati equations arise in the study of linear-quadratic distributed parameter problems [23, 25]. They are evolutionary differential equations with a quadratic right-hand side that act on an infinite-dimensional Banach space X. Operator Riccati equations can be considered as equations on the infinite-dimensional Grassmann manifold (or on the infinite-dimensional Lagrange–Grassman manifold if the space X is endowed with the symplectic structure [104]).

Isomonodromic deformations of complex linear systems of ordinary differential equations are described by the Schlesinger equation [8, 93], the system of partial differential equations whose right-hand side depends quadratically on unknown matrix functions. Quadratic terms in the Schlesinger equation have the form of commutators of the unknown matrices and are skew symmetrical in this sense; at the same time, the partial quadratic equations, which are considered in this book, have symmetrical quadratic terms.

A number of integrable Hamiltonian systems (for example, a Toda lattice) are reduced to an equation with a double commutator, i.e., to the Lax-type equations with the quadratic nonlinearity of the form

$$\dot{H} = [H, [H, N]],$$

where N is a constant operator. Equations with a double commutator were studied in works by R. Brockett and his school using Lie group and Lie algebra theory [7, 10].

One of the leitmotivs of this book is the connection of the Riccati equations with homogeneous spaces on which the group of generalized linear-fractional transformations act (exactly in the same way that linear equations are connected with the group of linear transformations). The theory of equations with a quadratic right-hand side is far from complete. It awaits the younger researchers to whom this book is devoted.

Appendix to the English Edition

To obtain all the relations between Plücker coordinates, a little more advanced technique of tensor calculus is needed.

Let V be n-dimensional linear space with the basis $e_1, \ldots, e_n$ and V^* be the space of linear functionals on V. The value of the functional $f \in V^*$ on the element $e \in V$ is denoted by $\langle f, e \rangle$. Let $e_1^*, \ldots, e_n^*$ denote the basis dual to $e_1, e_2, \ldots, e_n$, i.e. $\langle e_i^*, e_j \rangle = \delta_{ij}$. The exterior product of the vectors $v_1, \ldots, v_k$ (denoted by $v_1 \wedge \cdots \wedge v_k$) is the antisymmetrized tensor product. For example, if $v_1, \ldots, v_k$ is a basis of the plane W, then coordinates of the tensor $v_1 \wedge \cdots \wedge v_k$ are Plücker coordinates of W. Let $\Lambda^k V$ denote the exterior k-power of the space V, $\Lambda^k V = \mathrm{Span}\{e_{i_1} \wedge \cdots \wedge e_{i_k}\}$, and let $\Lambda^* V = \oplus_k \Lambda^k V$. The element $\omega \in \Lambda^k V$ is called *simple* if there exist $v_i \in V$, $i = 1, \ldots, k$, such that $\omega = v_1 \wedge \cdots \wedge v_k$.

We consider the following operators:

1. Exterior multiplication by e_i is the operator E_i defined by the relations

$$E_i : \Lambda^k V \to \Lambda^{k+1} V, \qquad E_i \omega = e_i \wedge \omega.$$

2. Contraction with e_i^* is the operator E_i^* defined by the relations

$$E_i^* : \Lambda^k V \to \Lambda^{k-1} V,$$

$$E_i^*(v_1 \wedge \cdots \wedge v_k) = \sum_{j=1}^{k} (-1)^{j-1} v_1 \wedge \cdots \wedge \langle e_i^*, v_j \rangle \cdots \wedge v_k.$$

3. The trace operator Ω of the tensor product of these operators is defined by the relations

$$\Omega : \Lambda^* V \otimes \Lambda^* V \to \Lambda^* V \otimes \Lambda^* V, \qquad \Omega = \sum_{i=1}^{n} E_i \otimes E_i^*.$$

It is easy to see that Ω is independent of the choice of the basis $e_1, \ldots, e_n$ of V.

Theorem A.1. *The element $\omega \in \Lambda^k V$ is simple iff*

$$\Omega(\omega \otimes \omega) = 0. \tag{A.1}$$

Proof. Let ω be simple, i.e., let there exist linear independent elements v_i, $i = 1, \ldots, k$, such that $\omega = v_1 \wedge \cdots \wedge v_k$. We complete v_i to a basis of V: $V = \mathrm{Span}\{v_1, \ldots, v_k, v_{k+1}, \ldots, v_n\}$. We write Ω in this basis,

$$\Omega(\omega \otimes \omega) = \sum_{i=1}^{k}(E_i \otimes E_i^*)(\omega \otimes \omega) + \sum_{j=k+1}^{n}(E_j \otimes E_j^*)(\omega \otimes \omega).$$

The first sum equals zero because the decomposition of ω contains v_i, $i \leq k$, and hence $v_i \wedge \omega = 0$, $i \leq k$. The second sum equals zero because the decomposition of ω does not contain v_j, $j > k$, and hence $\langle v_j^*, \omega \rangle = 0$, $j > k$. Equation (A.1) is proved.

The proof of the converse is somewhat more complicated. We must prove that $\Omega(\omega \otimes \omega) = 0$ implies ω is simple. We use induction on $\dim V$.

Let $V_{n-1} = \mathrm{Span}\{e_1, \ldots, e_{n-1}\}$ and $V = V_{n-1} \oplus e_n$. For any $\omega \in \Lambda^k V$, we can decompose

$$\omega = e_n \wedge \omega' + \omega'', \tag{A.2}$$

where $\omega' \in \Lambda^{k-1} V_{n-1}$ and $\omega'' \in \Lambda^k V_{n-1}$. Then

$$\omega \otimes \omega = e_n \wedge \omega' \otimes e_n \wedge \omega' + e_n \wedge \omega' \otimes \omega'' + \omega'' \otimes e_n \wedge \omega' + \omega'' \otimes \omega''. \tag{A.3}$$

We represent Ω as the sum $\Omega = E_n \otimes E_n^* + \Omega_{n-1}$, where $\Omega_{n-1} = \sum_{i=1}^{n-1} E_i \otimes E_i^*$. We consider the equation

$$\begin{aligned} \Omega(\omega \otimes \omega) &= 0 \\ &= e_n \wedge e_n \wedge \omega' \otimes \langle e_n^*, e_n \rangle \omega' + e_n \wedge e_n \wedge \omega' \otimes \langle e_n^*, \omega'' \rangle \\ &\quad + e_n \wedge \omega'' \otimes \langle e_n^*, e_n \rangle \omega' + e_n \wedge \omega'' \otimes \langle e_n^*, \omega'' \rangle \\ &\quad + \Omega_{n-1}(e_n \wedge \omega' \otimes e_n \wedge \omega') + \Omega_{n-1}(\omega'' \otimes e_n \wedge \omega') \\ &\quad + \Omega_{n-1}(e_n \wedge \omega' \otimes \omega'') + \Omega_{n-1}(\omega'' \otimes \omega''). \end{aligned} \tag{A.4}$$

Some of the terms in Eq. (A.4) equal zero. We only need the following corollaries of Eq. (A.4):

$$\Omega_{n-1}(e_n \wedge \omega' \otimes e_n \wedge \omega') = 0 \tag{A.5}$$

because all other terms in (A.4) do not contain e_n in both parts of the tensor product and cannot cancel with the left-hand side of (A.5);

$$\Omega_{n-1}(\omega'' \otimes \omega'') = 0 \tag{A.6}$$

because it does not contain e_n and all other terms in (A.4) do contain e_n; and

$$\Omega_{n-1}(e_n \wedge \omega' \otimes \omega'') + e_n \wedge \omega'' \otimes \omega' = 0 \tag{A.7}$$

because only these terms contain e_n in the first multiplier of the tensor product.

It follows from (A.6) and the induction hypothesis that ω'' is simple:

$$\omega'' = v_1 \wedge \cdots \wedge v_k. \tag{A.8}$$

We use the notation $\omega_i'' = (-1)^{i-1} v_1 \wedge \cdots \hat{v}_i \cdots \wedge v_k$, where the hat over a letter means that this letter must be omitted. We choose $v_1, \ldots, v_k, v_{k+1}, \ldots, v_{n-1}$ as a new basis of V_{n-1}. Then $v_1, \ldots, v_k, v_{k+1}, \ldots, v_{n-1}, e_n$ is a basis of V. Writing Eq. (A.7) in this basis yields

$$\sum_{i=1}^{k} v_i \wedge e_n \wedge \omega' \otimes \omega_i'' + e_n \wedge \omega'' \otimes \omega' = 0. \qquad (A.9)$$

Here, the sum is only taken from 1 to k because all other terms of Ω_{n-1} equal zero in view of (A.8). It follows from (A.9) that ω' cannot depend on v_j, $j > k$; otherwise, the corresponding terms cannot cancel with those of $e_n \wedge \omega'' \otimes \omega'$. Consequently,

$$\omega' = \sum_{j=1}^{k} a_j \omega_j'', \quad a_j \in \mathbb{R}. \qquad (A.10)$$

We demonstrate that all coefficients a_j, except at most one, equal zero. It is easy to see from (A.5) that

$$0 = \sum_{p=1}^{k} v_p \wedge \omega' \otimes \langle v_p^*, \omega' \rangle$$

$$= \sum_{p=1}^{k} \left(\left(\sum_{i=1}^{k} a_i v_p \wedge \omega_i'' \right) \otimes \left(\sum_{j=1}^{k} a_j \langle v_p^*, \omega_j'' \rangle \right) \right)$$

$$= \sum_{i,j=1}^{k} a_i a_j \sum_{p=1}^{k} \left(v_p \wedge \omega_i'' \otimes \langle v_p^*, \omega_j'' \rangle \right)$$

$$= \sum_{i,j=1}^{k} \pm a_i a_j \left(\omega'' \otimes \omega_{ij}'' \right). \qquad (A.11)$$

In the right-hand side of (A.11), $i \neq j$, and all tensors that are coefficients of $a_i a_j$ are distinct from each other. Hence, $a_i a_j = 0$ for any $i \neq j$. Consequently, at most one coefficient a_i can be nonzero. We can assume without loss of generality that $a_1 \neq 0$. It is conclusively proved that

$$\omega'' = v_1 \wedge \cdots \wedge v_k, \qquad \omega' = a_1 v_2 \wedge \cdots \wedge v_k,$$

and hence

$$\omega = (a_1 e_n + v_1) \wedge v_2 \wedge \cdots \wedge v_k$$

by (A.2). Theorem A.1 is proved.

Relation (A.1) gives the system of quadratic equations on the coefficients of a tensor ω that is the necessary and sufficient condition for this tensor to be the Plücker coordinates of a plane W.

Corollary A.1. *The image of the Plücker embedding ϕ (see p. 116) is described by Eq. (A.1).*

Exercise. Write Eq. (A.1) in terms of the Plücker coordinates of the manifold $G_2(\mathbb{R}^4)$.

References

1. Alekseev, V.M., Tikhomirov, V.M., Fomin, S.V. (1980): Optimal Control. Moscow: Nauka. English transl.: Contemp. Sov. Math. New York: Consultants Bureau 1987. Zbl. 689.49001[1]
2. Andrianov, A.N, Zhuravlev, V.T. (1990): Modular Forms and Hecke Operators. Moscow: Nauka. English transl.: Transl. Math. Monogr. **145**. Providence, R.I.: Am. Math. Soc. 1995. Zbl. 888.11032
3. Arnol'd, V.I. (1974): Mathematical Methods of Classical Mechanics. Moscow: Nauka. 3rd ed. 1989. English transl.: Grad. Texts Math. **60**. New York–Heidelberg–Berlin: Springer-Verlag 1978. Zbl. 386.70001
4. Arnol'd, V.I. (1984): Ordinary Differential Equations. Moscow: Nauka. English transl.: Berlin–Heidelberg–New York: Springer-Verlag 1992. Zbl. 744.34001
5. Atiyah, M. (1990): The Geometry and Physics of Knots. Cambridge Univ. Press. Zbl. 729.57002
6. Bliss, G.A. (1946): Lectures on the Calculus of Variations. Univ. of Chicago Press. Zbl. 063.00459
7. Bloch, A.M., Brockett, R.W., Ratiu, T. (1990): A new formulation of the generalised Toda lattice equations and their fixed point analysis via the momentum map. Bull. Am. Math. Soc., New Ser. **23**, No. 2, 477-485. Zbl. 715.58033
8. Bolibrukh, A.A. (1994): The 21st Hilbert problem for linear Fuchsian systems. Trudy Mat. Inst. Akad. Nauk. **206**. English transl.: Proc. Steklov Inst. Math. **206**. Providence, R.I.: Am. Math. Soc. Zbl. 844.34004
9. Brioscki, F. (1879): Sulla equazione dell ot taedro. Transunti Accad. Nar. Lincei. **3**
10. Brockett, R.W. (1988): Dynamical systems that sort lists, diagonalize matrices, and solve linear programming problems. Proc. 27th IEEE Conference on Decision and Control. Austin, Tex., 779-803. Linear Alg. Appl. **146**, 79-91 (1991). Zbl. 719.90045
11. Bucy, R.S. (1975) Structural stability for the Riccati equation. SIAM J. Control Optimiz. **13**, 749–753. Zbl. 301.34062
12. Cantor, M. (1901): Vorlesungen über Geschichte der Mathematik. Vol. 4. Leipzig and Bibliotheca Mathematica Teuneriana, New York: Johnson Reprint Corp. (1965). Zbl. 192.00701
13. Caratheodory, C. (1929): Über die Variationsrechnung bei mehrfachen Integralen. Acta Szeged Sect. Scient. Math. **4**, 193–216
14. Caratheodory, C. (1935): Variationsrechnung und Partielle Differentialsgleichungen Erster Ordnung. Berlin: Teubner. Zbl. 011.35603
15. Cartan, E. (1913): Les groupes projectifs qui ne laissent aucun multiplicité plane. Bull. Soc. Math. France. **41**, 53–96
16. Cartan, E. (1936): Sur les domaines bornés homogènes de l'espace de n variables complexes. Abh. Math. Sem. Hansischen Universität. **11**, 116 –162 (French). Zbl. 011.12302
17. Cartan, E. (1952): Oeuvres completes. Paris: Gauthier-Villars.
18. Chevalley, C. (1946): Theory of Lie Groups I. Proc. First Canad. Congr., Univ. of Toronto Press. English transl.: Princeton, N.J.: Princeton Univ. Press 1946. Zbl. 063.00842

---

[1] For the convenience of the reader, references to reviews in Zentralblatt für Mathematik (Zbl.), compiled using the MATH database, have, as far as possible, been included in this bibliography.

19. Clebsch, A. (1859): Über die zweite Variation vielfacher Integrale. J. Reine Angew. Math. **56**, 122–148

20. Clifford, V. (1955): Common Sense of the Exact Sciences: Origin of the Doctrine on Number and Space. Reprint New York: Dover Publ. 1955. Zbl. 066.00103

21. Clifford, W.K. (1882): Mathematical Papers. London

22. Courant, R. (1950): Dirichlet's Principle, Conformal Mapping, and Minimal Surfaces. With an appendix by M. Shiffer. New York–London: Interscience Publ.

23. Curtain, R.F., Pritchard, A.J. (1976): The infinite-dimensional Riccati equation for systems defined by evolution operators. SIAM J. Control Optimiz. **14**, 951–983. Zbl. 352.49003

24. Dao, Chong Thi, Fomenko, A.T. (1987): Minimal Surfaces and the Plateau Problem. Moscow: Nauka. English transl.: Transl. Math. Monogr. **84**. Providence, R.I.: Am. Math. Soc. 1987. Zbl. 716.53003

25. Deligne, P. (1970): Equations differentielles a points singuliers reguliers. Lect. Notes Math. **163**. Zbl. 244.14004

26. Dirac, P.A.M. (1927): The quantum theory of the electron. Proc. Roy. Soc. London A. **117**, 610–625; The quantum theory of the electron: Part II. Proc. Roy. Soc. London A. **118**, 351–361 (1928)

27. Douglas, J. (1931): Solution of the problem of Plateau. Trans. Amer. Math. Soc. **33**, 263–321. Zbl. 001.14102

28. Dubrovin, B.A., Novikov, S.P., Fomenko, A.T. (1979): Modern Geometry. Moscow: Nauka. English transl.: Grad. Texts Math. **93**, 2nd ed. 1992 and **104** 1985, Berlin–Heidelberg–New York: Springer-Verlag. Zbl. 751.53001, 565.57001

29. Dunford, N., Schwartz, J. (1988): Linear Operators. Part I: General Theory. Reprint of the orig., publ. 1959 by Wiley & Sons, Wiley & Sons 1988. Zbl. 635.47001

30. Edwards, H.M. (1964): A generalized Sturm theorem. Ann. Math. Ser. 2. **80**, No. 1, 22–57. Zbl. 127.03805

31. Federer, G. (1969): Geometric Measure Theory. Berlin–Heidelberg–New York: Springer-Verlag. Zbl. 176.00801

32. Fomenko, A.T. (1982): Variational Methods in Topology. Moscow: Nauka. (Russian) Zbl. 526.58012

33. Funk, P. (1962): Variationsrechnung und Ihre Anwendung in Physik und Technik. Berlin–Göttingen–Heidelberg: Springer-Verlag. 2nd ed. 1970. Zbl. 188.17504

34. Gauss, K.F. (1801): Disquisitiones arithmeticae

35. Giaquinta, M., Hildebrandt, S. (1996): Calculus of Variations. Grundlehren Math. Wiss. **310**, **311**. Berlin–Heidelberg–New York: Springer-Verlag. Zbl. 853.49001, 853.49002

36. Gibson, J.S. (1979): The Riccati integral equations for optimal control problems on Hilbert spaces. SIAM J. Control Optimiz. **17**, 537–565. Zbl. 411.93014

37. Gilbert, D. (1906): Zur Variationsrechnung. Math. Ann. **62**, 351–370

38. Giusti, E. (1984): Minimal Surfaces and Functions of Bounded Variation. Monogr. Math. **80**. Boston–Basel–Stuttgart: Birkhäuser Verlag. Zbl. 545.49018

39. Goursat, E. (1936): Cours d'analyse mathematique. Tome II. Reprint of the 4th ed. 1925. Gauthier-Villars. English transl.: New York: Dover Publ. Zbl. 144.04501

40. Grugnetti, L. (1992): Sur Carteggio Jacopo Riccati – Nicola 2 Bernulli. J. Riccati e la Cultura della Marca nel Settecento Europeo. Firence

41. Hadamard, J. (1905): Sur quelque question de calcul des variations. Bull. Soc. Math. France. **33**, 77–80

42. Hamilton, W.R. (1931, 1940): The Mathematical Papers of Sir William Rowan Hamilton. Part I and II. Cambridge Univ. Press 1931, 1940

43. Harnad, J., Saint-Aubin, Y., Shnider, S. (1984): Quadratic pseudopotential for GL(N, $\mathbb{C}$) principal sigma models. Physica D. **10**, No. 3, 394–412. Zbl. 584.58016

44. Helgason, S. (1962): Differential Geometry and Symmetrical Spaces. Pure and Appl. Math. **12**. New York–London: Academic Press. Zbl. 111.18101

45. Hermann, R., Martin, C.F. (1977): Applications of algebraic geometry to system theory: I. IEEE Trans. Automat. Control. **AC-22**, 19–25. Zbl. 355.93013

46. Hermann, R., Martin, C.F. (1982): Lie and Morse theory for periodic orbits of vector fields and matrix Riccati equations I: General Lie-theoretic method. Math. System Theory **15**, 277–284. Zbl. 508.58034

47. Hermite, S. (1865): Sur l'equation du cinquieme degre. C. R. Acad. Sci. **1**; (1866): **2**

48. Hodge, V., Pido, D. (1952): Methods of Algebraic Geometry. Vol. II. Cambridge Math. Library. Cambridge Univ. Press. Reprint 1994. Zbl. 796.14002

49. Hua, L.K. (1944): On the theory of automorphic functions of a matrix variable. II. The classification of hypercircles under the symplectic group. Amer. Journal Math. **66**, 470–488, 531–563. Zbl. 063.02920

50. Hua, L.K. (1965): Harmonic Analysis of Functions of Several Complex Variables in Classical Domains. Peking: Academic Press, Revised ed. 1965 (Chinese). English transl.: Transl. Math. Monogr. **6**. Providence, R.I.: Am. Math. Soc. 1979. Zbl. 507.032025

51. Hurewicz, W. (1948): The Foundations of the Theory of Algebraic Invariants. Moscow: GITTL (Russian). [Also Weitzenböck, R. Invariantentheorie. Gröningen 1923]

52. Hurwitz, A. (1923): Über die Komposition der quadratischen Formen. Math. Ann. **88**, 1–25

53. Husemoller, D. (1993): Fibered Bundles. 3rd ed. Berlin–Heidelberg–New York: Springer-Verlag. Zbl. 794.5501

54. Kirillov, A.A. (1972): Elements of Representation Theory. Moscow: Nauka (Russian)

55. Klein, F. (1968): Vorlesungen über Nicht-Euklidische Geometrie. Springer-Verlag 1928. Reprint 1968

56. Klein, F. (1989): Vorlesungen über das Ikosaeder und die Auflösung der Gleichungen vom fünften Grade. Basel: Birkhäuser Verlag, Stuttgart: Teubner Verlagesellschaft. Reprogr. Nachdruck der Ausg. Leipzig: Teubner 1884. Zbl. 803.01037. English transl.: Lectures on the Ichosahedron and the Solutions of Equations of the Fifth Degree. 2nd rev. ed. New York: Dover Publ. Zbl. 072.25901

57. Klingen, H. (1956): Über die analytischen Abbildungen verallgemeinerter Einheitskreise auf sich. Math. Ann. **132**, 134–144. Zbl. 071.07701

58. Klötzler, R. (1971): Mehrdimensionale Variationsrechnung. Berlin: Deutscher Verlag der Wiss., Basel–Stuttgart: Birkäuser Verlag. Zbl. 199.42901

59. Kobayashi, Sh., Nomizu, K. (1963): Foundations of Differential Geometry. Vol. 1. Wiley Classics Library

60. Koecher, M. (1976): Über eine Gruppe von rationalen Abbildungen. Invent. Math. **3**, 136–171. Zbl. 163.03002

61. Lagrange, J.L. (1779): Sur la construction des cartes géographiques. Nouveaux Mémories de l'Académie de Berlin

62. Lang, S. (1965): Algebra. Reading, Mass.: Addison-Wesley Publ. Co., 3rd ed. 1993. Zbl. 848.13001

63. Lee, E.B., Marcus, L. (1967): Foundation of Optimal Control Theory. New York–London–Sidney: Wiley & Sons

64. Lepage, J. T. (1941): Sur le champs geodesiques des integrales multiples. Bull. Acad Roy. Belg., Class de sciences **27**, 27–46. Zbl. 027.10702

65. Levin, J.J. (1959): On the matrix Riccati equation. Proc. Amer. Math. Soc. **10**, No. 4, 519–524. Zbl. 092.07803

66. Lie, S. (1891): Vorlesungen über Differentialgleichungen mit bekannten infinitesimalen Transformationen. Leipzig

67. Lie, S. (1893): Berührungstransformationsgruppen. Leipzig

68. Lion, G., Vergne, M. (1980): The Weil Representation, Maslov Index, and Theta-Series. Progress in Mathematics **6**. Boston–Basel–Stuttgart: Birkhäuser Verlag. Zbl. 444.2205

69. Matyukhin, D.V., (1995): On positive-definite biquadratic forms irreducible to sums of squares of bilinear forms. Vestn. MGU. Ser. 1. No. 2, 29–33. English transl.: Mosc. Univ. Math. Bull. **50**, No. 2, 28-32 (1995). Zbl. 886.11022

70. Milnor, J. (1963): Morse Theory. Ann. Math. Stud. **51**. Princeton, N.J.: Princeton Univ. Press. Zbl. 108.10401

71. Milnor, J., Stashef, J. (1979): Characteristic Classes. Ann. Math. Stud. **76**. Princeton, N.J.: Princeton Univ. Press. Zbl. 298.57008

72. Mishchenko, A.S., Fomenko, A.T. (1980): A Course in Differential Geometry and Topology. Moscow: MGU (Russian). Zbl. 283.35019

73. Miura, R.M., Gardner, C.S., Kruscal, M.D. (1968): Korteweg–de Vries equation and generalizations: II. J. Math. Phys. **9**, 1204–1209. Zbl. 283.35019

74. Morrey, C.B. (1966): Multiple Integrals in the Calculus of Variations. Berlin–Heidelberg–New York: Springer-Verlag. Zbl. 142.38701

75. Narasimhan, R. (1985) Analysis on Real and Complex Manifolds. Amsterdam–New York–Oxford: North Holland, 3rd printing

76. Nikol'skii, S.M (1975): A Course of Mathematical Analysis. Vol. 2. Moscow: Nauka. English transl.: Moscow: MIR Publ. 1977. Zbl. 397.00003

77. Petrovskii, I.G. (1961): Lectures on Partial Differential Equations. Moscow: Fizmatgiz. English transl.: London: Iliffe Books Ltd. Zbl. 163.11706

78. Pontryagin, L.S. (1961): Ordinary Differential Equations. Moscow: Fizmatgiz. English transl.: London–Paris: Pergamon Press. Zbl. 112.05502

79. Pontryagin, L.S. (1984): Continuous Groups. Moscow: Nauka. English transl.: Selected Works: Vol. 2, Topological Groups. New York: Gordon and Breach Publ. 1986. Zbl. 882.01025

80. Prasolov, V.V., Solov'ev, Yu.P. (1997): Elliptic Functions and Algebraic Equations. Moscow: Faktorial (Russian)

81. Pressley, E., Segal, G. (1988): Loop Groups. Oxford: Clarendon Press, 2nd ed. Zbl. 638.22009

82. Pyatetskii–Shapiro, I.I. (1961): Automorphic Functions and the Geometry of Classical Domains. Moscow: Fizmatgiz. English transl.: New York–London–Paris: Gordon and Breach Publ. Zbl. 196.09901

83. Pyatetskii-Shapiro,I.I. (1966): Geometrie des domaines classiques et theorie des fonctiones automorphes. Paris: Dunod. Transl. from the Russian (1961). Zbl. 142.05101

84. Rado, T. (1933): On the Problem of Plateau. Berlin–Heidelberg–New York: Springer-Verlag. Reprint 1971. Zbl. 211.13803

85. Radon, J. (1922): Lineare Scharen ortogonaler Matrizen. Abhandlungen aus dem Math. Sem. der Hamburg Univ. **1**, 1–14

86. Radon, J. (1927): Über die Oscillationstheoreme der konjugierten Punkte beim Problem von Lagrange. Münchener Sitzungberichte, 243–257

87. Radon, J. (1928): Zum Problem von Lagrange. Abhandlungen aus dem Math. Sem. Hamburg Univ. **6**, 273–299

88. Rashevskii, P.K. (1956): A Course of Differential Geometry. Moscow: GITTL, 4th ed. (Russian). Zbl. 071.14407

89. Reid, W.T. (1963): Riccati matrix differential equations and non-oscillation criteria for associated linear differential systems. Pacific J. Math. **13**, No. 2, 665–685. Zbl. 119.07401

90. Riccati J. (1724): Animadversationes in aequationes differentiales secundi gradus. Actorum eruditorum quae Lipsiae publicantur. Supplementa 8, 66–73

91. Rokhlin, V.A., Fuks, D.B. (1977): A Beginner's Course in Topology: Geometric Chapters. Moscow: Nauka. English transl.: Berlin–Heidelberg–New York: Springer-Verlag 1984. Zbl. 562.54003

92. Sandor, S. (1959): Sur l'equation differentielle matricielle du type Riccati. Bull. Soc. Sci. Math. Phys., R. P. Roumainie (New Ser.). **51**, No. 3, 229–249. Zbl. 098.05701

93. Schlesinger, L. (1905): Über Lösungen gewisser Differentialgleichungen als Functionen der singularen Punkte. J. Reine Angew. Math. **129**, 287–294

94. Schwarz, H.A. (1872): Über dienige Falle, in welchen die Gaussische hypergeometrische Reihe eine algebraische Function ihres vierten Elements darstellt. Borhardt's J. **75**, 292–335

95. Serre, J.-P. (1965): Lie Algebras and Lie Groups. New York–Amsterdam: W. A. Benjamin, Inc. Zbl. 132.27803, 2nd ed. Lect. Notes Math. **1500**, 1992. Zbl. 742.17008

96. Shabat, B.V. (1985): Introduction to Complex Analysis. Vols. 1, 2. 3rd ed. Moscow: Nauka. English transl.: Transl. Math. Monogr. **110**, Providence, R.I.: Am. Math. Soc. 1992. Zbl. 799.32001

97. Shayman, M.A. (1986): Phase portrait of the matrix Riccati equation. SIAM J. Control Optimiz. **24**, 1–65. Zbl. 594.34044

98. Shiffer, M. (1950): Some new results in conformal mapping theory. Appendix in Courant, R., op. cit.

99. Siegel, C.L. (1942): Note on automorphic forms of several complex variables. Ann. of Math., II. Ser. **43**, 613–616. Zbl. 138.31402

100. Siegel, C.L. (1943): Symplectic geometry. Amer. J. Math. **65**, 1–86. Zbl. 138.31401

101. Stepanov, V.V. (1953): A Course of Differential Equations. Moscow: GITTL (Russian)

102. Stiefel, E. (1936): Richtungsfelder und Fernparallelismus in n-dimensionalen Mannigfaltigkeiten. Comment. Math. Helv. **8**, 305–353. Zbl. 014.41601

103. Suprun, D.G. (1984): A generalization of the Caratheodory transformation. Vestn. MGU. Ser. 1. No. 3, 82–85. English transl.: Mosc. Univ. Math. Bull. **39**, No. 5, 104-108 (1984). Zbl. 568.49018

104. Swanson, R.C. (1980): Linear symplectic structures on Banach spaces. Rocky Mt. J. Math. **10**, 305–317. Zbl. 449.58007

105. Terpstra, F.J. (1938): Die Darstellung biquadratischer Formen als Summen von Quadraten mit Anwendungen auf die Variationsrechnung. Math. Ann. **116**, 166–180. Zbl. 019.35203

106. Tyurin, A.N. (1978): On periods of quadratic differential forms. Uspekhi Mat. Nauk. **33**, No. 6, 149–195 (Russian)

107. Van Hove, L. (1947): Sur l'extension de la condition de Legendre du calcul des variations aux integrales multiples a plusiers fonctions inconnues. Nederl. Akad. Wetensch. **50**, 18–23. Zbl. 029.26802

108. Warner, P. (1971): Foundations of Differentiable Manifolds and Lie Groups. Glenview, Ill.: Scott, Foresman & Co. 1971. Reprint New York–Berlin–Heidelberg: Springer-Verlag 1983. Zbl. 516.58001

109. Weyl, H. (1935): Geodesic fields in the calculus of variations for multiple integrals. Ann. Math. **36**, 607–629. Zbl. 013.12002

110. Whyburn, W.W. (1934): Matrix differential equations. Amer. J. Math. **54**, No. 1, 587–592. Zbl. 010.11101

111. Wilczinski, E.J. (1906): Projective Differential Geometry of Curves and Ruled Spaces. B. G. Teubner

112. Winternitz, P. (1983): Lie groups and solutions of nonlinear differential equations. Lect. Notes Phys. **189**, 263–331. Zbl. 571.34002

113. Wolf, J. (1963): Geodesic spaces in Grassmann manifolds. Illinois J. Math. **7**, No. 3, 425–446. Zbl. 114.37002

114. Wolf, J. (1963): Elliptic spaces in Grassmann manifolds. Illinois J. Math. **7**, No. 3, 447–462. Zbl. 114.37101

115. Wolf, J.A. (1967): Spaces of Constant Curvature. New York: McGraw-Hill. Zbl. 162.53304

116. Wong, Y.C. (1961): Isoclinic n-planes in Euclidean $2n$-space, Clifford parallels in elliptic $(2n-1)$-space, and Hurwitz matrix equations. Mem. Amer. Math. Soc. **41**, 1–112. Zbl. 124.13401

117. Young, L. (1969): Lectures on the Calculus of Variations and Optimal Control Theory. Philadelphia: W. B. Saunders Co. Zbl. 177.37801

118. Zakhar-Itkin, M.Kh. (1973): Matrix Riccati differential equation and the semigroup of linear fractional transformations. Uspekhi Mat. Nauk. **28**, No. 3, 83–120. English transl.: Russ. Math. Surv. **28**, No. 3, 89–131 (1973). Zbl. 293.34056

119. Zakharov, V.E., Faddeev, L.D. (1971): Korteweg–De Vries equation: A completely integrable Hamiltonian system. Funkts. Anal. Prilozh. **5**, No. 4, 18–27. English transl.: Funct. Anal. Appl. **5**, 280–287 (1971). Zbl. 257.35074

120. Zelikin, M.I. (1991): To the theory of the matrix Riccati equation. Mat. Sb. **182**, No. 7, 970–984. English transl.: Russ. Acad. Sci., Sb., Math. **73**, No. 2, 341–354 (1992). Zbl. 774.49003

121. Zelikin, M.I. (1992): To the theory of the matrix Riccati equation: 2. Mat. Sb. **183**, No. 10, 87–108. English transl.: Russ. Acad. Sci., Sb., Math. **77**, No. 1, 213–220 (1994). Zbl. 786.49023

122. Zelikin, M.I. (1997): On the connection generated by the problem of minimizing a multiple integral. Mat. Sb. **188**, No. 1, 59–72. English transl.: Russ. Acad. Sci., Sb., Math. **188**, No. 1, 61–74 (1997). Zbl. 980.03744

Index

Druck: Strauss Offsetdruck, Mörlenbach
Verarbeitung: Schäffer, Grünstadt